INTRODUCTION TO AIR TRANSPORT ECONOMICS

Reviews of the first edition of
Introduction to Air Transport Economics

This is a landmark textbook in airline economics as, for the first time, disparate information from various economic disciplines is integrated with key institutional factors that affect the industry. Basic material is presented in an easy-to-read and understand format with plenty of real-world situations. Much of the material in this book cannot be found in other sources. I highly recommend this book.
Darryl Jenkins, Ohio State Airline Operations Center, USA

This book represents a thorough and comprehensive coverage of the key topics in aviation economics by applying fundamental theories and methods available in micro-economics and industrial organization economics. This book also treats excellently the current policy and strategic issues facing policy makers, airlines, airports and other professionals in aviation. This is a rare book that combines both theory and applications in a very meaningful way. As such, I recommend it very strongly to graduate students, policy makers, managers and researchers in aviation.
Tae H. Oum, Air Transport Research Society (ATRS) and University of British Columbia, Canada

The aviation industry provides a good backdrop for understanding a variety of economic principals. Introduction to Air Transport Economics offers valuable insight into how the fundamentals of economics apply to aviation. The book covers a broad range of topics from underlying passenger demand to aviation infrastructure. It offers an overview into the dynamic industry environment while using practical industry applications to explain general economic concepts—a great introduction to the world of aviation and explanation for what keeps airlines in the air.
Zane Rowe, Senior Vice President Network Strategy, Continental Airlines

In helping to establish a new international university with a focus on aviation, I have recently had the opportunity to review many books and materials that might be critical as a resource for our future students. The authors' new book, Introduction to Air Transport Economics, is just such a resource. I was privileged to see this book early on, and I was immediately struck with its scope and timeliness. Most importantly, it provides a look at the industry from the keen perspective of a distinguished economist. So, rather than another survey of industry trends, this book takes rigorous, well established economic principles and applies them to the most important aspects of the international aviation industry. Even in areas such as safety and security, the authors provide clear insight to show how it is economics that drives the successful airlines to adhere to the highest standards of safety and security. I believe that this book will be used by faculties and students, worldwide, in their effort to understand, and then manage, the activities of the aviation industry.
Irwin Price, Dubai Aerospace Enterprise University

Dr Vasigh has produced a detailed introduction to the subject of airline economics, covering both the theory and its application to the modern air transportation industry. Students and those already engaged in the industry will find this an illuminating guide to an ever more complex business.
Barry Humphreys, Virgin Atlantic Airways

Introduction to Air Transport Economics: From Theory to Application *provides a comprehensive treatment of the economics of the commercial air transport industry. Written in easy-to-read, jargon-free English, the book first reviews the fundamentals of economics, using air transport examples, to illustrate concepts related to demand, costs, and competition. Specific chapters then cover, again from an economic perspective, key topic areas, such as the emergence of low cost carriers, international air transport, and aviation infrastructure. The book will be beneficial to those seeking a deeper understanding of the workings of the major players in the air transport industry, including airlines, airports, and aircraft manufacturers. Readers will learn, for example, how airlines use revenue management to maximize their profits or how airlines and airports forecast passenger traffic. Since the book does cover economic principles, it can be read as a standalone text by undergraduates and graduate students with limited backgrounds in economics. Those students with a deeper understanding of economics, at the intermediate level or higher, can read quickly through the principles chapters and delve directly into the airline applications.*
Martin Dresner, University of Maryland, USA

Introduction to Air Transport Economics

From Theory to Applications
Second Edition

BIJAN VASIGH

Embry-Riddle Aeronautical University, Florida, USA

KEN FLEMING

ACG Aviation Consulting Group LLC, Florida, USA

&

THOMAS TACKER

Embry-Riddle Aeronautical University, Florida, USA

Routledge
Taylor & Francis Group

LONDON AND NEW YORK

First published 2013 by Ashgate Publishing

Published 2016 by Routledge
2 Park Square, Milton Park, Abingdon, Oxon OX14 4RN
711 Third Avenue, New York, NY 10017, USA

Routledge is an imprint of the Taylor & Francis Group, an informa business

British Library Cataloguing in Publication Data
Vasigh, Bijan.
  Introduction to air transport economics : from theory to
  applications. -- 2nd ed.
  1. Aeronautics, Commercial--Finance. 2. Airlines--
  Deregulation. 3. Airlines--Cost of operation.
  4. Aeronautics and state. 5. Aeronautics, Commercial--
  United States.
  I. Title II. Fleming, Ken. III. Tacker, Tom.
  387.7'1-dc23

  ISBN: 978-1-4094-5486-1 (hbk)
       978-1-4094-5487-8 (pbk)

Library of Congress Cataloging-in-Publication Data
Vasigh, Bijan.
  Introduction to air transport economics : from theory to applications / by Bijan Vasigh, Ken
Fleming, and Thomas Tacker.
     p. cm.
  Includes bibliographical references and index.
  ISBN 978-1-4094-5486-1 (hardback) -- ISBN 978-1-4094-5487-8 (ebook)
  1. Aeronautics, Commercial--United States--Finance. 2. Airlines--Deregulation--United States.
  3. Airlines--United States--Cost of operation. 4. Aeronautics and state--United States. I. Fleming,
Ken. II. Tacker, Tom. III. Title.
  HE9803.A5V37 2013
  387.7'10973--dc23
                                                                                    2012030097

Printed and bound in Great Britain by
TJ International Ltd, Padstow, Cornwall

Contents

List of Figures

List of Tables

Glossary of Terms

A

AEA: *Association of European Airlines.*

Aircraft Utilization: *Aircraft utilization is calculated by dividing aircraft block hours by the number of aircraft days assigned to service on air carrier routes and presented in block hours per day.*

Air Carrier: *Any airline that undertakes directly, by lease, or other arrangement to engage in air transportation.*

Aircraft Crews Maintenance Insurance (ACMI): *A lease between two parties where the first party is a lessor with an Air Operator's Certificate (AOC) responsible for the aircraft crews, maintenance, and insurance and the second party is the lessee, usually with an AOC, who is responsible for schedules, flight charges, cargo handling, crew support, flight operations, ramp handling, and aircraft servicing and fueling.*

Aircraft Daily Utilization: *Aircraft hours flown (block-to-block) divided by aircraft days available.*

Alliance: *Several airlines participating in a commercial relationship or joint venture.*

Available Seat Kilometer (ASK): *A measure of a passenger airline's carrying capacity that is calculated as follows:*

$ASK = Number\ of\ seats\ \times\ number\ of\ kilometers\ flown$

Available Seat Mile (ASM): *A measure of a passenger airline's carrying capacity that is calculated as follows:*

$ASM = Number\ of\ seats\ \times\ number\ of\ miles\ flown$

Available Ton Mile (ATM): *A measure of a cargo airline's carrying capacity that is calculated as follows:*

$ATM = weight\ in\ non\text{-}metric\ tons\ x\ number\ of\ miles\ flown.$

Average Stage Length (ASL): *The ASL is the average distance flown per aircraft departure.*

$$ASL = \frac{Plane\ Miles}{Departures}$$

B

Bankruptcy: *The inability of an airline to pay its creditor is called bankruptcy.*

Block Hour: *Block hours are the airline industry basic measure of aircraft utilization. Block hour is the time from the minute the aircraft door closes at departure of a revenue flight until the moment the aircraft door opens at the arrival gate.*

Break-even: *It is the volume of goods or services that have to be sold in order for the business to make neither a loss nor a profit.*

Break-even Load Factor: *The load factor that covers the necessary operating costs for scheduled traffic revenue.*

$$\text{Break-Even}_{LF} = \frac{CASM}{R/RPM} = \frac{CASM}{Yield}$$

C

Cabotage: *Cabotage is the transport of passengers between two points in the same country by a an aircraft registered in another country.*

Certified Air Carrier: *An air carrier that is certified by the US Department of Transportation (DOT) to conduct scheduled or nonscheduled services interstate. The certificate issued to the air carrier by the DOT is the Certificate of Public Convenience and Necessity.*

Code-sharing: *An arrangement where an airline may place its own code to another carrier's flight. The airline that is actually operating the flight is called the operating carrier, and the airline that is marketing the flight is called the marketing carrier. Both carriers may sell tickets for the flight.*

Commercial Service Airport: *Airport receiving scheduled passenger service and having 2,500 or more enplaned passengers per year.*

Commuter Air Carrier: *A passenger air carrier operating aircraft with 30 seats or less and performs at least five scheduled roundtrips per week. It operates for hire or compensation under FAR Part 135.*

Concentration Ratio: *Measures the proportion of an industry's output accounted for by several of the largest firms.*

Consumer Surplus: *Consumer surplus is the difference between the maximum that a consumer is willing and able to pay for a good or service and the total amount that he actually pays.*

Cost per Available Seat Mile (CASM): *It is represented in cents and is calculated as follows:*

$$CASM = \frac{Operating\ Costs}{ASM}$$

CRS: *Computer Reservation System.*

D

Deregulation: *The term refers to the Airline Deregulation Act of 1978, which ended US Government regulation of airline routes and charges.*

Derived Demand: *It is the demand that is generated as the result of the demand for other goods or services.*

Duopoly: *Any market that is dominated by two firms. Commercial aircraft industry may be considered as a duopoly.*

Duopsony: *Two major buyers of a good or service in a market.*

E

Economies of Density: *Cost reductions that result when a company utilizes a bigger plant size in the production a single product.*

Economies of Scale: *The decrease in unit cost of a product, or the increase in efficiency of production as the number of goods being produced increases.*

Economies of Scope: *Cost reductions that result when a company provides a variety of products rather than specializing in the production a single product.*

Enplanement: *The boarding of scheduled and nonscheduled service aircraft by domestic, territorial, and international revenue passengers for intrastate, interstate, and foreign commerce and that includes in transit passengers.*

European Aviation Safety Agency (EASA): *The European Aviation Safety Agency is an agency of the European Union (EU) which has been given regulatory and executive tasks in the field of civilian aviation safety.*

Extended-range Twin-engine Operations (ETOPS): *This rule allows twin-engine aircraft (such as the Airbus A300, A310, A320, A330 and A350, the Boeing 737, 757, 767, 777, 787, the Embraer E-Jets, ATR) to fly longer distance routes that were previously off-limits to twin-engine aircraft.*

F

Federal Aviation Administration (FAA): *A US government agency responsible for air safety and operation of the air traffic control system.*

Federal Air Regulation (FAR): *Title 14 of the US Government's Code of Federal Regulations. The FAR covers all the rules regarding aviation in the US.*

Financial Leverage: *A measure of the amount of debt used in the capital structure of the airlines. An airline with high leverage is more vulnerable to downturns in the business cycle because the airline must continue to service its debt regardless of how bad business is.*

Flight Stage: *The operation of an aircraft from take-off to its next landing.*

Form 41 Data: *Information collected from airline filings with the Bureau of Transportation Statistics. Airline financial data is filed with the Bureau of Transportation Statistics (BTS) quarterly; traffic and employment numbers are filed monthly.*

Freight: *Any commodity other than mail and passenger baggage transported by air.*

Freight-ton Mile: *A ton-mile is defined as one ton of freight shipped one mile.*

Frequent Flyer Programs: *A service in which airline customers accrue points corresponding to the distance flown on an airline. These points can be used for free air travel, increased benefits such as airport lounge access, or priority bookings, and other products or services.*

G
Gross Domestic Product (GDP): *GDP is a measure of the total market value of all final goods and services produced in our country during any quarter or year.*

H
Hub and Spoke: *Many airlines designate an airport as a hub through which they transit passengers from spoke (origin) to spoke (destination).*

Herfindahl–Hirschman Index (HHI): *A standard measure of industry concentration, expressed as the sum of the squares of the market shares of the firms in the same industry.*

Homogeneous Good: *The outputs of different firms are identical and indistinguishable.*

I
ICAO: *International Civil Aviation Organization.*

Inflation Rate: *The percentage change in the price level from one period to the next.*

Intransit Passengers: *Revenue passengers on board international flights that transit an airport for nontraffic purposes domestically.*

International Air Transport Association (IATA): *An international organization which regulates many of the world's scheduled airlines.*

J
The Joint Aviation Authorities (JAA): *JAA was an associated body of the European Civil Aviation Conference (ECAC) representing the civil aviation regulatory authorities of a number of European States implementing safety regulatory standards and procedures.*

L
Lease: *A lease is a contract granting use or occupation of property during a specified period in exchange for specified lease payments.*

Lessee: *A person who leases a property from its owner (lessor).*

Lessor: *The owner of an asset who grants another party to lease the asset.*

Leverage: *The use of debt to supplement investment.*

Load Factor: *Load factor is the ratio of revenue passenger miles over available seat-miles, representing the proportion of aircraft seating capacity that is actually sold and utilized.*

$$LF = \frac{RPM}{ASM} \times 100$$

M

MACRS: *Allows for depreciation towards the beginning of the life of the capital asset and allowing the tax deductible depreciation expense to be taken sooner.*

Major Airlines: *Airlines earning revenues of $1 billion or more annually in scheduled service.*

Marginal Cost: *The marginal cost is the cost of producing an additional output.*

Market Failure: *Market failure happens when free markets fail to deliver an efficient allocation of resources.*

Maximum Certificated Take-off Weight (MCTOW): *The maximum weight at which the pilot of the aircraft is allowed to attempt to takeoff, because of the aircraft's structural limitation.*

Maximum Zero Fuel: *The total weight if the airplane and all its contents, minus the total weight if the fuel on board.*

N

Narrow-body Aircraft: *An aircraft with a single aisle.*

Net Profit Margin: *Net profit after interest and taxes as a percent of operating revenues.*

Nonscheduled Service: *Revenue flights not operated as regular scheduled service, such as charter flights, and all nonrevenue flights incident to such flight.*

O

Operating Expenses: *Expenses incurred in the performance of air transportation, based on overall operating revenues and expenses.*

Operating Lease: *A short-term lease, for example, an aircraft which has an economic life of 30 years may be leased to an airline for four years on an operating lease.*

Operating Leverage: *A measure of the extent to which fixed assets are utilized in the business firm.*

Operating Profit Margin: *Operating profit (operating revenues minus operating expenses) as a percent of operating revenues.*

Operating Revenues: *Revenues from air transportation and related incidental services.*

P

Passenger Load Factor: *Load factor is the ratio of revenue passenger miles over available seat-miles, representing the proportion of aircraft seating capacity that is actually sold and utilized.*

Passenger Revenue per Available Seat Mile (PRASM): *The average revenue received by the airline per unit of capacity available for sale.*

$$PRASM = \frac{Revenue}{ASM}$$

Pax: *Abbreviation of passenger.*

Payload: *The part of an aircraft's load that generates revenue (freight and passengers).*

Penetration Pricing: *A pricing policy used to enter a new market, usually by setting a very low price.*

Price Discrimination: *Price discrimination happens when an airline charges different prices to different passengers for an identical service, for reasons other than costs.*

Price Ceiling: *A legally established maximum price.*

Price Floor: *A legally established minimum price.*

Primary Market: *The market for raising of new capital for the first time.*

Profitability Ratios: *They are a group of ratios that are used to assess the return on assets, sales, and invested capital.*

R

Return on Investment (ROI): *The percentage amount that is earned on a company's total capital, calculated by dividing the total capital into earnings before interest, taxes, or dividends.*

Revenue Passenger Enplanement: *The total number of revenue passengers boarding aircraft including origination, stopover, or connecting passengers.*

Revenue Passenger Mile (RPM): *RPM is computed by the summation of the products of the revenue aircraft miles flown by the number of revenue passengers carried on that route. RPM is a principal measure of an airline's turnover.*

Revenue per Available Seat Mile (RASM): *The revenue in cents received for each seat-mile offered. This is computed by dividing operating income by Available Seat Miles and is not limited to ticket sales revenue.*

Revenue Ton Mile (RTM): *One nonmetric ton of revenue traffic transported one mile.*

S

Secondary Market: *The markets for securities that have already been issued and traded among investors with no proceeds go to the company.*

SEC: *US Securities and Exchange Commission.*

SLOT: *The scheduled time of arrival or departure allocated to an aircraft movement on a specific date at an airport.*

Small Certificated Air Carrier: *An air carrier holding a certificate issued under section 401 of the Federal Aviation Act of 1958, as amended, that operates aircraft designed to have a maximum seating capacity of 60 seats or fewer or a maximum payload of 18,000 pounds or less.*

Sunk Costs: *These costs cannot be recovered.*

T

Tariff: *Import tax.*

Trade Deficit: *Imports of goods and services exceed exports of goods and services.*

Trade Surplus: *Exports of goods and services exceed imports of goods and service.*

Treasury Bill: *A short-term bond issued by the US Government.*

U

US Flag Carrier: *One of a class of air carriers holding a Certificate of Public Convenience and Necessity issued by the US Department of Transportation (DOT) and approved by the President, authorizing scheduled operations over specified routes between the United States and one or more foreign countries.*

V

Variable Costs: *A variable cost is a cost that changes in proportion to a change in a company's activity or business.*

Variance: *A measure of how much an economic variable varies across the mean.*

W

Wet Lease: *Refers to the leasing of an aircraft and includes the provision of crew and supporting services such as fuel, airport fees, and insurance.*

Wide-body Aircraft: *A commercial aircraft with two aisles.*

Y

Yield: *Average revenue per revenue passenger mile or revenue ton mile, expressed in cents per mile.*

List of Abbreviations

AFC	Average Fixed Cost
ANSP	Air Navigation Service Provider
ASK	Available Seat Kilometer
ACI	Airports Council International
ASEAN	Association of Southeast Asian Nations
AFTA	ASEAN Free Trade Area
AU	African Union
APEC	Asia-Pacific Economic Cooperation
ATM	Available Ton Mile
ATC	Average Total Cost
AOC	Air Operator's Certificate
AOPA	The Aircraft Owners and Pilots Association
ASL	Average Stage Length
ASM	Available Seat Miles
A4A	Airlines for America
ATC	Air Traffic Control
ATI	Air Transport Intelligence
AVC	Average Variable Cost
BAA	British Airport Authority
BEA	Bureau of Economic Analysis
BLS	Bureau of Labor Statistics
BLF	Breakeven Load Factor
BOP	Balance of Payments
BRIC	Brazil, Russia, India & China
BTS	Bureau of Transportation Statistics
CAA	Civil Aviation Authority
CAB	Civil Aeronautics Board
CACM	Central American Common Market
CAFTA	Central American Free Trade Agreement
CANSO	Civil Air Navigation Services Organization
CARICOM	Carribean Community
CASM	Cost per Available Seat Mile
CAST	Commercial Aviation Safety Team
CFIT	Controlled Flight Into Terrain
CPD	Cumulative Probability Distribution
CPI	Consumer Price Index
DOC	Direct Operating Cost
DOL	Degree of Operating Leverage
DOT	Department of Transportation
DRVSM	Domestic Reduced Vertical Separation Mimum
EASA	European Aviation Safety Agency

EBITA	Earnings Before Interest, Tax, Depreciation and Amortization
ECAC	European Civil Aviation Conference
ECJ	European Court of Justice
EMSR	Expected Marginal Seat Revenue
ETOPS	Extended Twin-engine Operations
EU	European Union
FAA	Federal Aviation Administration
FAR	Federal Aviation Regulation
FBO	Fixed-base Operator
FC	Fixed Cost
FFP	Frequent Flyer Programs
GAAP	Generally Accepted Accounting Principles
GAFTA	Greater Arab Free Trade Area
GAO	Government Accountability Office
GATT	General Agreement on Tariffs and Trades
GDP	Gross Domestic Product
GDS	Global Distribution Systems
GECAS	GE Commercial Aviation Services
GNP	Gross National Product
GSE	Government Sponsored Entity
HHI	Herfindahl-Hirschman Index
IATA	International Air Transport Association
IASB	International Accounting Standards Board
ICAO	International Civil Aviation Organization
IFR	Instrument Flight Rules
ILFC	International Lease Finance Corporation
ILS	Instrument Landing System
IMF	International Monetary Fund
IOC	Indirect Operating Cost
IRR	Internal Rate of Return
JFTA	Jordan Free Trade Agreement
JAA	The Joint Aviation Authorities
LCC	Low-cost Carrier
LOC	Loss of Control
MAD	Mean Absolute Deviation
MC	Marginal Cost
MR	Marginal Revenue
MP	Marginal Profit
MSE	Mean Squared Error
MTOW	Maximum Take Off Weight
NAFTA	North American Free Trade Agreement
NBFU	National Board of Fire Underwriters
NGO	Non-governmental Organization
NMAC	Near Midair Collision
NOC	Non Operating Cost
NTSB	National Transportation Board Safety
O&D	Origin and Destination
OAG	Official Airline Guide

OC	Operating Cost
OFT	Office of Fair Trading
OLS	Ordinary Least Square
OP	Operating Profit
OPEC	Organization of the Petroleum Exporting Countries
PPC	Production Possibility Curve
PPI	Producer Price Indexes
PPP	Purchasing Power Parity
QIZ	Qualifying Industrial Zones
R&D	Research and Development
RASM	Revenue per Available Seat Mile
RJ	Regional Jet
ROA	Return On Assets
ROE	Return On Equity
ROI	Return On Investment
RPK	Revenue Passenger Kilometer
RPM	Revenue Passenger Mile
RRPM	Revenue per Revenue Passenger Mile
RTA	Regional Trade Agreement
RTK	Revenue Ton Kilometers
RTM	Revenue Ton Miles
SAARC	South Asian Association for Regional Cooperation
SAFTA	South Asian Free Trade Area
SDR	Special Drawing Rights
SEC	Securities Exchange Commission
TC	Total Cost
TCAS	Traffic Collision Avoidance System
TFC	Total Fixed Cost
TOC	Total Operating Cost
TP SEP	Trans-Pacific Strategic Economic Partnership
VC	Variable Cost
VFR	Visual Flight Rules
VIF	Variance Inflation Factors
VLA	Very Large Aircraft
VOR	Very-high-frequency Omnidirectional Range
WTO	World Trade Organization

About the Authors

Bijan Vasigh is Professor of Economics and Finance in the College of Business at Embry-Riddle Aeronautical University, Daytona Beach, Florida. He received a PhD in Economics from the State University of New York, and has published many articles concerning the aviation industry. The articles were published in numerous academic journals including the *Journal of Economics and Finance, Journal of Transportation Management, National Aeronautics and Space Administration (NASA) Scientific and Technical Aerospace Reports, Transportation Quarterly, Airport Business, Journal of Business and Economics*, and *Journal of Travel Research*. He has been quoted in major newspapers  and magazines around the world. In 2006, his paper, "A Total Factor Productivity Based for Tactical Cluster Assessment: Empirical Investigation in the Airline Industry," was awarded the Dr Frank E. Sorenson Award for outstanding achievement of excellence in aviation research scholarship. Bijan is the author of *Aircraft Finance: Strategies for Managing Capital Costs in a Turbulent Industry* and *The Foundations of Airline Finance: Methodology and Practice*.

Dr Ken Fleming received his PhD from the University of California at San Diego, August 1978 with specialties in Econometrics and Public Finance. He received his MA in Economics from the University of California at Los Angeles in June 1969. A former military pilot with over 3,000 hours in 11 different aircraft, he holds an FAA Commercial Pilot Airplane Single and Multi-engine Instrument Rating. Dr Fleming was the leader of a group of 15 research analysts and computer programmers who actively participated in applied aviation research projects with Boeing, NASA, and the FAA and he has been the principal author or co-author of over 20 reports and articles including two textbooks during the past 15 years; these publications have dealt with all aspects of the economics of the aviation industry.

Thomas Tacker, PhD is Professor of Economics and the Chair of the Department of Economics, Finance and Information Systems at Embry-Riddle Aeronautical University. He received his doctoral degree from the University of North Carolina, Chapel Hill and his BS in Business Administration from Embry-Riddle. He won the National Federation of Independent Businesses Award for Most Outstanding Paper on Entrepreneurship and Public Policy and the prestigious Leavey Award for Excellence in Private Enterprise Education. Funded research includes NASA grants for study on: The Future Supply of SATS Pilots, The Economic Benefits of SATS in US Labor Markets, and The Future Supply of SATS Aircraft. He is author of numerous academic papers and conference proceedings publications as well as several publications in the popular press. Areas of research expertise include Airline Anti-trust Issues, Regulation, Privatization and Macroeconomic Policy.

Acknowledgements

The development and writing of any textbook is a long and trying experience—as all aspiring authors know! Our effort has been no exception. And, as with any other text, we have received a great deal of assistance and encouragement from a number of individuals whose efforts we would like to acknowledge.

Many graduate students from Embry-Riddle Aeronautical University helped in the preparation of this manuscript. We owe a special debt to the graduate assistants: Brian Sherman, Pascal Lawrence, Liam Mackay, Christian Vogel, James Cirino, Chris Weeden, and Duane Miner who participated in data collection, editing graphs, tables, and figures. This necessary preparation was often difficult and time-consuming work. Other student assistants who were very helpful in preparing this text were: Megan Praschak, Sean Ross, Charles Nipper, Brian Merzlak, Giby Abraham Jacob, Bryan Lange, Prince Nudze, Stephen P. Luxion, Isabella Salim, and Andrew Taylor. We would also like to thank the following members of Thomas Tacker's class, who were the first to use the text and made many helpful suggestions: James Cirino, Lavnish Budhiraja, Chih-Han Chang, Constantin Corneaunu, Tom Derosiers, Thomas Hilgers, Kevin Hyatt, Virginie Janneteau, Ryan Johnson, Mahmoud Khatib, Niles Klungboonkrong, Brian Lech, Vaida Maleskis, Mary Mankbadi, Nathan Monroe, Jae-Woo Park, Rushi Patel, Naef Saab, Worachat Sattayalekha, Paul Smith, Heli Trivedi, and Odartey Williams.

Ashgate Publishing has done an admirable job in being patient with the text and turning the raw manuscript into an actual textbook, with special thanks to Charlotte Parkins, Gillian Steadman, and Guy Loft for all their help finalizing the second edition. An additional thank you is also extended to VISIVO Design Studio for the book cover design for this second edition.

Thanks also go to many other people who have helped with the book, and especially Hossein Tavana. Many individuals have read all or part of the manuscript and made helpful suggestions. Among these are Tae Oum, Irwin Price, Martin Dresner, Zane Rowe, Barry Humphreys, and Darryl Jenkins.

We would also like to acknowledge the significant contribution of Professors Frank van der Zwan of the Delft University of Technology and G. Rod Erfani of Transylvania University who read the entire first printing of the manuscript very carefully and pointed out some typos and errors in a few of the formulas.

Foreword

This is the book for anyone who wants to know what drives the behavior in air transport markets, of customers, suppliers, the airlines themselves, and even the governments that regulate the industry. Bijan Vasigh, Ken Fleming, and Thomas Tacker showed, in the first edition of this book, that looking at the air transport industry with an economist's perspective gives the reader important insights into many of the issues that make aviation so fascinating. We are given the opportunity to see through the eyes of three distinguished economists, from one of the leading universities specializing in aviation. That opportunity will be well worth taking in this second edition.

Economics matters. The authors describe the "economics way of thinking;" understanding how people in various parts of the air transport market respond to incentives, in ways that economics can generally predict. This is not about forecasting the future, though they do have a chapter about the techniques for those brave enough to attempt this. It is about how the perishability of the airline product (empty seats are worthless once the plane takes off) and the volatility of consumer demand (with peaks during the day and occasional one-direction flows) have led to sophisticated pricing techniques. It is about how the delivery of the service is controlled by air traffic control companies who have little interaction or response to the airline customers, but how resulting issues like congestion impose costs just as real as financial payments. It is about how market structure is so important in driving the efficiency and behavior of companies, and how the labyrinthine rules governing airline services often hampers the normal operation of market forces.

This is a fascinating industry. There are not many that have expanded three to four times as much as the global economy over the past 40 years; where efficiency improvements have more than halved the cost, in real terms, of providing the product; where investment is constantly attracted; and yet where companies (the airlines) routinely destroy billions of dollars of shareholder value each year. Yet the wider economic benefits are clear. Air transport has both enabled and benefited from the globalization of the world economy. Connecting cities and countries around the world has brought tremendous value not just to air travelers and shippers but also to the wider economy, as connectivity has opened new markets and facilitated the benefits of specialization in the businesses that cluster in well-connected cities. Understanding the economic forces that drive this industry will give students both intellectual satisfaction and, hopefully, the wisdom required from future regulators and CEOs to make sure these economic benefits continue to flow.

The authors make good use of the economist's basic tool of supply and demand curves to illuminate what drives behavior throughout the industry, from customer demand, the behavior of infrastructure providers, to the impact of the structure of costs and the market. Analytical tools are also taken from public choice theory, to throw light on the behavior of governments, and from game theory, to give some insight into the difficult issue of why running an airline is usually so unprofitable. But the emphasis throughout the book is to use the economist's tool kit with simplicity and clarity. Complex mathematics and detailed footnotes are kept to a minimum. The analysis is leavened with a multitude of good examples and a lot of the institutional detail the reader will need to know to fully understand the issues. There is even a chapter arguing for a more explicit assessment of the costs and benefits of safety and security decisions. For anyone in the air transport industry, or intending to join it, this is a great textbook for understanding how economics drives so many of the important issues. I would also recommend this as a very good guide for economists from other fields, who want to understand what makes air transport special.

There are a lot of books about air transport. There are few that use the "economics way of thinking" to understand why the industry operates as it does today, as effectively as this book.

Brian Pearce
Chief Economist, International Air Transport Association, Geneva
Visiting Professor, Cranfield University

Preface to the Second Edition

The second edition features new coverage of macroeconomics for managers, and expanded analysis of modern revenue management and pricing decisions. The macroeconomic section covers basic trends in economic growth and the business cycle and includes a short overview of issues such as sovereign debt crises that are likely to impact the future of air travel and the economy overall. All material is extensively updated, particularly the numerical examples and industry data. Sections on market structure and anti-trust issues are comprehensively revised, reflecting the many recent mergers that have occurred since the publication of the first edition. Although the authors have preserved their basic pedagogical approach, a synthesis of solid economic principles and theories with aviation industry practices and institutional structures, the presentation of material is substantially refined. The authors' own experiences teaching from the text as well as suggestions from numerous readers have shaped revisions that make the new edition much easier to read and comprehend, without any sacrifice of the rigor and comprehensiveness that made the first edition a bestseller in its field. Instructors will find this modernized edition easier to use in class, and suitable to a wider variety of undergraduate or graduate course structures while industry practitioners and all readers will find this latest edition more intuitively organized and more user friendly.

While it is undoubtedly true that there are many books that cover different aspects of various economic problems, it is also true that there comes a time when there is a requirement for a textbook that brings together the disparate elements of analysis that are covered in other separate areas. For example, there are textbooks on labor economics, monetary economics, international economics, comparative economic systems, industrial organization, and numerous other specialized fields. It is our conviction that the time has come to bring together the numerous and informative articles and institutional developments that have characterized the field of airline economics in the previous two decades. While some might argue that this is too specialized an area in economics, we would contend that the unique nature of the economics of this industry make it particularly appropriate for a separate text. Included here, and as covered in the text, we would suggest the perishable nature of the product and the consequent elasticity of demand and pricing complications, the control of the method of delivering the service by a disinterested third party (namely, air traffic control), the presence of only two major suppliers of the means of providing the service, the unique dominance of this form of transportation for long-haul passenger traffic, the interesting and complicated financial arrangements that are used to provide the service, the existence of quasi-monopolistic entities to jointly deliver the service (airports) and last, but by no means least, the international legal aspects of the industry. All of these areas are covered in one place or another, but there is as yet no single text or article that brings them together in such a way that the critically important underlying economics of the industry is made clear to the interested reader.

Moreover, the air transport industry continues to undergo significant restructuring. From the early days of an industry regulated by governments to an increasingly privatized industry as it is today, the industry continues to evolve. After the period of government regulations and control, deregulation, begun first by the United States government in 1978, was followed by a series of continued liberalization movements of the European air transportation industry. Since the early period of national carriers was justified by many governments who desired to have absolute control over their airspace for military reasons, the natural result was a single state-owned carrier as the norm. Thus, the system evolved from government owned and operated carriers (flag carriers) to privatized companies that operated on the basis of economic principles and theories and were more concerned with performance. Under this ongoing restructuring, national carriers continue to be privatized, including the 1987 privatization of British Airways. Other fully privatized carriers are Lufthansa and Iberia. Another segment of the air transport industry that is experiencing continued privatization efforts is airports. Airport privatization began in 1987 with the sale of the British Airport Authority (BAA); since then, major airports around the world have been fully or partially privatized including Paris Charles de Gaulle (partial), Rome Fiumicino (fully), Bangkok (partial), and Beijing (partial). All of these developments are covered using basic economic theory and relevant examples and statistics from the industry itself.

Therefore, the underlying foundation of the text is the idea that the reader should be introduced to the economic way of thinking and approaching problems in aviation rather than the more traditional institutional and governmental regulatory approach. In the early chapters the reader is introduced to the elementary ideas of demand and supply and market equilibrium. This is followed by an in-depth presentation of costs and their key applicability to managerial decision-making. The basic economic principles are then applied to a unique analysis of the effect of air traffic control and the governmental ownership of airports on the industry. Following this, there is a thorough discussion of market structures and how they affect the industry. In particular, this section introduces the idea of contestability theory which appears to be particularly applicable to this industry. The international aspects of the industry and global alliances are then discussed in detail.

Chapters 10 through 13 are devoted to what might be called applications of the earlier theoretical chapters. There is an elementary overview chapter on the various types of forecasting that are prevalent in the industry. The next chapter ties together the basic principles of demand (that were covered earlier) in a somewhat more sophisticated presentation of the critically important topic of revenue management. Clear numerical examples are presented tying this mainstay of the industry to the theoretical idea of elasticity of demand. This discussion is followed by another unique chapter that is entirely devoted to the phenomenon of so-called low cost carriers. Finally, the text presents a decidedly non-conventional approach to the controversial topic of safety within the industry. That is, rather than the conventional safety at any cost approach (which is in reality not followed anyway), the text adopts a more balanced cost-benefit approach to this important topic.

As economists in a university that specializes in the aviation industry, our preferred approach is to apply economic principles to the industry, and here is the area where we see the unique need. As such, we deemed it appropriate for the second edition to add an introductory section on macroeconomic principles and therefore, we feel that this text will bring all of these areas together in one book. In summary and as discussed above, our approach will follow a more or less standardized format. That is, we first present the necessary economic principles that will be used to analyze the industry. We follow this with a discussion of institutional arrangements, particularly in the international area, that make the aviation industry a truly global enterprise. Finally, the last chapters of the text are devoted to practical applications and comparisons within the industry and conclude with a contextual perspective on macroeconomics. It is our hope that the text will appeal to interested readers within the industry as well as students who intend to enter the industry.

1

The Evolving Air Transport Industry

If you want to be a millionaire, start with a billion dollars and open an airline. Soon enough you will be a millionaire.

Sir Richard Branson, Founder, Virgin Atlantic Airways

As the comment above implies, in the last 30 years the airline industry's earnings have fluctuated wildly (mostly downward). As the industry fluctuated a series of consolidations and bankruptcies began to occur. New carriers such as JetBlue and AirTran in the US, EasyJet and Ryanair in Europe, Gol and Volaris in Latin America and a few others entered the industry, but many others such as Eastern, Pan Am, and Midway declared bankruptcy and ultimately ceased operations. Other carriers like Delta and Northwest, and United and Continental survived bankruptcy but merged post-bankruptcy, further consolidating the industry. In more recent times, a new airline paradigm called low-cost carriers (LCCs) were created to accommodate the economic changes that were happening. These LCCs have rapidly gained market share and currently hold approximately 24 percent of global capacity. In response to the LCCs, legacy carriers trimmed services and boosted ancillary means of revenue generation.

With relatively minimal profit margins, the financial condition of the aviation industry is highly dependent on the global economic conditions. During times of economic boom, profits soar and in times of distress carriers are forced to cut back capacity. The purpose of this chapter is to describe the evolution of the air transport industry including airlines and airports. The topics include the following:

- The Airline Industry
- Financial Condition of the Airline Industry
- Consolidation and Bankruptcies
- Factors Affecting World Air Traffic Growth
- Economic Impact of Air Transport Industry
  - Direct Impact
  - Indirect Impact
  - Induced Impact
  - Total Impact
- Outlook for the Air Transport Industry
- Summary
- Discussion Questions

THE AIRLINE INDUSTRY

It is estimated that airline travel will nearly double over the next 20 years. The number of passengers in the US will grow from 732 million in 2012 to 1.2 billion in 2032. Since the US Airline Deregulation Act of 1978, the US airline industry (and to a certain extent the global airline industry) has been characterized by volatility. This volatility produces airline bankruptcies, large layoffs or employee pay cuts, loss of shareholder wealth, and great uncertainty in the market. Periods of high revenues are followed by periods of economic drought. The most recent economic "troughs" followed the September 11 terrorist attacks in 2001 and the economic recessions of 2008 and 2009. Prior to deregulation, the airline industry was relatively stable with minimal losses and healthy profits; however, it was also clear that this state of affairs was due mainly to government regulation that virtually eliminated any meaningful competition between airlines and certainly prevented new competitors from entering the market. The biggest loser in all of this was the passenger who had to pay ticket prices that were set to cover average airline costs with no competitive discounts permitted. Therefore, while deregulation may seemingly have caused huge financial losses, it also reflected the fact that airlines were faced with their first bout of meaningful competition and some did not measure up. On the other hand, deregulation also opened up the opportunity for some airlines, such as Southwest Airlines and Ryanair, to post some of the greatest profits in the history of the industry. Figure 1.1 graphically displays this trend of volatile profitability for the global airline industry.

> "People Express is clearly the archetypical deregulation success story and the most spectacular of my babies. It is the case that makes me the proudest."
>
> Alfred Kahn, Professor of Political Economy, Cornell University

The major reason why the deregulation of the US airline industry had such a large impact on the global airline industry is that the North American airline industry has historically been the most dominant player in the global aviation industry. Table 1.1 shows that although the North American market still currently holds the distinction of being the largest market in terms of aircraft movements (34.8 percent), it has lost market share for passengers and cargo. In 2011, the European region surpassed North America in terms of passengers with 30.6 percent of global passenger market, with Asia-Pacific closing the gap with 25.4 percent share of the market. This growth has been a direct result of the explosive economic expansion in those regions, particularly China and India (expected to continue at 6.7 percent over the next 20 years).[1] Boeing and Airbus market forecasts both point out that the Asia-Pacific region is expected to surpass the North American market in the next 20 years in absolute terms of growth and market share with over half of the world's air traffic growth driven by travel to, from, or within the Asia-Pacific region.

It is interesting to note that in terms of international passengers, both the Asia-Pacific and Europe markets have far surpassed North America (Table 1.2). Europe's dominance in international air transportation is mainly a result of its historical ties to former colonial

---

1   Boeing Current Market Outlook 2011–2030.

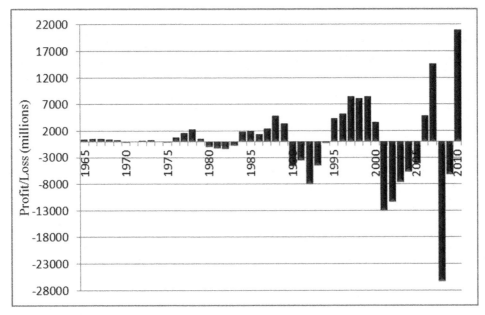

Figure 1.1   World Airline Operating Profits

Source: Compiled by Authors

Table 1.1   World Airlines Passenger and Cargo Traffic (2011)

	Movements	% Change	Passengers	% Change	Cargo	% Change
Africa	2,223,682	(1.0)	133,558,845	(5.9)	1,453,797	0.3
Asia-Pacific	10,636,150	5.2	1,254,317,321	5.7	30,207,926	(1.5)
Europe	17,965,635	3.4	1,508,974,011	7.1	17,376,134	1.2
Latin American	5,829,835	5.5	388,084,263	8.6	4,706,559	6.1
Middle East	1,558,943	2.9	178,862,214	8.4	5,603,482	1.6
North America	25,911,039	(0.7)	1,467,455,324	1.8	24,695,659	(0.7)
Total	**74,454,642**	**2.0**	**4,931,251,978**	**4.9**	**84,043,557**	**(0.1)**

Source: Airports Council International

countries and its relatively small geographical area. Because of this small area and significant government support for other surface transportation (mainly railroads), most of the domestic aviation industries in Europe are relatively small; therefore, European airlines survive on international travel. The growth in economies of Eastern Europe will also undoubtedly increase its share of international travel.

Another trend in the air transportation industry is the growth and expansion of the cargo industry. Table 1.1 and Table 1.2 quote cargo (in metric tons) for each region and for international cargo. The Asia-Pacific region is the dominant region for cargo (especially international cargo) as it currently holds a 40 percent share of the international cargo market, even with a 3 percent decline in 2011. Much of this growth is a result of China's burgeoning economy and the large and growing amount of exports that come from the region.

Table 1.2   International Passengers and Cargo Traffic (2011)

	Passengers	% Change	Cargo	% Change
Africa	81,372,791	(9.4)	1,300,233	5.9
Asia-Pacific	458,035,610	6.3	20,993,996	(3.0)
Europe	1,091,153,140	8.0	13,218,501	1.8
Latin American	109,557,990	7.1	3,216,553	6.2
Middle East	156,149,157	8.6	5,391,681	1.8
North America	203,780,170	4.2	9,307,266	(0.1)
Total	2,100,048,858	6.4	53,428,230	(0.1)

Source: Airports Council International

Another way of looking at the global distribution of air transportation is the number of in-service aircraft. Table 1.3 divides aircraft into two categories based on the aircraft's age: new and middle, or old. While the North American market contains the highest number of commercial aircraft, it also contains a relatively high percentage of mid-generation and older aircraft (41 percent). For aircraft manufacturers, this is attractive as many of these older aircraft will have to be replaced in the coming years. Europe, on the other hand, does not have very many mid-generation to older aircraft (25 percent) as this is probably a result of the stringent noise regulations implemented by the European Community that banned many older generation aircraft. Finally, some correlation can be made between the number of new aircraft and the region's economic growth. For example, Africa has the highest percentages of older aircraft (44 percent) while more prosperous regions like Europe and Asia have low ratios (25 percent and 21 percent respectively).

Table 1.3   Aircraft in Service by Region (2010)

	New	Mid	Old
North America	59%		41%
Europe	75%		25%
Asia	79%		21%
Latin America	67%		33%
Africa	56%		44%
Middle East	72%		28%
World	69%	27%	4%

Source: Compiled by authors using Boeing 2011 market forecast

A final method to analyze the composition of the air transport industry is to analyze traffic data on an airport-by-airport basis. Tables 1.4 and 1.5 provide a list of the top 15 airports in terms of total passengers, total international passengers, and total cargo volume. The rankings of the top airports mirror the distribution of passengers by regions as evidenced by the two largest airports in terms of passengers. Since North America is the second largest market in terms of passengers, it is not surprising that Atlanta is the top airport. Conversely, since Europe is the top region for international passengers, it should come as no surprise that the top two airports, in terms of international passengers, are in Europe. Also, many of the airports on the international passenger traffic list are airports located in countries that have small or non-existent domestic air travel markets. Similar to the airline market, the Asia-Pacific region is experiencing tremendous growth in airports. For example, Beijing Capital International Airport has seen a more than 50 percent increase in passenger traffic since 2006. Several Asian airports that were already on the list saw phenomenal increases in passenger traffic from 2010 to 2011; most notably, Jakarta saw a 19.3 percent increase in passenger traffic.

Much of the airport development in the Asia-Pacific region is still taking place in China and India, though there are notable projects in Vietnam, the Philippines, and Indonesia. The division of airports by region in terms of cargo volume is not as clear cut, but both North America and Asia-Pacific airports are well represented in the top 15.

Table 1.4   Total Passenger Traffic

Total Passenger Traffic				International Passenger Traffic			
Rank	Airport	Total	% Chg	Rank	Airport	Total	% Chg
1	Atlanta (ATL)	92,365,860	3.4	1	London (LHR)	64,687,737	6.20
2	Beijing (PEK)	77,403,668	4.7	2	Paris (CDG)	55,674,880	4.80
3	London (LHR)	69,433,565	5.4	3	Hong Kong (HKG)	52,749,262	6.00
4	Chicago (ORD)	66,561,023	(0.5)	4	Dubai (DXB)	50,192,013	8.40
5	Tokyo (HND)	62,263,025	(2.9)	5	Amsterdam (AMS)	49,680,283	10.10
6	Los Angeles (LAX)	61,848,449	4.8	6	Frankfurt (FRA)	49,477,184	6.90
7	Paris (CDG)	60,970,551	4.8	7	Singapore (SIN)	45,429,263	11.00
8	Dallas/Fort Worth (DFW)	57,806,152	1.6	8	Bangkok (BKK)	35,009,002	11.40
9	Frankfurt (FRA)	56,436,255	6.5	9	Seoul-Incheon (ICN)	34,537,845	4.80
10	Hong Kong (HKG)	53,314,213	5.9	10	Madrid (MAD)	32,449,857	4.70
11	Denver (DEN)	52,699,298	0.9	11	London (LGW)	29,923,391	7.50
12	Jakarta (CGK)	52,446,618	19.3	12	Munich (MUC)	27,879,045	10.10
13	Dubai (DXB)	50,977,960	8.1	13	Tokyo (NRT)	26,331,010	(18.10)
14	Amsterdam (AMS)	49,754,910	10.1	14	Kuala Lumpur (KUL)	25,915,723	10.70
15	Madrid (MAD)	49,644,302	(0.4)	15	Rome (FCO)	24,448,925	5.00

Source: Compiled by authors using Airports Council International
Data is on the 12 months preceding and including December 2011

Table 1.5   Cargo Traffic at Major International Airports

Rank	Airport	Cargo (metric tonnes)	% Change
1	Hong Kong (HKG)	3,968,397	(4.7)
2	Memphis (MEM)	3,916,535	—
3	Shanghai (PVG)	3,103,030	(4.3)
4	Anchorage (ANC)	2,625,201	0.5
5	Incheon (ICN)	2,539,222	(5.4)
6	Dubai (DXB)	2,269,768	—
7	Frankfurt (FRA)	2,215,181	(2.6)
8	Louisville (SDF)	2,187,766	1.0
9	Paris (CDG)	2,095,773	(4.0)
10	Tokyo (NRT)	1,945,110	(10.3)
11	Singapore (SIN)	1,898,850	3.1
12	Miami (MIA)	1,840,231	0.2
13	Los Angeles (LAX)	1,688,351	(7.2)
14	Beijing (PEK)	1,668,751	7.7
15	Taipei (TPE)	1,627,461	(7.9)

Source: Compiled by authors using Airports Council International
Data is on the 12 months preceding and including December 2011

Hong Kong has replaced Memphis as the busiest cargo airport in the world. In the Asia-Pacific region, the large amount of export trade has spurred cargo growth, especially in Hong Kong and Shanghai (numbers 1 and 3 respectively).

FINANCIAL CONDITION OF THE AIRLINE INDUSTRY

> I don't think JetBlue has a better chance of being profitable than 100 other predecessors with new airplanes, new employees, low fares, all touchy-feely… all of them are losers. Most of these guys are smoking ragweed.

> Gordon Bethune, CEO Continental Airlines, June 2002

The US Airline Deregulation Act of 1978 dramatically changed the global financial condition of the airline industry. Soon after the US, other countries also began to deregulate their own industries. And, as mentioned earlier, prior to 1978, the industry was relatively stable based mainly on the government's enforcement of non-competitive regulation and pricing. In the post-deregulation era, the industry took on the more cyclical nature of a competitive industry, where periods of robust financial profitability could be followed by periods of severe economic distress. Like other competitive industries, the financial condition of the airline industry is highly related to economic growth, so it is therefore not surprising that the airline industry suffered when the economy stalled.

In the early 1980s, shortly after US deregulation, the airline industry suffered a minor crisis as the economy slowed and competition soared. More specifically, the US domestic industry experienced overcapacity, as the many new airlines that were formed out of deregulation either went bankrupt or merged with other carriers. The result was four years of global net losses for the industry, largely based on the situation in the US. A similar situation occurred in the early 1990s as the economy once again experienced a downturn, but this downturn was aggravated by political uncertainty from the first Gulf War and increased fuel costs.

While the early parts of each decade following deregulation have proved to be troublesome for the airline industry, the industry has recovered to post record short-run profits in the late 1980s and again in the late 1990s. This was partly as a result of the overall improvement in the global economy, but financial distress and competition also caused airlines to be more innovative and conscious of controlling costs. Tools such as revenue management and frequent flier programs were created and developed during these periods. Additionally, technological innovations allowed the airlines to improve their profit margins. For example, simpler cockpit design has been able to reduce the number of flight crewmembers, better engine design has reduced the number of engines required to fly long distances, and fuel costs have been reduced with more fuel-efficient engines. All of these technologies have enabled airlines to reduce their costs and/or increase revenue. A more recent technological innovation has been e-ticketing, which allows airlines to reduce their ticket distribution costs.

The post-deregulation airline profitability cycle continued into the new century with the global industry experiencing its worst downturn in the history of commercial aviation. While the September 11, 2001 terrorist attacks were the proximate cause of the global airline industry's financial problems, the root cause of the problem was a slowing economy that reduced passenger yields. Added to this were rising jet fuel costs, increased airline operating costs stemming from overcapacity in domestic markets, and increased security costs at commercial airports. As a point in fact, the airline industry was in trouble before the September 11 disaster, with many airlines losing money and with no significant initiatives to reduce costs and increase productivity. The result of this situation was net losses for the entire airline industry from 2001 to 2005 until an economic resurgence returned the industry to profitability in 2006 and 2007. These profits were short lived however, as 2008 brought about arguably the worst year in the history of commercial aviation, with a $26.1 billion net loss as shown in Table 1.6. Again, an overall economic recovery has somewhat stabilized the industry's return to profitability in 2010 and 2011.

However, the road to recovery has been slow for the airline industry as a result of political instability (in various parts of the world), rising fuel prices, and persistent competition between network and LCCs. This situation has been most evident in the North American market where the high-profile bankruptcies of US Airways, United, Delta, Northwest, and now American have highlighted the increasing effects of fierce competition from lower-cost airlines and the bloated cost structures of the more traditional airlines. Globally, we see some different patterns. Table 1.7 presents the international air transport market's net profits based on the different geographical regions. After a year of severe losses across the board, except for Asia-Pacific and Latin America carriers in 2009, the industry recovered in 2010. The Asian-Pacific carriers also enjoyed the highest level of profitability followed by North America for 2011.

Table 1.6   Scheduled Airlines' Financial Performance

	2003	2004	2005	2006	2007	2008	2009	2010	2011F
REVENUES, $ billion	322	379	413	465	510	570	476	547	596
Passenger	249	294	323	365	399	444	374	425	469
Cargo	40	47	48	53	59	63	48	66	66
Other	33	38	42	47	52	63	54	56	61
Sched. passengers, millions	1,849	2,064	2,211	2,325	2,518	2,507	2,479	2,681	2,840
Freight tonnes, millions	38	41	42	45	47	45	41	46	46
Passenger yield, %	2.4	2.6	2.7	7.8	2.7	9.5	(14.0)	6.1	4.0
Cargo yield %	2.0	7.4	2.4	5.9	5.5	7.4	(14.2)	15.0	-
EXPENSES, $ billion	323	376	409	450	490	571	474	525	583
Fuel	44	65	91	117	135	189	125	139	178
Crude oil price, Brent, $/b	29	38	55	65	73	99	62	79	112
Jet kerosene price, $/b	35	50	71	82	90	127	71	91	128
Non-fuel	279	311	318	333	355	382	349	386	405
Cents per ATK (non-fuel cost)	39	40	39	39	39	42	40	42	41
Breakeven weight load factor, %	61.1	61.9	62.0	61.2	60.9	63.2	62.3	63.1	63.1
Weight load factor achieved, %	60.8	62.5	62.6	63.3	63.4	63.1	62.6	65.7	64.5
Passenger load factor achieved, %	71.5	73.4	74.9	76.1	77.7	76.0	76.0	78.4	78.2
OPERATING PROFIT, $ billion	(1.4)	3.3	4.4	15.0	19.9	(1.1)	1.9	21.7	13.2
% margin	(0.4)	0.9	1.1	3.2	3.9	(0.2)	0.4	4.0	2.2
NET PROFIT, $ billion	(7.5)	(5.6)	(4.1)	5.0	14.7	(26.1)	(4.6)	15.8	6.9
% margin	(2.3)	(1.5)	(1.0)	1.1	2.9	(4.6)	(1.0)	2.9	1.2

Source: Compiled by authors from ICAO and IATA airline financial data

Table 1.7   Financial Performance of Airline Industry

	Industry Net Profits: US$ Billion		
	2009	**2010**	**2011**
Global	(4.6)	15.8	6.9
North America	(2.7)	4.1	2.0
Europe	(4.3)	1.9	1.0
Asia-Pacific	2.6	8.0	3.3
Middle East	(0.6)	0.9	0.4
Latin America	0.5	0.9	0.2
Africa	(0.1)	0.1	-

Source: Compiled by the authors using IATA, 2012

There have been some bright spots in the industry as bankruptcy protection has enabled several carriers to restructure their costs and receive wage concessions from labor groups. Moreover, the overcapacity issue has been addressed with carriers not only reducing capacity as a whole, but also shifting capacity to international markets that are less competitive. Several LCCs have also remained successful and profitable by continuing to expand while keeping costs relatively constant. Innovations such as e-ticketing and fleet rationalization have been instrumental in helping airlines achieve cost reductions, and these reductions narrowed the cost gap between network airlines and LCCs. Figure 1.2 provides a comparison between the network carriers and LCCs in terms of Cost per Available Seat Mile (CASM).

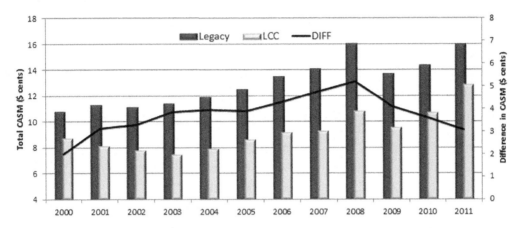

Figure 1.2   Cost Comparisons between Low-Cost and Legacy Carriers

Source: Compiled by the authors using Form 41

Among network carriers, the most profitable airline was JAL. This highlights a trend in the global airline industry where the cargo industry is thriving because of the increased globalization of the business marketplace. In terms of operating profit, network carriers did well especially in Europe and Asia Pacific as shown in Table 1.8. Also, LCCs, all from different continents, ranked among the highest commercial airlines based on operating profit margin (Table 1.9), and net profit margin (Table 1.10).

The global airline industry is well on the road to recovery with the International Air Transport Association (IATA) forecasting a global net profit in 2012 of $3 billion. This global profit is largely spurred by a forecast that Asia-Pacific carriers are expected to post large profits along with North American carriers. European carriers are expected to make net losses as Europe's sovereign debt crisis remains a threat to earnings. Other dangers to profitability are ever present, with fuel costs constituting the largest threat. In fact, for many airlines, fuel costs are now larger than labor costs, and airlines have few options in dealing with fuel costs. Moreover, since higher fuel costs affect all airlines in a somewhat similar manner, there is little or no competitive advantage to be gained in this area. Therefore, as this extra cost is passed on to the consumer, the relative price structure should remain proportionately the same. In this case, the ultimate question of

Table 1.8   Top 25 Airlines by Operating Profitability (2009 and 2010)

	2009			2010	
Rank	Airline	USD million	Rank	Airline	USD million
1	Emirates Airline	971	1	JAL Group	2,274
2	Air China	834	2	Delta Air Lines	2,217
3	FedEx Express	794	3	United Continental Holdings	1,818
4	Cathay Pacific	576	4	Air China	1,658
5	Ryanair	567	5	Lufthansa Group	1,642
6	Turkish Airlines (THY)	470	6	Emirates Airline	1,482
7	LAN	436	7	Cathay Pacific	1,420
8	Lufthansa Group	386	8	FedEx Express	1,127
9	TAM	316	9	SIA Group	1,007
10	JetBlue Airways	285	10	Korean Air	989
11	Aeroflot Group	278	11	Southwest Airlines	988
12	China Eastern Airlines	276	12	United	976
13	Republic Airways	271	13	China Southern Airlines	954
14	Alaska Air Group	267	14	China Eastern Airlines	864
15	AirAsia Berhad	263	15	ANA Group	818
16	Southwest Airlines	262	16	US Airways Group	781
17	UPS Airlines	259	17	Thai Airways Int'l.	757
18	South African Airways	258	18	Ryanair	689
19	GOL Linhas Aereas	248	19	Continental Airlines	660
20	Copa Holdings	240	20	LAN	623
21	China Southern Airlines	218	21	TAM	609
22	SkyWest Inc.	212	22	Asiana Airlines	541
23	WestJet	211	23	British Airways	529
24	AirTran Airways	177	24	China Airlines	508
25	Qantas Group	174	25	Aeroflot Group	499

Source: ATW World Airline Report

Table 1.9   Top 25 Airlines by Operating Profit Margin (2009 and 2010)

	2009			2010	
Rank	Airline	Percent	Rank	Airline	Percent
1	Mahan Air	35.6%	1	Mahan Air	35.2%
2	AirAsia Berhad	25.5%	2	AirAsia Berhad	33.9%
3	Allegiant Air	21.9%	3	Midex Airlines	29.9%
4	Middle East Airlines	21.4%	4	Shuttle America	26.2%
5	Abu Dhabi Aviation	20.7%	5	Abu Dhabi Aviation	25.1%
6	SA Express	19.8%	6	Astar Air Cargo	24.7%
7	Copa Holdings	19.1%	7	Cebu Pacific Air	22.2%
8	Vensecar Int'l.	17.7%	8	Copa Holdings	20.5%
9	Holdings	16.5%	9	Lynden Air Cargo	20.2%
10	Spirit Airlines	15.9%	10	Alpine Air Express	19.8%
11	Royal Jordanian	15.7%	11	Skymark Airlines	19.3%
12	Pullmantur Air	15.2%	12	Sabah Air Aviation	19.0%
13	Atlas Air W.W. Holdings	14.1%	13	Deraya Air Taxi	19.0%
14	Air Arabia	14.0%	14	Ryan Int'l Airlines	18.3%
15	Cebu Pacific Air	13.6%	15	Atlas Air W.W. Holdings	17.0%
16	Ryanair	13.5%	16	Vensecar Int'l.	16.7%
17	S7 Airlines	13.2%	17	Allegiant Air	15.8%
18	Alpine Air Express	12.5%	18	Thai AirAsia	15.6%
19	North American Airlines	12.4%	19	Thomas Cook Belgium	15.5%
20	Mongolian Airlines	12.3%	20	Gulf & Caribbean Air	14.9%
21	REX—Regional Express	12.3%	21	USA Jet Airlines	14.7%
22	LAN	11.9%	22	Middle East Airlines	14.5%
23	Vueling Airlines	11.9%	23	Amsterdam Airlines	14.3%
24	Amsterdam Airlines	11.8%	24	GoJet	14.2%
25	LAM Mozambique	11.2%	25	JAL Group	13.8%

Source: ATW World Airline Report: World Airline Financial Results 2010

Table 1.10   Top 25 Airlines by Net Profit Margin (2009 and 2010)

	2009			2010	
Rank	Airline	Percent	Rank	Airline	Percent
1	Air Arabia	22.9%	1	AirAsia Berhad	31.4%
2	Skymark Airlines	19.9%	2	Cebu Pacific Air	23.8%
3	Copa Holdings	19.8%	3	Thai AirAsia	23.0%
4	Middle East Airlines	19.4%	4	IndiGo Airlines	20.6%
5	Vensecar Int'l.	17.1%	5	Lynden Air Cargo	19.1%
6	AirAsia Berhad	15.9%	6	Ryan Int'l Airlines	18.3%
7	GOL Linhas Aereas	14.8%	7	Indonesia AirAsia	17.2%
8	Cebu Pacific Air	14.0%	8	Copa Holdings	17.0%
9	Allegiant Air	13.7%	9	Cathay Pacific	15.9%
10	TAM	12.8%	10	Nok Air	15.6%
11	SA Express	12.3%	11	Thomas Cook Belgium	15.5%
12	AirAsia X	12.1%	12	SA Express	15.0%
13	Spirit Airlines	12.0%	13	Air China	15.0%
14	Nok Air	11.6%	14	Air Arabia	14.9%
15	Ethiopian Airlines	11.0%	15	Xiamen Airlines	14.5%
16	Ryanair	10.2%	16	Sabah Air Aviation	14.3%
17	REX—Regional Express	10.0%	17	Astar Air Cargo	14.3%
18	Hawaiian Holdings	9.9%	18	Hainan Airlines	13.9%
19	Abu Dhabi Aviation	9.6%	19	Abu Dhabi Aviation	13.9%
20	Air China	9.3%	20	Juneyao Airlines	13.9%
21	Tyrolean Airways	9.1%	21	Middle East Airlines	13.4%
22	Air Astana	8.6%	22	Omni Air Int'l.	13.3%
23	Nouvelair Tunisie	8.4%	23	Kalitta Air	13.3%
24	Emirates Airline	8.1%	24	Everts Air Cargo	12.4%
25	Turkish Airlines (THY)	7.9%	25	Centurion Air Cargo	12.3%

Source: ATW World Airline Report: World Airline Financial Results 2011

profitability depends to a large extent on the elasticity of demand for the product and the cost containment ability of the airline's management.

CONSOLIDATION AND BANKRUPTCIES

The airline industry has been affected by economic recession, rising fuel costs, political uncertainty, and stiff competition. These factors have caused many major carriers, such as Eastern Airlines, Pan American, and Piedmont into liquidation and American Airlines, US Airways, United Airlines, Delta Air Lines, and Northwest

Airlines into bankruptcy protection. The period immediately following deregulation (1980s) saw one of the most turbulent periods for commercial aviation in the US and the greatest rate of airline bankruptcies. Since 1990, more than 189 airline bankruptcy filings have occurred. In more recent times, Delta and Northwest filed in 2005 and later merged, shortly followed by United and Continental in 2010.

> "If you look at the history of mergers, the assumption was that you couldn't do them successfully. Everybody had come to the conclusion that these things are too big, too complex and too unwieldy to manage."
>
> Richard Anderson, Delta's Chief Executive

Table 1.11 displays the market share for various US carriers for the domestic market. In 2000, Southwest Airlines passed Delta Air Lines to become the largest domestic carrier (in terms of passengers flown) in the US, and it has maintained its ranking through 2011. This highlights the fact that LCCs are capturing more of the domestic market share while legacy network carriers are losing theirs. This is mainly due to the LCCs' continual expansion and the advantages they possess because of their lower cost structure. Another major trend (in terms of market share) is the emergence of regional carriers. In 1998 Expressjet, American Eagle, and Skywest had less than 1 percent combined market share, yet in early 2006 they had acquired 7.5 percent of the total US domestic market. In this case, the reason was the increased use of regional jets by legacy carriers to open up new markets and to combat LCCs.

Table 1.11   US Airline Industry Domestic Market Share

Airlines	2004	2005	2006	2007	2008	2009	2010	2011
Southwest Airlines	16.9%	17.5%	18.8%	19.8%	20.7%	21.5%	21.5%	21.6%
American Airlines	11.3%	11.5%	11.2%	11.0%	10.9%	10.1%	9.8%	9.5%
Delta Air Lines	12.5%	11.5%	9.0%	8.1%	8.0%	8.0%	12.8%	12.9%
United Airlines	9.7%	8.7%	8.3%	8.1%	7.7%	7.0%	6.4%	5.9%
US Airways	6.9%	6.5%	5.3%	5.6%	7.5%	7.2%	7.2%	6.9%
Northwest Airlines	7.3%	7.3%	7.0%	6.9%	5.6%	5.1%	0.0%	0.0%
Continental Airlines	5.1%	5.1%	5.5%	6.3%	5.8%	5.8%	4.7%	4.7%
JetBlue	2.6%	3.0%	3.6%	4.0%	4.3%	4.4%	4.6%	4.9%
AirTran Airways	2.4%	2.7%	3.3%	4.0%	4.4%	4.5%	4.5%	4.4%
Alaska Airlines	2.6%	2.6%	2.5%	2.4%	2.5%	2.5%	2.5%	2.7%
American Eagle	2.2%	2.5%	2.7%	2.6%	2.4%	2.3%	2.3%	2.4%
Skywest Airlines	0.0%	0.6%	2.6%	2.7%	2.8%	3.0%	3.3%	3.2%
Atlantic Southeast Airline	1.5%	1.7%	1.5%	1.5%	1.6%	1.7%	1.8%	1.9%
Frontier Airlines	1.2%	1.2%	1.5%	1.6%	1.6%	1.6%	1.6%	1.8%
Industry Herfindahl Index	852.0	833.1	807.5	836.8	869.8	873.2	930.3	926.5

Source: Compiled by authors using OAG Data

Historically, mergers rapidly increased following deregulation in 1978, and for the ten years following deregulation, there were 51 airline mergers and acquisitions (Dempsey, 1990). The result of these mergers and acquisitions was the creation of six legacy carriers from the 15 independent carriers that had close to 80 percent US market share in 1987 (Dempsey, 1990).

While the number of mergers reduced during the 1990s, critics argue that most mergers still were part of well-planned strategies to lessen competition in various markets. As a result, starting in 1985 the Department of Transportation (DOT) assumed approval authority for all airline mergers. To approve the merger, the DOT must now balance the consumer benefits resulting from mergers against the possibly negative effects of increasing concentration (Dempsey, 1990). On the other hand, the extraordinary financial problems of legacy carriers suggest that reductions in capacity, whether through mergers or alliances, may be inevitable. Some economists argue that less intense competition, through consolidation and coordination, can actually benefit consumers by allowing airlines to build more efficient networks with greater economies of scale, scope, and density. Figure 1.3 provides a framework of the major airline mergers that have occurred in the US since deregulation.

The industry Herfindahl-Hirschman index (HHI) is a measure of US market consolidation. As Table 1.11 shows, since 1998 the industry has become less consolidated.[2] This spreading out of competition usually equates to lower fares and increased service. In fact, an American Express travel survey has shown that average US domestic airfares have steadily declined since 2000 (Amex, 2006). While mergers and acquisitions have slowed in the US, the relatively low HHI and the poor financial condition of the legacy carriers indicate that there is still the potential for additional mergers and acquisitions within the industry. Moreover, airline mergers are not limited to the US and have played a large part in international aviation. A few recent international mergers include:

- Air France and KLM
- Cathay Pacific Airways acquired full ownership over Dragonair
- Cimber Air acquired bankrupt Sterling
- Lufthansa and Swiss Air
- Air Canada acquiring Canadian
- Japan Airlines purchase of Japan Air System
- British Airways and Iberia.

Historically, mergers have not been overly successful in the aviation industry. Many mergers do not obtain the envisioned benefits, and one-time merger costs such as aircraft painting and IT harmonization end up being far more costly than planned. Airline mergers also have difficulty in dealing with labor groups, especially with regard to issues such as merging seniority lists. Corporate culture can also be a much underestimated barrier to successful mergers as different companies' cultures may impede merger success. Finally, one of the greatest challenges a merger faces is managing multiple and powerful stakeholders; these can include but are not limited to politicians, regulators, labor leaders, and consumers (McKinsey & Company, 2001). Many of these stakeholders are suspicious of the mergers because they fear lessened competition and increased travel prices (McKinsey & Company, 2001). Many potential mergers have been thwarted by

---

2   For more information on HHI see Chapter 9.

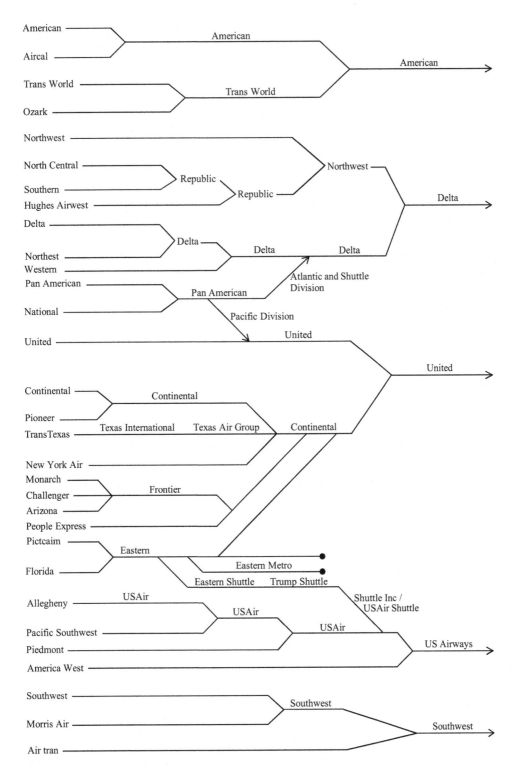

Figure 1.3   Evolution of US Airline Industry

regulators, and one of the key measures regulators use in analyzing potential mergers is the planned mergers' effect on the HHI.

However, there are, as mentioned, potential benefits from airline mergers. McKinsey & Company estimate that a merger of two mid-sized carriers could unlock synergies in excess of 7 percent (McKinsey & Company, 2001). The major benefit of mergers is cost rationalization. Since the airline industry exhibits large economies of scale, merged airlines are able to spread their high fixed costs over a greater network. Additionally, the new merged carrier can increase its bargaining power with key suppliers and merge such functions as parts inventories, back office functions, and sales forces (McKinsey & Company, 2001).

Another major benefit of mergers is network harmonization. Network harmonization can include a variety of things, and using the last analysis as an example, the merged airline's route network is greater than that of the individual airlines. A good example of this is the America West/US Airways merger. America West was predominantly a west-coast airline while US Airways was primarily an east-coast airline, but the merged airline had a strong route network on both coasts. Without a merger, both carriers would have had a difficult time increasing their presence on the opposite coast. The economies of scope that resulted from the merger allowed US Airways to widen its customer base and strengthen its market power throughout the US.

Another way to look at consolidation in the domestic US industry is to look at it from the airport level. After deregulation, the major carriers adopted a hub-and-spoke system which funneled passengers through a few airports (Dempsey, 1990). This in turn led to some carriers holding dominant positions at certain hub airports throughout the US. Table 1.12 depicts the consolidation of carriers by enplanements and operating carrier at the ten largest airports in the US:

The general trend in airport consolidation from 2004 through 2011 is one where the largest carrier has become less dominant. This has occurred for all the airports presented in Table 1.12, except for Dallas and Las Vegas. In Dallas, Delta's withdrawal left American Airlines as the only major airline still operating, while simultaneously leaving Dallas as the most consolidated airport of the top ten domestic US airports between 2004 and 2011.

The general reduction in consolidation at US airports can largely be attributed to two factors: 1) increased competition, particularly from LCCs; and 2) major carriers pushing more flying to regional affiliates. LCCs such as Frontier and AirTran have situated themselves in the dominant hubs of Denver and Atlanta, and have been successful at taking away market share in those airports. With the emergence of regional jets, major carriers have been pushing capacity toward regional carriers in an effort to reduce costs. As an indication of this trend, in May of 2007, Bombardier Aerospace introduced the next generation versions of its CRJ700, CRJ900, and CRJ1000 regional jets. These new CRJ NextGen aircraft have featured significant operating cost improvements and the increased use of composite materials.

Since the Form41 data used in Table 1.12 breaks data down by operating carriers, regional carriers are treated separately. For example, ExpressJet operations at Houston are separate, even though the flights are marketed by Continental. This could potentially distort the level of consolidation at airports with large regional carrier presence.

As shown in Table 1.12, the HHI indicates a decline in consolidation for the US domestic airline industry when consolidation is analyzed on an airport basis. However, it is important to remember that the level of consolidation at major US airports is much greater than the level of consolidation of the airline industry. The least consolidated US airport in the list, Los Angeles, still has a HHI score well above the airline industry's level of consolidation.

Table 1.12   Major US Airports Concentration (with Enplanements by Operating Carrier)

		2004	2005	2006	2007	2008	2009	2010	2011
Atlanta (ATL)	Largest Carrier %	68.4%	66.8%	60.0%	56.8%	56.7%	56.1%	61.1%	65.6%
	Largest Carrier	DL	DL	DL	DL	DL	DL	DL	DL
	Herfindahl Index	4916	4794	4063	3771	3764	3707	4196	4682
Chicago (ORD)	Largest Carrier %	40.7%	36.5%	36.4%	36.2%	36.4%	34.2%	31.0%	28.8%
	Largest Carrier	UA	UA	UA	UA	UA	UA	UA	UA
	Herfindahl Index	2716	2421	2352	2339	2320	2091	1847	1686
Dallas (DFW)	Largest Carrier %	65.8%	73.8%	73.8%	74.1%	74.6%	74.6%	74.4%	73.7%
	Largest Carrier	AA	AA	AA	AA	AA	AA	AA	AA
	Herfindahl Index	4521	5607	5612	5665	5730	5708	5689	5566
Denver (DEN)	Largest Carrier %	48.8%	44.5%	44.0%	41.6%	38.0%	32.8%	28.1%	24.7%
	Largest Carrier	UA	UA	UA	UA	UA	UA	UA	UA
	Herfindahl Index	2729	2432	2468	2340	2140	1838	1673	1596
Detroit (DTW)	Largest Carrier %	67.4%	64.6%	63.7%	62.4%	58.4%	52.2%	51.5%	47.5%
	Largest Carrier	NW	NW	NW	NW	NW	NW	DL	DL
	Herfindahl Index	4684	4338	4220	4086	3640	3028	2957	2596
Houston (IAH)	Largest Carrier %	66.7%	64.6%	63.7%	65.1%	65.6%	65.1%	64.1%	64.0%
	Largest Carrier	CO	CO	CO	CO	CO	CO	CO	CO
	Herfindahl Index	4855	4711	4715	4653	4695	4712	4655	4469
Las Vegas (LAS)	Largest Carrier %	33.9%	33.7%	36.5%	37.1%	38.8%	41.5%	43.0%	42.1%
	Largest Carrier	WN	WN	WN	WN	WN	WN	WN	WN
	Herfindahl Index	1705	1701	1835	1753	1982	2097	2199	2110
Los Angeles (LAX)	Largest Carrier %	21.9%	20.5%	21.1%	19.4%	18.3%	17.6%	17.1%	15.7%
	Largest Carrier	UA	UA	UA	UA	UA	UA	UA	UA
	Herfindahl Index	1268	1290	1303	1214	1191	1180	1226	1148
Minneapolis (MSP)	Largest Carrier %	69.4%	67.1%	67.4%	67.2%	62.0%	54.3%	53.9%	51.2%
	Largest Carrier	NW	NW	NW	NW	NW	NW	DL	DL
	Herfindahl Index	4930	4625	4663	4632	3997	3170	3098	2844
Phoenix (PHX)	Largest Carrier %	38.1%	37.5%	35.8%	30.4%	29.6%	29.9%	30.5%	31.5%
	Largest Carrier	HP	HP	HP	WN	WN	WN	WN	WN
	Herfindahl Index	2387	2411	2360	1893	2555	2604	2716	2759

Source: Compiled by authors using Form41 Data

Therefore, in at least a few markets, potential airline mergers could have a much greater and controversial effect on the level of consolidation at some major airports.

FACTORS AFFECTING WORLD AIR TRAFFIC GROWTH

The factors that affect world air traffic growth are numerous, complex and occur on global, national, regional, and civic levels. This complexity helps explain why air travel can grow significantly in one country or city and why it can stagnate or flounder in another; chief among these factors is the level of prosperity in the region. This amount of economic prosperity is measured by such indicators as Gross Domestic Product (GDP) or Gross National Product (GNP). GDP is the total market value of all final goods and services produced in a country in a given year. Increased prosperity derives increased demand for air travel in two separate but concurrent ways. First, increased economic activity helps generate employment, which ultimately causes an increase in business travel, the most important segment of travelers for airlines. Business travel is the primary reason why world financial centers such as London and New York have experienced strong air traffic growth. Additionally, increased economic activity will also spur air cargo growth.

The second result of economic prosperity is a decrease in unemployment and a concurrent increase in household income. People have more discretionary income and are more able to afford more leisure travel trips. A good example of this has been in China, where a growing middle class has fueled a large expansion of air travel within the country.

A decrease in the real cost of air travel will also create air traffic growth. This was first experienced in the 1970s when deregulation resulted in a dramatic decline in the cost of air travel. Air travel was now affordable to a greater number of people and they took advantage of the opportunities. LCCs generated increased air travel with low fares, and airports experienced tremendous growth in their passenger statistics once a low-cost airline initiated service. This phenomena has been coined the "Southwest Effect." Ryanair is accomplishing similar feats in Europe where weekend getaways are now affordable to almost everyone.

Another factor influencing world air traffic growth is population growth rates. Strong population growth rates in developing countries such as India and China have helped spur air travel growth. However, population growth must generally be accompanied by income growth for this factor to significantly affect air travel.

Economic liberalization is another major factor impacting air transportation. Government restrictions on an economy, such as wage and price controls or excessive regulation, ultimately constrain demand. When such artificial barriers are lifted, the marketplace dictates demand for goods and services and increased air travel is almost always the result. The reason for this is the fact that government regulation in the aviation industry usually involves ticket prices and market access. That is, favored airlines (usually a national airline) are granted monopoly access with some sort of a fare structure that is structured to cover average costs. This effectively eliminates competition and restricts the growth of air traffic. A good example of economic liberalization is the US itself. Following deregulation, airfares plummeted and air traffic growth increased significantly. Moreover, the freedom for airlines to fly to whatever destination they wished made flying more convenient for passengers by providing more non-stop flights with greater frequency. Recent air transport liberalization in Europe and India has led to a tremendous growth in air traffic in these countries.

Politics and political stability also play a role in air travel. It is not surprising that countries that choose extremely protectionist and radical policies do not experience great air traffic growth; in these cases the government restricts air travel as a matter of political policy. Political instability can also greatly influence air travel, since people do not want to travel to regions where they feel unsafe. It is likely that political instability can be blamed for the poor air traffic growth rates in those parts of Africa where governments are in constant turmoil. Finally, political instability reduces and/or restricts business activities within the country.

Terrorist attacks can also affect air travel. Following the tragic events of September 11, air travel dropped off drastically as passengers no longer felt safe traveling. Additionally, many felt unsafe to travel to international destinations in the event that some other terrorist attack would occur. Finally, the amount of leisure time people have can impact the demand for leisure flights. Typically, individuals who possess greater discretionary free time have a greater demand for leisure and/or vacation flights. Tourism promotion can also help spur an increased demand for air travel to a particular destination. For example, Walt Disney World has turned Orlando into the number one destination airport in the US.

ECONOMIC IMPACT OF THE AIR TRANSPORT INDUSTRY

Commercial aviation is comprised of two primary segments: large commercial air carriers and regional/commuter air carriers. Since deregulation of the commercial airline industry in 1978, both the large commercial and regional/commuter air carriers have enjoyed more robust growth than the domestic economy. Generally, the commercial airline industry has closely followed the movement of the domestic economy. After deregulation, the large US commercial air carriers have averaged an annual revenue growth of 4.8 percent, compared to a 2.6 percent average growth rate of the US gross domestic product. During this same time period, US regional/commuter air carriers grew at an annual growth rate of 14.3 percent.

More recently, the Federal Aviation Administration (FAA) forecasts long-term growth in enplanements for large US carriers to average 2.7 percent domestically and 4.2 percent internationally through 2030, while growth will average 2.9 percent for regional/commuter airlines. And, as has been mentioned, international aviation continues to grow with IATA forecasting higher growth rates than the US industry (Pearce, 2006). This growth has been largely spurred by the soaring economies in the Asia-Pacific region. The following section explores the economic impact that the growth of air transportation is likely to have on the economy.

Economic impact can be divided into three categories: direct, indirect, and induced. Direct impact represents economic activities that would not have occurred in the absence of air transportation. In the air transportation industry, both airlines and airports provide the economy and local communities with a direct economic impact. Examples of direct economic impacts include the salaries of airline personnel, fuel purchased, landing fees, salaries of airport personnel, and other similar purchases and expenditures. Indirect economic benefits include the financial benefits that are attributed to airport/airline activities. Examples of indirect economic impacts for air transportation include hotels, restaurants, and other retail activities. There is usually a causal relationship between the industry and indirect impacts. For example, if there were a reduction in air travel for a community, the hotel industry in that community would most likely realize a fall in occupancy rates as well. Finally, induced economic impacts are the multiplier effects of the direct and indirect impacts. Induced impacts account for the increased employment and salaries that come from secondary spending that is a result of the direct and indirect economic impacts. The total of these economic impacts measure the importance of an industry in terms of the employment it provides and the goods and services it consumes. The following sections explore the effect of air transportation on each of these economic impacts.

Direct impact

Direct economic impacts are the consequences of what might be termed first-tier economic activities carried out by an industry in the local area. In the air transportation industry, airports provide the greatest direct impact to local economies. The reason for this is the more or less obvious fact that the economic activities that take place at the airport directly involve the local economy. Most direct impacts, like airport employment and fixed-based operations, occur at the airport; others, like local production of goods and services for use at the airport, may occur off site. In 2011, Over 56 million people are employed worldwide in aviation and related tourism. By 2026, it is forecast that aviation will contribute $1 trillion to world GDP.[3]

Expenditures by airlines, fixed-based operators, and tenants also generate direct impacts, but only those expenditures that lead to local business activity are relevant for a regional economic assessment. For this reason, it is important to distinguish between the local value-added component of expenditures and the regional import component. Thus, airline expenditures on fuel generate local fuel storage with distribution systems and they also contribute to the importation of fuel into the region. In most parts of the country, only the former component is relevant for any local economic impact analysis. Therefore, the direct economic impacts of air transportation for a community are usually measured based on the airport's immediate economic activity. Additionally, large aircraft manufacturers can have a huge direct economic benefit by locating their production facilities in a given community or state. For example, the direct economic impact of the Boeing 787 Dreamliner project on the state of Washington in 2006 has been estimated at approximately 11,470 jobs with an economic output of $2.268 billion (Deloitte, 2004). There are of course numerous other examples of large direct economic impacts that are provided by the air transportation industry. If aviation were a country, it would rank 19th in the world in terms of GDP by generating about $540 billion worth of product and services per year.[4]

Indirect impact

Indirect impacts derive from off-site economic activities that are attributable to air transportation activities. For example, indirect economic impacts include services provided by travel agencies, hotels, rental car companies, restaurants, and retail establishments. These enterprises have a strong relationship to the air transportation industry and, like airport businesses, employ labor, purchase locally produced goods and services, and invest in capital projects. Indirect impacts differ from direct impacts because they originate entirely off-site. Typically, indirect economic impacts are generated by visitors to the area who are traveling by air. A good example of an industry that has a strong indirect economic impact relationship with air transportation is the hotel industry. Airlines provide economic benefits to the hotel industry by requiring hotel rooms for passengers who have business in, or are vacationing in a city. This increased demand for hotel accommodation in the city creates employment and may require construction of more hotels, thereby creating more economic impact. The large demand for hotel accommodation caused by air transportation is one of the main reasons why areas around major airports almost always contain many hotels.

Induced impact

As mentioned earlier, induced economic impacts are the multiplier effects that are caused by the increases in employment and income generated from the direct and indirect economic

---

3   Air Transport Action Group (ATAG), 2012.
4   Air Transport Action Group (ATAG), 2012.

impacts of air transportation. A simple example will help make this concept clear. Imagine a new airline employee who purchases a house in the local community. The builder of the house then uses this income to purchase other goods and services and the income to the suppliers of these goods and services is also spent. This framework of expenditures is the basis behind the multiplier effect; that is, one transaction leads to multiple economic transactions.

More economically self-sufficient regions tend to have higher multipliers than do regions that are more dependent on regional imports, since more of the spending and re-spending is done within the region. Therefore, the larger the region under consideration, the higher the multiplier will be.

Total impact

Total economic impact is defined as the sum of direct, indirect, and induced impacts. Total impact is usually expressed in terms of economic output, earnings, or employment (sometimes full-time equivalents). The basic formula for total economic impact is:

Total Impact = Direct Impacts + Indirect Impacts + Induced Impacts

Table 1.13 provides a comparison of the total economic impact in terms of employment for 11 airports located in the US. The report for each airport was done independently and at different times, but the methodology used for each is similar. While the 11 airports vary in size, they all provide strong economic impacts for their communities. When normalized in terms of commercial departures, Memphis generates one job for every departure, or in other words, one additional daily flight would generate approximately 365 new jobs for the region. Wichita's extremely high ratio of three is probably attributable to the large manufacturing and maintenance facilities for Cessna and Bombardier. The presence of Federal Express in Memphis explains its high economic impact to departure ratio. And finally, much of Seattle's total economic impact can be attributed to the simultaneous indirect economic impact of the presence of aircraft manufacturing giant Boeing and of tourism.

Table 1.13   The Economic Impact of Selected Airports

Airport	Year of Report	Total Jobs	Total Jobs per Commercial Departure
Cincinnati/Northern Kentucky	2000	78,573	0.651
Wichita	2002	41,634	3.184
Seattle-Tacoma	2003	160,174	0.964
Greenville-Spartanburg	2003	5,787	0.246
Memphis	2004	165,901	1.010
Minneapolis-St Paul	2004	153,376	0.630
TF Green Rhode Island	2006	21,857	0.781
Portland International	2007	38,571	0.429
Central Wisconsin Airport	2007	981	0.075
Southwest Florida International	2010	41,588	0.979
Anchorage International	2011	15,500	0.115

Source: See chapter references for all source documents

These disparate examples highlight the diversity (cargo operations, manufacturing, and tourism) and strength of the economic impact of the aviation industry.

OUTLOOK FOR THE AIR TRANSPORT INDUSTRY

Over 1,700 airlines operate a fleet of 23,000 aircraft by serving 3,750 airports around the world.[5] Since demand for the air transport industry is highly correlated with overall economic growth, it is not surprising that the global outlook for the air transport industry mirrors the global economic outlook. The airline industry was hit hard in 2011, by the persistent financial problems, and global financial problems after a strong rebound in 2010.[6] Therefore, the air transport industry is expected to grow significantly in regions where economies are developing, such as Asia-Pacific, while other regions' air transport outlook is expected to be steady. GDP and economic growth are strong leading indicators of the air transport industry's growth, so in the short-term these measures can be used to assess the industry.

However, direct correlations between GDP and air transport growth are never exact due to a variety of issues. For example, structural barriers in the air transport industry can cause drastic differences between economic growth and the growth of the air transport industry. A good example of this was the effect of deregulation in the US; deregulation was a major structural change that caused a rapid increase in the air transport industry's growth compared to overall economic growth.

Airport capacity and, in the US, antiquated air traffic control, are also potential structural barriers. Major international airports in the US and Europe have severe capacity issues with relation to the number of aircraft that they are capable of handling. As these capacity limits are reached, delay at these airports tends to increase in an exponential fashion. These delays, especially if they are on an ongoing basis, discourage demand and constrain growth.[7] Similar capacity issues could plague airports in the Asia-Pacific region, especially Indian, Chinese, and Japanese airports. This capacity barrier to air transport growth is a prime reason why Airbus embarked on the creation of its new super-jumbo A380 aircraft.

The two major sources for the long-term air transport outlook are Boeing and Airbus. Each aerospace giant has published forecasts for the future of the aviation industry. They have similar growth estimates for world air traffic growth with Boeing forecasting that world revenue passenger kilometres (RPK) will grow at 5.0 percent per annum for the next 20 years. Table 1.14 summarizes the average regional growth rate forecasts between Boeing and Airbus.

Both Airbus and Boeing also forecast worldwide demand for new aircraft for the next 20 years. Not surprisingly, each company's forecasts vary slightly, highlighting each company's strategic plan and product offerings. Boeing estimates that there will be demand for 33,500 aircraft seating over 90 passengers in the next 20 years, while Airbus forecasts a worldwide demand for 26,900 similar-sized aircraft over the same period. While both companies agree that roughly 70 percent of the demand for new aircraft will be for single-aisle aircraft, Airbus predicts a greater demand for large wide-body aircraft, while Boeing believes the remainder of aircraft demand will be for small and medium

---

5   Air Transport Action Group (ATAG), 2012.
6   Zacks Equity Research, January 10, 2012.
7   See Chapter 5 for an analytical discussion of this issue.

Table 1.14   Regional Economic Growth Forecast

2011–2030 Estimated Growth		
Region	GDP	RPK
North America	2.7%	3.2%
Latin America	4.2%	6.9%
Europe	2.0%	4.3%
CIS	3.4%	5.0%
Middle East	4.1%	8.2%
Asia	4.7%	6.7%
Africa	4.4%	5.7%
World	3.3%	5.0%

Source: Compiled by authors using Airbus Global Market Forecast 2011–2030 and Boeing Current Market Outlook 2011–2030

wide-body aircraft. Airbus forecasts demand for 1,331 very large aircraft (747s and A380s) while Boeing only forecasts 820 aircraft in this segment. Additionally, the firms differ on where demand for new aircraft will be. Boeing still forecasts that North America will be the largest market for new aircraft (mostly narrow-body aircraft); while Airbus forecasts the Asian-Pacific market will order the most aircraft in the next 20 years. Additionally, Airbus foresees greater LCC growth in this region to spur narrow body sales.

One other sector of the air transport industry that should be mentioned is the air cargo market. Both Boeing and Airbus forecast world air cargo to grow by about 6 percent per year for the next 20 years. This worldwide forecast growth outstrips passenger growth forecasts, and this situation is especially true in international markets where the air cargo industry has not developed to the extent of the passenger industry. As a result, demand for cargo aircraft (new or second-hand) is expected to be strong, especially for wide-body aircraft. China is expected to lead the way in air cargo growth, both domestically and internationally. The US domestic air cargo market appears to be mature with Airbus forecasting a modest 2.8 percent annual air cargo growth and Boeing forecasting a 4.8 percent growth rate

SUMMARY

With relatively minimal profits margins, the financial condition of the aviation industry is highly dependent on the global economic conditions of the day. During times of economic boom, profits soar and in times of distress carriers are forced to cut back capacity. This chapter introduces the reader to the present state of the aviation industry with a representative data set that covers the volume of traffic, the existing finances, mergers, bankruptcies, and levels of concentration within the industry. The chapter then covers the economic impact of the industry along with forecasts for future growth. The purpose of this chapter is to describe the evolution of the air transport industry including airlines and airports. As the preceding discussion and statistics amply demonstrate, the chapter introduces the industry as a large and growing segment of the domestic and international economies. As such it is an important area for economic analysis. Although the industry is similar in some ways to other large industries, it has some peculiar characteristics that can best be understood in the context of standard economic analysis. This text aims to apply economic analysis to the industry to explain and illuminate those characteristics. To that end the first four chapters of the text will introduce the reader to basic economic theory including demand, supply, costs, and production analysis. These ideas will be presented in the context of the aviation industry and will be presented with applicable examples from the industry.

DISCUSSION QUESTIONS

1. What are the factors influencing world air traffic growth?
2. The Airline Deregulation Act of 1978 practically removed government control over fares, routes and market entry from commercial aviation.
    a   Identify some of the characteristics of the US airline industry before de-regulation.
    b   How did this era affect airlines and passengers?
    c   Was deregulation successful?
3. What are direct and indirect economic impacts related to air transportation and how do they differ? Provide an example of each.
4. Which regions serve the largest number of passengers? Movements? Cargo?
5. Does regional aviation activity reflect in average age of aircraft by region?
6. What are the top five busiest airports in terms of passengers? International Passengers? Cargo?
7. What are some trends with respect to consolidation in the industry? At airports? How is this demonstrated in the HHI?
8. Mergers have been an important part of the airline industry. Have they been successful?
9. Which regions are forecasted to have the highest growth in aviation over the next 20 years?

REFERENCES

Amex. (2006). *2005 US Domestic Airfares for American Express Business Travel Clients Drop to Six-Year Low*. Retrieved on August 31, 2006 from http://home3.americanexpress.com/corp/pc/2006/4q05_monitor_print.asp.

Center for Economic Development and Business Research. (2003). *Wichita Mid-Continent Airport Economic Impact*. Retrieved on September 13, 2006 from http://webs.wichita.edu/cedbr/AirportImpact.pdf.

Deloitte. (2004). *Employment and Income Analysis of the Boeing 7E7 Project*. Retrieved on September 13, 2006 from http://www.aia-aerospace.org/stats/resources/Boeing_EmploymenAndIncomeAnalysis.doc.

Dempsey, P. (1990). *Flying Blind: The Failure of Airline Deregulation*. Washington, DC: Economic Policy Institute.

Economics Research Group. (1999). *The Cincinnati/Northern Kentucky International Airport Economic Impact Analysis*. Retrieved on September 13, 2006 from www.cba.uc.edu/econed/1998-2011impact. pdf.

John C. Martin Associates. (2005). *The Local and Regional Economic Impacts of the Minneapolis/St Paul International Airport*. Retrieved on September 13, 2006 from www.mspairport.org/msp/docs/ misc/mspimp04_FINAL.pdf.

Martin Associates. (2005). *The 2003 Economic Impacts of the Port of Seattle*. Retrieved on September 13, 2006 from http://www.portseattle.org/downloads/business/POS2003EIS_Final.pdf.

McKinsey & Company. (2001). Making mergers work. *Airline Business*. Retrieved on August 31, 2006 from Air Transport Intelligence.

Pearce, B. (2006). New Financial Forecast. *IATA Industry Financial Forecast, September*. Retrieved on August 31, 2006 from www.iata.org/economics.

Sparks Bureau of Business and Economic Research. (2005). *The Economic Impact of Memphis International Airport*. Retrieved on September 13, 2006 from www.memphisairport.org/EcImpactFinal.pdf.

Wilbur Smith Associates. (2003). *The Economic Impact of Greenville-Spartanburg International Airport — Update 2003*. Retrieved on September 13, 2006 from www.gspairport.com/images/downloads/ AirImpact.pdf.

2

Principles of Economics with Applications in Aviation

Economics is haunted by more fallacies than any other study known to man. This is no accident. The inherent difficulties of the subject would be great enough in any case, but they are multiplied a thousand fold by a factor that is insignificant in, say, physics, mathematics, or medicine—the special pleading of selfish interests.

Henry Hazlitt, American economist and journalist (1894–1993)

This chapter introduces students to the "economic way of thinking" primarily through the study of incentives and prices. Economics is a social science and similar to other social sciences, it has its own concepts, presumptions, and rules. Throughout this chapter we will examine the basic concepts of scarcity, choice, and opportunity cost in economics. The subject of economics is generally divided into two major sections: microeconomics and macroeconomics.

Microeconomics is a subdivision of economics that studies how individual households and businesses make decisions to distribute limited productive resources to maximize profit. Microeconomics focuses on pricing strategy and production policy and how to assign the products and services among the competing customers. Macroeconomics is the aggregate of microeconomics and examines the economy as a whole, and deals with the subjects of unemployment, inflation, and economic growth. We will look at these two fields of economic thought through an in-depth analysis of the characteristics and importance of each. The basic economic framework can then be applied to issues in aviation including the role of government in aviation, deregulation, operating costs, foreign operations, and certification. We then present economics as a discipline for informing and critiquing political policy.

- Basic Economics
- Scope of Economics
  - Microeconomics
    - Pricing decisions
    - Output decisions
    - Choice and opportunity cost
  - Macroeconomics
    - Inflation
    - Unemployment

- ○ Natural unemployment
- ○ Cyclical unemployment
- ○ Frictional unemployment
- The Role of Economic Systems
  - – Government and aviation
- Government Failures and Market Failures
- Summary
- Discussion Questions

BASIC ECONOMICS

Economics may be defined as the science of decision-making and resource allocation under scarcity. Many decisions carried out throughout the aviation industry are prime examples of economic decisions where scarce or limited resources have to be allocated. An example of this was the decision to construct a $1.28 billion fifth runway at Atlanta's Hartsfield-Jackson airport (ATL) to increase the operational capacity of the airport. Broadly speaking, every resource is scarce, and the allocation of resources under a variety of incentives and decision-making parameters forms the core of economic analysis. An important feature of economic analysis is the assumption that people understand and can act in their own best interest. That is, people respond predictably to a given set of choices in order to maximize their benefits or minimize their losses. Though this is sometimes referred to as the fundamental assumption of economics, economists believe it is not an assumption at all but a simple fact confirmed by common empirical reality. For instance, individuals will engage in a search process to find the lowest ticket price for a given itinerary and set of requirements like flexibility, refundability, and service level. However, there is a limit to the amount of time they will spend in such market research since the time spent searching for the best price has a cost, as it is time that could be spent in doing something else.

> "The only way that has ever been discovered to have a lot of people cooperate together voluntarily is through the free market. And that's why it's so essential to preserving individual freedom."
>
> Milton Friedman

This implicit cost of resources is known as opportunity cost, and is yet another fundamental concept in economics. Opportunity cost is defined as the alternative cost of using a resource; that is, the benefits that would accrue if the resource were being utilized in its next best allocation. In other words, the next best use of the resource is the economic cost of using it in the current allocation. For instance, in the above example, the opportunity cost of the time spent in searching for the best ticket price is the next best use of that time, perhaps in running the company, generating greater sales, or spending time with one's family. For an airline, the opportunity costs of acquiring a new aircraft could be an alternative use of that money, perhaps in acquiring another type of aircraft, opening a new route, or restructuring.

The opportunity costs of a resource determine the eventual allocation of those resources. To continue with the example above, as long as the opportunity costs of time do not exceed the potential cost savings that can be achieved, the individual will continue the search.

When the opportunity cost of time exceeds the potential cost savings, the individual will cease the search and take the best price that can be found.

Continuing the example, travel aggregators like Expedia.com and Travelocity have massively reduced the search costs associated with booking airline tickets. By aggregating the various ticket prices into a convenient form, time spent in searching for the best fare is reduced, and the opportunity costs of travel arrangements are reduced. Effectively, travel has been made cheaper due to information provided by a knowledgeable middleman.

Yet another fundamental of economic thinking is the concept of prices. Prices constitute the central allocating mechanism of economics, and are the decision-making parameters by which individuals organize their actions. For example, consider the price of oil; literally millions of people use the price of oil to make decisions affecting both business and personal consumer interests. Large corporate firms use the price of oil to make important corporate decisions regarding their own pricing (consider the airlines for example), exploration and development activities depend upon the present price of oil, numerous transportation decisions are related to the price of oil, and of course, consumers alter their behavior based directly on the price of gasoline and oil. In many instances, individuals are hardly aware of the impact that prices have on their decision-making process since prices can be implicit as well as explicit. In the case of the oil prices mentioned above, prices are explicitly given in the marketplace; however, there are also implicit prices that must be considered in the economic way of thinking. As an example the price of flying from Orlando, FL to Houston, is $254 by Southwest airlines, and the cost of driving there is $99.91.[1] On the surface, it might appear far cheaper to drive. However, the true price of driving does not take into account the cost of 14 hours and 53 minutes of the passenger's time. Assuming a median income of $53,207 in 2011 (US Census Bureau, 2011), which translates roughly into $26.60 per hour, the true price of driving is $372.40 + $99.91 = $472.31, which makes it 84.8 percent more expensive than flying. Therefore, when the implicit price of driving is considered, flying would be the more efficient alternative.

SCOPE OF ECONOMICS

If we read different books on economics, we would see many different definitions for the subject, but they all share two factors. First, there is the notion of scarcity and second, the idea of unlimited wants. So we can define economics as the art or science of using limited productive resources such as land, capital, labor, and technology to produce different goods and services to satisfy unlimited human wants. At its core, economics involves choices and tradeoffs; choices matter because resources are limited while human wants are unlimited. The economic method of thinking centers on analyzing the decisions of individuals and the movements of markets in light of incentives and choices. Economics has many subdivisions and specialties, but the broadest distinction has been between "macro" and "micro" economics. These respective fields are covered in the following pages.

---

1   Retrieved from Southwest.com on October 11, 2009, for travel on October 19, 2009 assuming fuel burn of 24 miles per gallon, and a gas price of $2.50 per gallon.

Microeconomics

Microeconomics deals with the behavior of individual households and businesses (decision-making units). It depends heavily on the concepts of supply and demand; that is, the way in which the market determines the price of goods and services and the level of production. Airlines provide particularly good examples for the decision-making aspect of microeconomics since they respond swiftly to changing market conditions by altering supply and/or prices by utilizing sophisticated pricing techniques. Capacity (quantity) management is a good example of microeconomic decisions. Some cases follow: during the Gulf War airlines experienced a significant decline in traffic so they reduced their capacity (supply) by 10 percent over a period of about one year. In India, the low-cost carrier (LCC) JetLite Airways, a subsidiary of Jet Airways, slashed fares on all sectors by up to 40 percent to increase quantity demand.[2] Airline revenue management is another good example of a microeconomic decision-making. Airbus and Boeing are two tough rivals in the aircraft industry; they compete with each other to capture a bigger market share by introducing more efficient aircraft.

Pricing decisions   Based on pricing, businesses can forecast what a consumer may buy, and how many units of that product or service will be sold. Revenue management will be explored in greater detail in a subsequent chapter, but the basic concept is that airlines price seats differently according to several variables including service level, time before flight, and day of the week. The idea is to charge each consumer the maximum he/she is willing to pay based on his/her personal characteristics. The business traveler with little flexibility and last-minute travel needs would intuitively be willing to pay a great deal more than the casual vacationer who is much more price sensitive. Airlines separate the aircraft into cabins (economy, business, and first in the typical three-class system) that are further separated into classes, each with its own price point and often slightly different ticket characteristics like flexibility and refundability. Pricing the cabins according to classes allows the airline to maximize its revenue based on traveler preferences, since it can charge a higher price to those who are willing to pay it, and use them to cross-subsidize, in a sense, those who would be much more price sensitive. The entire exercise of revenue management is predicated on microeconomic decision-making—travelers with certain characteristics are likely to support a higher price than others, and pricing according to those characteristics benefits both the airline and the consumers.

Decisions including aircraft acquisition, fleet selection, and route planning are all microeconomic decision-making activities. Will this new aircraft have a justifiably high load factor if employed on this route? Given route characteristics, will adding this new aircraft to the fleet result in an optimal mix? Is there a large enough market on this route to support the entrance of a new airline? All these decisions are based on the behavior of individual consumers, and the effect that changing certain variables will have on individual choices.

When an airline purchases or leases an aircraft, the airline economists evaluates the opportunity cost. You buy one aircraft at the expense of another. Management usually purchase or lease the aircraft that will give them the most value for the available resources.

---

2   Oasis Hong Kong Airline was launched in 2006 and is a now-defunct long-haul low-cost airline.

Output decisions   Output decisions for an airline often come in the form of capacity decisions, aircraft size, aircraft type, and schedule selection. Capacity decisions fall under the umbrella of microeconomic decisions; in recent years airlines have shifted capacity away from domestic US markets to more profitable international markets. In April of 2008, Emirates Airline started New York service with a double-deck aircraft, but pulled it two months later and replaced it with the smaller Boeing 777. In response to soaring fuel costs, Frontier cut one-third of its daily departures from Milwaukee, from 67 to 45. In addition, non-stop service to six cities was suspended.[3] In May 2011, Ryanair announced a capacity cut by grounding 80 aircraft in the winter schedules between November 2011 and April 2012 due to the high cost of fuel and continuing weak economic conditions.[4] Ryanair clearly focused on a strategy that is low in cost structure, which allowed them to charge low fares.

Choice and opportunity cost   Choosing to utilize a resource has an implicit cost referred to as opportunity cost. As mentioned, opportunity cost can be defined as the true cost of using a resource, which is the benefits that could accrue if the resources were not being utilized in the current allocation. In other words, the next best use of the resource is the true cost of using it. This is why it may be efficient for executives to outsource their travel arrangements to their executive assistant. The opportunity costs of the assistant's time are less than the executive's—the latter could be better employed in running the company and generating profits. Based on their relative opportunity costs, executives can perhaps spend ten minutes searching for a ticket before the opportunity costs of their time exceed the potential benefits. The Executive Assistant can spend 30 minutes at the same task because of lower opportunity costs and therefore has a better chance of finding a better deal.

Macroeconomics[5]

In contrast, macroeconomics studies the decision-making process for the entire economy. Movements in gross domestic product (GDP), interest rates, inflation rates, exchange rates, balance of trades, and their interrelationships are all considered macroeconomics. In contrast to microeconomics, macroeconomics is a field that comes into play during the decision-making of airlines and aircraft manufacturers as a structural variable; that is, one which presents circumstances and parameters which they may not be able to influence, but need to take into account. For instance, movements in GDP and economic activity have implications for travel—the microeconomic decisions made by individuals are impacted by the overall economy, and the airline or airport manager has to take into account these circumstances while making decisions. In a slow economic climate for instance, introducing discounted fares or sales promotions might be more effective strategies at generating demand than introducing a new luxury class. Airport and airline managers have to deal with seasonality as well—travel is typically slow in the winter months, but picks up during the holiday season, and prices may have to be altered to account for the effects. Similarly, the impact of taxes and passenger fees on airline tickets is a macroeconomic variable the airlines have to contend with. According to the MIT

---

3   Airlines cut flights as jet-fuel costs climb. *Denver Post*, October 24, 2011.
4   *The Financial Times*, Ryanair to cut capacity for first time, May 23, 2011.
5   Macroeconomics will be discussed in much greater detail in Chapter 14.

Ticket Tax Project, 16 percent of the ticket price of an airline ticket is determined by macroeconomic variables that the airline has to take into account in terms of pricing and the price sensitivity of customers.

For aircraft manufacturers, macroeconomic variables become crucial in terms of predicting the demand for aircraft over the long term, since the financial health of airlines is highly dependent on economic cyclicality. Therefore, forecasting when the demand for aircraft is likely to fall off and pick up again will determine the timing of expected cash flows on aircraft, and consequently, the breakeven year and quantity that will determine project success or failure. In general, macroeconomic activity can be best understood in terms of business cycles.

Inflation   Inflation can be simply defined as the rise in prices of goods and services in an economy over time. In order to measure inflation, the Bureau of Labor Statistics (BLS) uses several indexes designed to measure the various components of inflation. The Consumer Price Index (CPI) is a measure of the average change over time in the prices paid by urban consumers for a market basket of consumer goods and services. CPI measures the cost of goods and services to a typical consumer, based on the costs of the same goods and services at a base period. The CPI is published monthly by BLS. Furthermore, the US BLS regularly reports the Producer Price Indexes (PPIs). PPIs are a family of indexes formerly called Wholesale Price Indexes that represent the change in the selling prices received by manufacturers of goods and services. These indexes are primarily used to carry out price adjustments.

Unemployment   Using the BLS's classifications, a person is considered unemployed if they do not have a job, have actively looked for work in the prior four weeks, and are currently available for work. Persons who were not working and were waiting to be recalled to a job from which they had been temporarily laid off are also included as unemployed. The unemployment rate represents the number unemployed as a percent of the labor force. The labor force is the number of persons in the economy who have jobs or are seeking a job, are at least 16 years old, are not serving in the military, and are not institutionalized. In other words, the labor force is all people who are eligible to work in the everyday US economy.

$$Unemployment\ Rate = \frac{Number\ of\ Unemployed}{Total\ Labor\ Force}$$

Assume 2,000 people are unemployed in a small city with a population of 60,000 people, 48,000 people in the labor force, and 46,000 people are employed. Hence, the unemployment rate is calculated as follows:

$$Unemployment\ Rate = \frac{2,000}{48,000}$$

$$Unemployment\ Rate = 4.12\%$$

Unemployment can be further classified as natural unemployment, cyclical unemployment, and frictional unemployment.

Natural unemployment   The natural rate of unemployment is the inherent rate that will occur in an economy, separate from unemployment occurring due to business cycles. The natural rate is determined by the rate at which jobs are simultaneously created and destroyed, the rate of turnover in particular jobs, and how quickly unemployed workers are matched with vacant positions. The Congressional Budget Office forecasts the natural unemployment rate of the US to be 5.0 percent through to 2017.

Cyclical unemployment   Cyclical unemployment is another class of unemployment, and refers to the unemployment that occurs as a result of business cycles. As businesses go through their cycles, cyclical unemployment occurs; when the business cycle is at its peak (maximum economic output), cyclical employment is low. Cyclical unemployment occurs when consumer expenditure is low and there is not enough demand for employers to hire everybody who wants a job.

Similarly, when the business cycle is low and economic output decrease, cyclical unemployment increases.

Frictional unemployment   Frictional unemployment is also referred to as transitional unemployment. This form of unemployment is always present in the economy as a result of persons transitioning between jobs and new workers entering the labor force. Frictional unemployment is therefore a function of the time it takes the labor market to match the available jobs with persons in the labor force. From an economic point of view, a business cycle is defined as the movement of economic activities such as unemployment, inflation, and economic growth and it is divided into four stages:[6]

- *Expansion or recovery*: Historically, the annual growth in air travel has been about twice the annual growth in GDP, with increased growth during periods of economic expansions.
- *Peak*: The peak occurs at the highest point between economic expansion and the start of economic contraction.
- *Contraction or recession*: Contractions (recessions) start at the peak of a business cycle and end at the trough and are marked by a significant decline in economic activity spread across the economy, lasting more than a few months, normally visible in real GDP, real income, employment, industrial production, and wholesale–retail sales.
- *Trough*: The trough signals the end of a period of declining business activity and the transition to expansion. This is the beginning of economic recovery.

THE ROLE OF ECONOMIC SYSTEMS

Broadly speaking, modern political systems can be characterized by the perceived role of the market and the prevalence of government in decision-making and resource allocation. At one end of the spectrum there lies hypothetical laissez-faire capitalism, which would be an economy driven purely by the market with very little government involvement and where the market dictates resource allocation.

---

6   Duration of periods and contractions can be found from a variety of sources including the National Bureau of Economic Research.

On the other end of the spectrum, there lies a pure command economy, where resources are allocated with reference to a central authority or a government, with no reference to a market. In the middle of the continuum is a mixed economy. The mixed economy is a combination of private enterprise with a larger government sector (usually involved with income transfer) and the existence of freedom of ownership, pricing policy, and profit earning.

Historically, no economies have ever achieved either extreme. The beginnings of the post-industrial revolution US in the nineteenth century was close to pure capitalism, although there was still some government involvement in the economy. The Soviet Union's communist economy for the better part of the twentieth century was very close to a complete command economy, although some market forces did exist in the form of black markets. Over time, however, every modern economy has come to some form of mixed system, some relying more heavily on government involvement than others.

Within every economy, there are industries that are subject to more or less government supervision. Industries like the financial sector have historically been subject to less regulation than manufacturing, pharmaceuticals, or aviation. Therefore, while the US might have less government involvement in economic activities than France, the aviation industry in the US might have more industrial regulation than is average for the country as a whole, although such regulation might be less than the regulation of the aviation industry in France.

Further, even within the aviation industry, there will be more or less regulation. For instance, the airline deregulation in the 1970s brought airlines firmly within the sphere of the free market. However, simultaneously, aircraft manufacturing and certification has grown more highly regulated over time, due to liability concerns. Further still, different parts of the aviation industry might be regulated differently. In Europe, for instance, airports are beginning to be privatized, while the same is not taking place to that extent in the US. Therefore arguably, airports in the US are subject to greater regulation than in Europe, while the opposite is true for operational restrictions due to environmental and noise concerns.

Increasingly, however, the push has been toward less government involvement in aviation. For instance, the various open skies agreements between the European Union (EU) and US allow any airline in the US to fly from any point in the US to any point in Europe. This EU–US Open Skies Agreement was first signed on April 2007 and became effective in March 2008. A second phase of the agreement aimed at reducing further barriers in the transatlantic aviation market was signed in June 2010. This reduces the involvement of governments in the decisions of individual carriers to offer transcontinental service, and allows for increasing competition between airlines in different countries. However, this agreement is by no means comprehensive, as it does not allow any European carrier to operate flights within two points in the US, while it allows American airlines to operate intra-European flights. There have been drives to create a completely open skies agreement between the two entities, which would represent a significant withdrawing of government influence in air transportation.

An economic incentive is best described as a force or circumstance that encourages an individual to engage in a particular activity. For instance, a tax credit on home ownership will incentivize a certain kind of economic activity, specifically, home ownership. Similarly, the emergence of LCCs and a competitive threat incentivized the legacy carriers like American Airlines to become more competitive and adopt practices like revenue management, which drove ticket prices lower. In contrast, a disincentive is an incentive that discourages an individual from engaging in a particular activity. Raising the tax on jet fuel would necessitate higher ticket prices for the consumer as a result of cost pass through, and give consumers a disincentive for air travel.

Incentives can be both economic and noneconomic in nature: the threat of litigation has incentivized the Federal Aviation Administration (FAA) to institute ever more stringent tests for aircraft certification, which has in turn dis-incentivized many general aviation manufacturers from major or frequent innovation. The Economic Way of Thought centers on analyzing incentives and predicting human behavior when dealing with clearly defined incentives.

Incentives are a powerful tool for analyzing public policy. Phrasing things in the language of incentives has the effect of throwing light upon the consequences of policy and regulations: while airline regulation allowed airlines to make a normal profit and enjoy stable operations, it incentivized inefficient operations since no matter how low the load factor on approved routes, they always generated a relatively predictable amount of costs and revenues. While airline deregulation may have caused airlines' operational structures to become unpredictable, their revenue to fall, and their costs to rise, it in turn incentivized competition, highly efficient operational practices like revenue management and route profitability analysis. As a result, airline consumers reaped the benefits of such operations in terms of new routes, new airline options, and vastly lower ticket prices. We will return to airline regulation and deregulation later in this chapter as an example of government involvement in aviation.

Government and aviation

Throughout the world, aviation has always been deeply intertwined with government. Since the beginning of commercial aviation in the 1920s and 1930s, governments on both sides of the Atlantic supported and subsidized the fledgling industry, encouraging innovation through lucrative mail contracts, and later through military contracts. The economics of early commercial aviation were extremely unattractive—without the possibility of night flights or flying through clouds or over weather phenomenon, with speeds of 90 miles an hour at best, commercial aviation in the early 1920s was hardly viable without government subsidies (Heppenheimer, 2001). In Europe, this took the form of direct government subsidies to aviation. Both France and Germany laid the foundations for their aviation industry with flag carriers Air France and the Deutsche Luft Reederei, which would later become Lufthansa.[7] The American and British governments were more cautious about direct government support—instead, they took the route of indirect subsidies through mail contracts. We see this pattern to this very day, with the styles of support afforded to Boeing and Airbus by the respective governments in America and Europe—both are subsidized, albeit through different mechanisms. Airbus is arguable subsidized more directly, while others argue that Boeing receives an indirect cross-subsidization through its defense contracts.

Aircraft and engine manufacturing, on the other hand, had been receiving government subsidies through direct contracts with the military since the First World War, with many of the developments in aircraft design conducted by Boeing, Lockheed Martin, McDonnell Douglas, Fokker, De Havilland, and Northrop. The Contract Air Mail Act of 1925, and

---

7   Deutsche Lufthansa timetable at timetableimages.com.

the subsequent Air Commerce Act of 1926, had two important effects—first, they put commercial aviation squarely into the hands of the private sector, but offered it massive government support and reserved the power of regulatory oversight in the determination and disposition of those contracts. It was under this umbrella that commercial aviation would truly come into its own—monoplane design for higher cruise speeds which would enable air mail to become an attractive competition to traditional mail routes. Lockheed's famous Orion and Vega, and Boeing's Monomail were all developed to maximize the profits that could be had by the efficient transportation of government mail. Since the airmail subsidies were based on weight, the greater the capacity of the airplanes and the greater their useful load, the more money an airline could potentially make from it. This had two effects—first, it shifted significant amounts of mail traffic from railroads onto airplanes. Secondly, it spurred manufacturer/airlines to focus on building and operating airplanes that could deliver the most weight with the highest speed, leading to innovations in airplane design that poised aircraft to take over high-speed passenger transportation in the future.

Yet another key piece of aviation legislation would take place with the 1930 Watres Act, championed by Congressman Walter Brown. Effectively, the Watres Act changed the nature of the subsidies for air mail from weight to mileage, which shifted the focus of aircraft manufacturers to build large planes with an extended range, and effectively shifted the subsidy from freight to passengers. If an airline was paid by the mile no matter what the weight of mail it transported, it would want to fly aircraft which had long-range capabilities, and incentivize intercontinental travel. Further, given that the demand for mail was fairly static, and longer range usually necessitated larger aircraft, it would have the incentive to use the extra space for the transport of passengers. The government hoped that over time, successful airlines would shift revenues from mail to passengers, and that the subsidies for the industry could slowly be phased out. Further, the Watres act awarded the government broad discretionary powers to award airmail contracts as they saw fit, removing some of the competitive bidding associated with the mail contracts. This was in line with Congressman Watres's vision, which saw the development of aviation in the US as being too chaotic. This consolidated the competition in the airline industry into a few large players, setting the stage for airlines that would eventually become known as the legacy carriers, and the creation of the Civil Aeronautics Board (CAB). Therefore, the very inception of aviation and the directions which the industry took were incentivized and controlled by governments. The development of the commercial aviation industry is an exercise in market responses to government incentives.

Increasing numbers of passengers and aircraft in the air created the need for extensive infrastructure and an air traffic management system. The Federal Airways Act of 1946 implemented key elements of infrastructure such as navigational aids like the Instrument Landing System (ILS), Very-High-Frequency Omnidirectional Range (the VORs), the designation of specific airways used for navigation, and the increasing use of radar (Fried and Myron, 1997). The implementation of direct pilot to controller communication took place in 1955. In 1959, airports were modernized under the direction of the Federal Aviation Authority (created from the Civil Aviation Authority in 1958). The tragic midair collision at the Grand Canyon in 1956, where a United Airlines DC-7 collided with a TWA Super Constellation while both were operating under visual flight rules in uncontrolled airspace, prompted the adoption of positively controlled airspace above 24,000 feet, and the abandonment of visual flight rules by commercial airliners. Further regulation was enacted in 1960 following a midair collision over Brooklyn that mirrored the Grand Canyon accident of 1956. Therefore, aviation infrastructure, navigational aids, airports, and aviation safety became tightly regulated, and airlines would remain under the regulatory umbrella until 1975. Before 1975, there were essentially four major airlines: United, American, Eastern, and TWA. Pan Am was the largest international carrier. These airlines were regulated by the CAB in terms of routes and fares, and new entrants had to apply for permission to carry out air transportation, which could be contested by the existing carriers. An airline wishing to expand service into a new route had to fill out a petition with the CAB which would be open to dispute by the existing carriers flying that route and could degenerate into full-blown court proceedings. Every aspect of airline operations was regulated, right down to new aircraft acquisitions, the type and disposition of freight carried, whether carriers could issue refundable tickets, whether a carrier certificated to operate a one-stop segment could change to non-stop service, whether the flight attendants of two financially affiliated airlines could wear similar uniforms, and so forth. Every operational detail of air transportation was under scrutiny and required approval—and while on one hand, no carrier ever went bankrupt under such regulation, neither did they have the flexibility to conduct any business on their own terms without extensive approval processes.

Arguably, it was airline deregulation that introduced the practice of revenue management, which was a byproduct of competition engendered by the emergence of LCCs. Revenue management, as discussed earlier, minimizes consumer surplus by charging each consumer what he/she is willing to pay, as opposed to a blanket single fare across consumer characteristics. Furthermore, the fall in ticket prices and the increase in route choices points to a distinct benefit. On the other side, airlines, especially legacy carriers, had to operate in an environment of much more heated competition, with thinner margins and potential price wars, with much the same cost and managerial structure they had utilized during the days of regulation (Heppenheimer, 2001).

This pattern is by no means restricted to the US—every country has a comparable aviation regulation framework, reflective of its economic and political development. Throughout the world and in any political or economic framework, aviation remains a tightly regulated industry with heavy government involvement. The impact of government on various aspects of aviation has shifted over time and tends to run in cycles. Recently, for example, there has been a push for privatization of airports, especially in Europe, which is triggering a similar call in America. Simultaneously, however, the costs of certifying non-experimental aircraft have steadily increased over the past few decades, although there has been an attempt to address this trend in the US through the introduction of a new Light Sport Aircraft category that bypasses most of the heavy certification and pilot licensure burden imposed on other aircraft.

With this degree of government involvement, no discussion of aviation economics would be complete without an analysis of the economics of government, its incentives, and impact on commercial aviation. Governments are a function of the political system and philosophy that each country has come to adopt—the development of the FAA, airline regulation and deregulation in the US and the historically unrestricted nature of general aviation; all of these developments are intimately linked to the political climate and prevalent philosophy at the time. In order to better understand the nature of government involvement in aviation, it is necessary to discuss a broader picture—market-driven political and economic systems in comparison to command economies.

GOVERNMENT FAILURES AND MARKET FAILURES

Broadly speaking, there are two kinds of economic failures; market failures and government failures. *Market failure* occurs when the market does not allocate resources to their most efficient use, or equity failures, in which the market does not allocate resources to their most—perceived—equitable distribution. In this case the market would allocate more or less than optimum. In economic market failures, the market is somehow not operating "as it should." The two main categories of market failures are externalities and a lack of competition.

Externalities are hidden costs and benefits associated with the production of a good or service that are not fully experienced by the individual producing or consuming it, but exist as a byproduct of such production or consumption. A negative externality is an undesirable consequence of production that is not experienced directly by the producer, while a positive externality is a desirable consequence that is a byproduct of production and also not experienced by the producer. A classical aviation example of externalities would be the congestion costs imposed by increased general aviation activity. In the US, the number of active general aviation aircraft increased from 131,743 to 223,370 from 1970 to 2010 (BTS, 2010). General aviation traffic does not pay user's fees for their share of consumption of airport and air traffic control resources. These are subsidized by the federal government, or in the case of large airports, by the landing and other fees imposed on commercial airlines. However, a high volume of general aviation traffic leads to increased congestion at large airports, ties up air traffic control resources, and imposes costs on all aircraft. If there were user's fees for general aviation, and the pilot experienced the full cost of his consumption of resources and the congestion he creates by being airborne, the volume of such activity would drop off sharply. Since the general aviation pilot only experiences a part of the true costs of flying, he *overproduces*, in this case by flying a lot more than he would if he were experiencing the full costs of flying.

Similarly, positive externalities exist when a remote city or rural area is connected to the rest of the country by air travel. Apart from the revenues generated by ticket sales to and from the area, the hidden benefits of air travel include job creation from the airline and airports in the area, increased possibilities for commerce and enterprise due to increased connectivity, the possibility of emergency relief and aid in times of natural disaster, and so forth. These externalities are not experienced by the airlines who seek to provide service to the region—all they experience are low revenues and high costs associated with operating in a thin market. Therefore, they *under produce*, by choosing not to operate that route since it will have a poor profitability. This is the role of the Essential Air Service Program, which subsidizes such unprofitable routes due to the presence of substantial positive externalities. The subsidies pass on some of the benefits of the positive externalities to airlines, allowing them to operate in such thin markets.

The second kind of market failure is a natural lack of competition. Some industries involve such a high level of fixed cost investment, high entry barriers, and great economies of scale that natural monopolies are the only sustainable firm structure. Large aircraft manufacturing is a good example of such an industry. It cost approximately $14 billion to develop the Airbus A380. The Boeing 777 cost approximately $5.5 billion to develop, and it had been based on an existing product platform. This establishes high barriers of entry into the industry and the economies of scale that are derived from mass-production make sustaining vigorous competition extremely difficult. Further, the uncertainty of the revenue stream is always extremely high. In other words, even though the manufacturer might collaborate with its customers in designing and building the aircraft, the volatile nature of the airline industry, the unstable competitive structure due to bankruptcies, changes in economic circumstances, and so forth make the end-revenues of a product extremely uncertain and raise the cost of investment. Therefore, the manufacturing industry is dominated by a few large players, and in order to keep the competition going, even if it devolves into a duopoly, there needs to be a stable source of income in order to effectively cross-subsidize the business of manufacturing large jets. In the case of Boeing and Airbus, this takes the shape of government defense and civilian contracts, as well as direct government subsidies. Many pundits have argued that the commercial airports enjoy significant monopoly power and to prevent monopoly profit, airport charges such as landing fees must be regulated. Globally, many airports face little or no competition and without regulation, they may charge monopolistic prices.

Government failure occurs when a government attempts regulation, but does so inefficiently and the resulting allocation of resources is inferior to that achieved by the free market. Airline regulation was an example of such a failure. Though a case may be made for its benefit, it has been shown in numerous studies like the US General Accounting Office studies (1996), Morrison and Winston (1995), and Goetz (2002) that average airline fares have declined through small, medium, and large airports since deregulation, due to competition engendered by the emergence of LCCs. The US General Accounting Office Study (1996) from their sample of 112 airports across the US noted that airfare declines have been observed in 79, or 70.5 percent of airports between 1978 and 1998. In real terms, passenger yield has fallen nearly 40 percent in small airports from 1979–2008, 12 percent in medium airports, and 20.6 percent in large airports. Economically, the decline in passenger yield has not been nearly so pronounced in large hub airports compared to small airports because of the hub premium. Most large airports are dominated by one major carrier, with monopoly or near-monopoly pricing power. Therefore, the premium associated with a hub airport is nearly always greater than the premium associated with smaller airports in terms of passenger yield. Arguably, small community airports have benefited the most from deregulation in terms of passenger yield. Further, the total number of scheduled departures in small, medium, and large hubs has also generally increased, as Figure 2.1 demonstrates.

However, studies like Brenner (1988) and Anderson, Gong and Lakshmanan (2005) have found decreasing competition in the airline market due to bankruptcies and consolidations, as well as deteriorating competition in major hub airports as a result of the dominance of major hub airlines. While this trend was indeed the case through the first 20 years of deregulation, the hub dominance of airlines like Delta at Atlanta was slowly eroded over time. Delta's market share at Atlanta (percentage of scheduled revenue passengers enplaned through Delta as a percentage of total scheduled revenue passengers) rose from 49.8 percent to 83.6 percent from 1977 to 1993, and has fallen to 63.2 percent in 2011 (see Table 2.1).[8] This is primarily due to the emergence of viable competitors like AirTran that is now responsible for over 17.1 percent of the market share at Atlanta. Therefore, it may be argued that in the long term, airlines return to a more or less competitive state in a deregulated environment, as a result of the threat of new entrants, the lowered entry barriers, and the accountability that is engendered through market forces.

---

8   Bureau of Transportation Statistics 2012–Airport Snapshots and Market Shares.

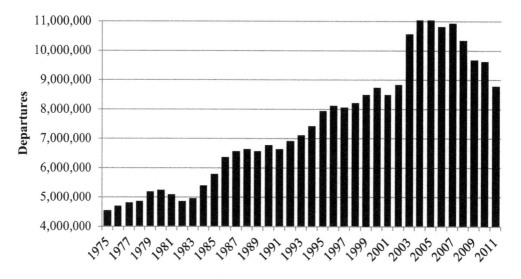

Figure 2.1   Scheduled Departures within the US

Source: Compiled by the authors using Form 41

Table 2.1   Hub Dominance Before and After Deregulation: 1977, 1993, and 2011

Airports	1977		1993		2011	
	Dominant Carrier	Market Share	Dominant Carrier	Market Share	Dominant Carrier	Market Share
ATL	DL	49.9%	DL	83.7%	DL	63.2%
CLT	EA	74.9%	US	94.8%	US	52.3%
CVG	DL	35.1%	DL	90.2%	DL	37.8%
DEN	UA	32.4%	UA	52.6%	UA	24.3%
DTW	DL	21.3%	NW	78.1%	DL	44.5%
MEM	DL	40.5%	NW	76.6%	DL	34.1%
MIA	EA	30.6%	AA	60.0%	AA	69.3%
MSP	NW	46.3%	NW	83.8%	DL	50.1%
PIT	US	45.7%	US	89.4%	SW	19.4%
SLC	WA	40.0%	DL	74.7%	DL	42.7%
STL	TW	39.5%	TW	61.4%	WN	45.4%

Source: Back Aviation O&D Lux data and Bureau of Transport Statistics 2012

Yet another argument against deregulation is that airline bankruptcies since deregulation have increased dramatically, since no airline was allowed to go bankrupt under the CAB's fares and competitive strategies. As of April 2012, there have been about 47 airline bankruptcies in the US, among these 17 airlines ceased operations. According to a US General Accounting Office Study in 1996, quality of service indicators have emerged with mixed results at various airports across the US. Therefore, was deregulation a government success or failure? It seems to depend on one's perspective. Air travelers today enjoy a greater choice of carriers, a greater number of routes from which to choose and lower fares, all of which point to greater competition in the airline industry. On the other hand, airline profitability has fallen considerably, as have the compensation of employees. However, taken in its entirety, airline deregulation may be said to be a government success—or rather a success of the free market over the organizational power of the government in aviation.

> "American will ground some planes and resize our network."
>
> Thomas Horton, AMR, CEO
> After Filing for Chapter 11 bankruptcy, 2012

A more unqualified government success in terms of effective regulation is the field of aviation safety. Since the creation of the FAA in 1958, accident rates have dropped almost 92.5 percent, from approximately one every 12.5 million aircraft miles to one in every 166.67 million aircraft miles flown. Figure 2.2 shows the dramatic decline in air traffic accidents between 1960 and 2010. Extensive regulations in terms of pilot certification, training, aircraft airworthiness, operational directives, the institution of controlled airspaces, navigational aids, sophisticated weather and flight planning tools, and the standardization of scheduled air carrier operations has contributed to this decline, which has contributed considerably to the development of aviation by establishing an excellent track record for safety and accountability.

It may also be argued that much of this decline in commercial aviation accidents and fatalities is attributable to factors other than government regulation—the introduction of jet engines, for instance, improvements in avionics instituted by the private sector, improvements in pilot training, the rise of simulators as a flight-training tool—these are factors that could have arguably arisen even without government regulation. However, regardless of the relative percentages of aviation safety attributable to the market or the government, the dramatic decline in aviation accidents and fatalities since the 1960s, coupled with strong regulatory pressures on airlines and aircraft manufacturers to emphasize safety presents a remarkable example of the government and the market working in tandem to achieve an optimal result.

SUMMARY

This chapter introduces the reader to the economic way of thinking in the context of aviation. We presented economics as a science of choice amidst the scarcity

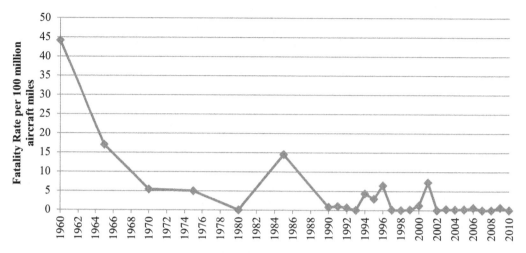

Figure 2.2   Accident Rates per Million Aircraft Miles Flown: 1960–2010

Source: BTS National Transportation statistics, Table 2.9: US Carrier Safety Data

of resources, and the concepts of incentives and opportunity costs which are fundamental to economic decision-making. We gave several examples of incentives and opportunity costs at work in the form of travel agents and aviation tax credits. We then presented the differences between microeconomics and macroeconomics, and how each might affect aviation managers in terms of variables they could control, and variables to which they had to devise an optimal reaction. We went on to present the role of government in an economy as a spectrum of government influence. On one extreme, we have *laissez-faire capitalism* with no government influence. On the other, we have a command economy in which the allocation of resources is entirely conducted by the government. Most modern economies throughout the world lie somewhere in between those two extremes; presenting a mixture of market-driven and government-regulated allocation of resources. We went on to show that even within a given economy, government influence is non-uniform, with some industries being more highly regulated than others, and even within an industry like aviation, some aspects—such as airline operations—are less highly regulated than aviation safety. We gave a brief history of government in aviation, noting that the industry has been intertwined with government since its inception and presenting macroeconomic and regulatory variables as extremely important decision-making inputs for the aviation manager. We analyzed government and market failures, examining the aircraft manufacturing industry as a market failure due to the high entry barriers that naturally restrict competition, and airline deregulation, as an example of a market success, where the FAA had removed restrictive regulation for the betterment of the industry. We mentioned aviation safety as an example of regulatory success, noting that while regulation might have had a significant impact on the falling accident rates, much of it may also be attributable to natural market forces such as emerging technologies. Having established the fundamentals of economic thought, the following chapter will focus on supply and demand, prices, and equilibrium in order to explore the implications and interpretations of this analysis in the context of aviation.

DISCUSSION QUESTIONS

1. Explain what the concept of opportunity cost means in economics.
2. How do individuals and firms generally make decisions concerning economic situations?
3. Give a standard definition of the science of economics.
4. What are the main differences between micro and macroeconomics?
5. How does the chapter classify government control of the economy?
6. List two types of perceived market failures.
7. What are some of the main problems associated with total government regulation of the aviation industry?

REFERENCES

Anderson, W. G. (2005). Competition in a Deregulated Market for Air Travel: The U.S. Domestic Experience and Lessons for Global Markets. *Research in Transportation Economics*, Vol. 13, 3–25.

Brenner, M. (1988). Airline Deregulation: A Case Study in Public Policy Failure. *Transportation Law Journal*, Vol. 16, No. 2, 179–227.

Bureau of Transportation Statistics (BTS) (2010). *Number of US Aircraft, Vehicles, Vessels, and Other Conveyances*, pp. Tables 1–11. Retrieved on May 13, 2012 from http://www.bts.gov/publications/national_transportation_statistics/html/table_01_11.html. Last accessed August 16, 2012.

Colander, D. (2004). *Microeconomics*. New York: McGraw-Hill.

Fried, W. and Myron, K. (1997). *Avionics Navigation Systems*, 2nd ed. New York: John Wiley & Sons.

Goetz, A. (2002). Deregulation, Competition, and Antitrust Implications in the US Airline Industry. *Journal of Transport Geography*, Vol. 10, No. 1, 1–19.

Heppenheimer, T. (2001). *Turbulent Skies: The History of Commercial Aviation*. New York: Sloan Technology Series.

Hirschey, M. (2006). *Managerial Economics (11th ed.)*. Mason, OH: South-Western.

Kane, T., Holmes, K. and O'Grady, M. (2007). *2007 Index of Economic Freedom*. Washington, DC: The Heritage Foundation.

McGuigan, J., Moyer, R. and Harris, F. (2008). *Managerial Economics: Applications, Strategies, and Tactics (11th ed.)*. Mason, OH: South-Western.

Morrison, S. (1997). Airline Deregulation and Fares at Dominated Hubs and Slot-Controlled Airports. *Hearing before the Committee on the Judiciary United States House of Representatives*.

Morrison, S. and Winston, C. (1995). *The Evolution of the Airline Industry*. Washington DC: The Brookings Institution.

US Census Bureau (2011). *Income, Poverty, and Health Insurance Coverage in the United States*. Retrieved on August 16, 2012 from http://www.census.gov/prod/2011pubs/p60-239.pdf .

US General Accounting Office (1996). *Airline Deregulation: Changes in Airfares, Services, and Safety at Small, Medium-sized, and Large Communities*. Washington DC: US General Accounting Office.

Vasigh, B. and Haririan, M. (2003). An Empirical Investigation of Financial and Operational Efficiency of Private Versus Public Airports. *Journal of Air Transportation*, Vol. 8, No. 1, 91–110.

3

Supply and Demand: Analysis in the Airline Industry

Economics has many substantive areas of knowledge where there is agreement but also contains areas of controversy. That's inescapable.

Ben Bernanke, Chairman of the Federal Reserve

Airline demand and supply analysis is concerned with understanding passenger behavior, measuring and characterizing the airline response to a change in ticket prices or incomes, and deriving the demand-side information necessary to make sound supply-side business decisions. In this chapter, we present supply and demand analysis in the context of airlines, airports, and aircraft manufacturers. We examine the characteristics of supply and demand such as price and income elasticities and we outline the factors that determine and shift the two market forces. We present market equilibrium analysis and analyze market disequilibrium in the context of supply or demand disruptions in aviation. We present examples of the efficiency of market allocation of resources, contrasting with a regulated allocation, using general aviation user fees and airport landing slot allocation as examples. We conclude by stressing the importance of a thorough understanding of the factors influencing supply and demand for an aviation manager.

- Basics of demand
  - Law of demand
    - Demand curve
    - Derived demand
    - Direct demand
    - Demand function
  - Determinants of demand for air transportation
  - Inverse demand function
  - Other demand functions
  - Characteristics of demand for air transportation
- Basics of Supply
  - Factors affecting supply of airline services
  - Characteristics of supply for airline services
- Market Equilibrium
  - Changes in equilibrium

- Equilibrium price maximizes consumer well-being
- Price controls
- Consumer and Producer Surplus
- Airport landing fees and airport congestion
- Price floors
- Price ceiling
- Disequilibrium
- Elasticity
- Price elasticity
- Cross-price elasticity
- Income elasticity
- Pricing and elasticity application
- Summary
- Discussion Questions

BASICS OF DEMAND

Demand may be defined as the ability and willingness to buy specific quantities of a good or a service at alternative prices in a given time period under *ceteris paribus* conditions.[1] Understanding demand theory and the demand function is one of the more important aspects for any business, since the characteristics of demand will dictate the patterns and characteristics of sales. For example, if an airline were to drop ticket prices, would the resulting increase in passenger demand cause an increase or decrease in passenger revenue? At what point would the increased sales generated by low ticket prices start to hurt passenger yield? What are the macroeconomic variables that could positively or negatively affect air transport demand? Demand analysis will shed light on these and other business-critical decisions.

> Revenue passenger miles (RPMs) are measures of traffic for an airline obtained by multiplying the number of revenue-paying passengers aboard the aircraft by the distance traveled.
>
> Airlines determine revenue ton kilometers (RTK) by multiplying the weight of paid in tonnage by the total number of kilometers it has been transported.

The quantification of demand varies from industry to industry. For instance in the airline industry, demand is usually expressed in terms of:

- the number of passengers (PAX);
- revenue passenger miles (RPMs), which normalizes passenger demand according to the number of miles traveled;
- revenue passenger kilometers (RPKs), which normalizes passenger demand according to the number of kilometers traveled;
- revenue ton miles (RTMs);

---

1   Ceteris paribus is a Greek term for all else being equal or for everything other than the variable of interest being held constant.

- revenue ton kilometers (RTKs).

The rationale behind normalizing by miles is that raw passenger numbers are only partially informative—a passenger traveling on a 100-mile route from New York, NY to New Haven, CT is not worth the same to an airline as the passenger who travels close to 9,500 miles from New York to Singapore. Therefore, expressing demand in RPMs gives a sense of airline traffic in terms of distance, and is the preferred metric for airline demand analysis throughout the industry. For aircraft manufacturers, demand would be represented as the number of aircraft sold. While demand varies from industry to industry, its characteristics remain similar, and its importance to business is always high. Therefore, it is critical to fully understand the nature of demand.

The law of demand

The law of demand states that, *ceteris paribus,* as price increases, the quantity demanded decreases. In other words, quantity demanded has a negative relationship with price.

In order to understand the law of demand's practicality, consider a transcontinental flight from New York to Los Angeles. For this round-trip flight, what would be the maximum a passenger would be willing to pay? $500? $1,000? $5,000? The fact is that at some price the passenger would consider it too expensive to fly and not take the trip. This decision to not fly is the law of demand in practice; that is, at some price the quantity demanded for the individual will decrease. The amount by which the quantity demanded decreases with ticket prices will vary by consumer characteristics and preferences—this responsiveness of quantity demanded to price is termed demand elasticity, and we will explore that at length later in the chapter, since it is the basis of airline pricing and revenue maximization. No matter what the extent of the response of demand, the law of demand states that the quantity demanded will fall with an increase in price.

When this decision is presented to all possible travelers, there will be different responses because people have different incomes and purposes for travel. These different responses help create a demand schedule, which is simply a table showing the quantities of a good that customers are *willing and able* to buy at alternative prices in a given time period, *ceteris paribus*. Such a table outlines the number of customers who would purchase a product or service at the given price. It is important to remember that demand is cumulative; that is, a consumer who is willing to pay $1,000 for the flight would certainly also be willing to pay $500 for the same flight. Table 3.1 provides a hypothetical demand schedule for the New York to Los Angeles flight.

The demand schedule contained in Table 3.1 highlights the law of demand since the quantity demanded (number of passengers) for the $200 airfare is significantly more than the quantity demanded for the expensive $5,000 airfare. At an airfare of $5,000, very few people are willing and able to pay for the ticket—perhaps only a few extremely wealthy and/or time-sensitive passengers.

Table 3.2 presents data from the BTS O&D Database on ticket prices of the four carriers along that route: American Airlines, US Airways, Delta Airlines, and Virgin America. The various observed fares were divided into 29 bins of $150 width. Fares lower than $100 were omitted from the dataset because of non-revenue passengers or airline-subsidized passengers like crewmembers or family who generally pay artificially low prices.

Graphically, the above data yields the demand curve shown in Figure 3.1:

Table 3.1   Hypothetical Demand Schedule: JFK–LAX

Ticket Prices New York–Los Angeles	Quantity Demanded Number of Passengers
$200	735
$500	690
$1,000	615
$1,500	540
$2,000	465
$2,500	390
$3,000	315
$3,500	240
$4,000	165
$4,500	90
$5,000	15

Table 3.2   Observed Average Fares for Route JFK–LAX, 2009 Quarter 2

Average Fare	Number of Passengers	Average Fare	Number of Passengers
$150	95,110	$2,100	210
$300	59,450	$2,250	340
$450	24,250	$2,400	1,130
$600	14,650	$2,550	50
$750	7,550	$2,700	30
$900	4,390	$2,850	20
$1,050	2,020	$3,000	10
$1,200	4,060	$3,150	20
$1,350	4,250	$3,300	10
$1,500	2,470	$3,450	10
$1,650	1,700	$3,600	10
$1,800	1,460	$3,750	10
$1,950	490	$3,900	10

Source: Compiled by the authors using Back Aviation O&D data
Note: American Airlines, Virgin America, US Airways, Delta Airlines.

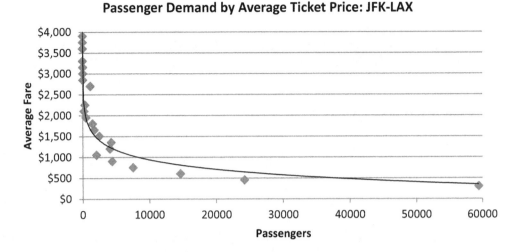

Figure 3.1   Observed Demand Curve for JFK–LAX Route

Source: Compiled by the authors using Back Aviation O&D data

As can be observed, the empirical demand curve is downward sloping, confirming the law of demand presented above, and demonstrating a negative relationship between quantity demanded and price. The curve is concave to the origin, suggesting that at some point, demand increases tremendously as a result of decreasing price, but above that point, demand is much less responsive to changes in price. For the JFK–LAX route, the abrupt change in demand elasticity occurs somewhere around $350.

Omitting the lowest four fare classes, and just graphing the demand curve at prices above the point of changing elasticity, the following graph shown in Figure 3.2 occurs:

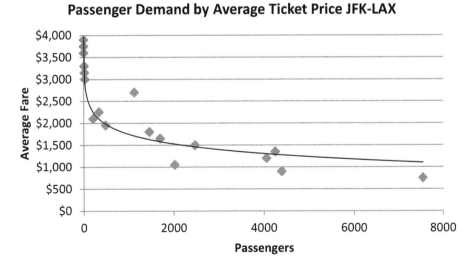

Figure 3.2   Observed Demand Curve for JFK–LAX Route: Omitting Four Lowest Fare Classes

Source: Compiled by the authors using Back Aviation O&D data

Here too, a negative relationship between price and quantity emerges, but the responsiveness of quantity demanded to changes in price is much lower. This clearly demonstrates the relative inelasticity of demand at lower ticket prices and is the basis for revenue management strategies in airlines.

> The law of demand states that, all else equal, consumers will buy less at a given time period when price of a product increases and buy more as price decreases.

Demand curve   From the demand schedules above, a demand curve can be constructed. A demand curve is a graphical description of the demand schedule, and the quantities of a good that customers are willing and able to buy at alternative prices in a given period of time. The law of demand can then be derived from this schedule and it shows that the demand curve will always be downward sloping.[2] It should be noted that the demand curve does not portray actual purchases, but only what consumers would be willing and able to purchase. Figure 3.3 provides the demand curve for the New York to Los Angeles trip. In this example, the demand curve is a linear negative sloping line.

The demand curve in this example is linear and downward-sloping. Both vertical and horizontal intercepts have an economic interpretation—the vertical intercept is called a "choke price," which is defined as the price at which demand shrinks to zero. In other words, this price is above what every consumer in the market would be willing to pay for the trip between JFK–LAX. The horizontal intercept is the limit of demand even if the travel were offered free of charge. This could represent the entire size of the market, since,

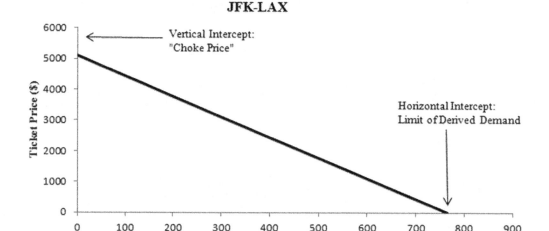

Figure 3.3   Abstract Demand Curve for JFK-LAX Flight

---

2   Stated exceptions to the law of demand usually involve confusion between the perceived quality and/or prestige that a high priced good confers on the purchaser. A simple thought experiment confirms this. Imagine that the exact same quality and/or prestige could be achieved at a lower price. Rational consumers would always select the lower priced good for the exact same quality and/or prestige. Therefore, it is the quality and/or prestige that the good has and not the high price that the consumer is responding to.

as discussed below, the demand for air transportation is a derived demand. The demand for air transportation is based on people's willingness to travel, which is itself contingent on their reasons for travel. Even if the trip between JFK and LAX were to be offered free of charge, demand might increase slightly due to an increase in vacationers and other leisure travelers, but would not increase greatly, as the underlying factors affecting the desire for travel have not changed.

Derived demand   An important aspect of air transportation demand analysis is the fact that the demand is *derived,* meaning that it is derived from the demand of consumption or utilization of another good or service. In other words, the demand for air transportation stems from factors other than the transportation itself. Individuals do not directly demand travel—their demand stems from the reason they desire to undertake the travel. Once they have decided to travel, they choose between different modes of transportation, and later, between different airlines. However, their core demand stems from their reasons to travel. A managerial implication of this analysis is that an *absolute* drop in prices might not necessarily cause an increase in travel if all the competing airlines on a particular route simultaneously drop their prices as well. In a derived-demand market, only *relative* price decreases matter, since the overall quantity demanded cannot necessarily be increased by a drop in prices. If every airline were to cut its prices for the JFK–LAX route by $100 tomorrow, travel might increase slightly due to the leisure segment, but overall, not many more people would necessarily want to go to JFK–LAX merely because it is cheaper to do so. However, if Virgin America alone were to cut its fares by $100, while American Airlines, Delta, and US Airways kept their prices unchanged, Virgin America could take customers from the other three airlines, thus increasing the demand for its product. However, derived demand markets tend to present zero-sum games in terms of decision-making equilibrium. One player gains at the expense of another, and this lends itself to price matching and price wars. The demand for that travel route is derived, and *ceteris paribus*, consumers will switch over to the competitor at the expense of the airline they are currently flying if the rival airline lowers its price. The derived nature of demand in the air transportation industry might also account for the slim operating and net profit margins in the industry; that is, zero-sum game markets tend to lend themselves to competition where each player earns zero economic returns.

Direct demand   In contrast to derived demand, direct demand refers to demand for goods and services meant for final consumption. For example, if you fly general aviation aircraft to enjoy flight, then your demand for aircraft is a direct demand. For commercial airlines, each aircraft has a unique cost/revenue analysis that makes it suitable for certain airlines. The pricing implication of this is that a drop in the price of aircraft will lead to an increased demand for aircraft in terms of market growth — airlines that had previously been unable to afford a certain type of aircraft due to fixed cost constraints will now be able to do so. Itineraries that were previously deemed uneconomical might suddenly become profitable, and the size of the market could grow. Therefore, a managerial implication is that for the manufacturer, demand and market size can be directly influenced by decisions about types of aircraft, and competition is not necessarily a zero-sum game. Add to this large degrees of technological differentiation and one gets a scenario where Boeing can create and capture an entirely different segment of the market than Airbus. Although both companies compete for the same set of customers, high levels of product differentiation and a direct demand for aircraft make them both able to segment the market and benefit from increasing numbers of customers.

Demand function   Using information from the demand schedule and/or the demand curve, a demand function can be constructed. A demand function is simply the functional relationship between the quantity demanded and factors influencing demand. There are numerous factors that influence the demand for air travel, for example, in the New York to Los Angeles market, and therefore price is not the sole determinant of demand. (However, when the other determinants are held constant, the relationship between price and quantity demanded is always negative.) If, however, some of those other characteristics are varied, the quantity demanded can change accordingly. For instance, if a drop in the price of airline tickets occurs in conjunction with a global influenza pandemic, the quantity of airline tickets might decrease, even though prices of tickets dropped simultaneously.

Demand functions may be classified into two categories based on their composition: *implicit* or *explicit.* Implicit demand functions simply state a general relationship between the quantities demanded and the factors affecting demand. Explicit demand functions are mathematical relationships between the quantity demanded and the various variables impacting demand. Implicit functions do not have the actual mathematical relationships but rather a more generalized statement of the factors affecting demand. For example, the implicit demand function for the New York to Los Angeles flight could be:

$$D_{NY-LA} = \text{Function } (P_X, P_Z, Y, A_{NY-LA}, PA_{NY-LA}, H)$$

Where:

- P_X is the own ticket price;
- P_Z is the competition's ticket price;
- Y is the annual income or state of the economy;
- A_{NY-LA} is the availability of other mode of transportation between JFK–LAX;
- PA_{NY-LA} is the price of other mode of transportation between JFK–LAX;
- H is a composition of other factors such as service, customer loyalty, safety, cabin amenities, and random factors.

The implicit demand function simply states that a relationship exists between dependent variable and the independent variables, but it does not state the extent to which the variables are related. The numerical relationship that displays the degree of influence that each factor has on the quantity demanded is the explicit demand function. Using the information obtained from the demand schedule and the demand curve, an explicit demand function for the New York to Los Angeles flight could be written as:

$$D_{NY-LA} = 15000 - 2P$$

Based on this linear demand function, two statements can be made about the nature of demand for a round-trip New York to Los Angeles flight. First, when the price of the ticket is $7,500, then demand for this trip drops to zero. In other words, demand for any product or service is always limited by the extent of the market demand (in this case any price above $7,500).

The other statement concerning the linear demand curve is that the negative price coefficient (or slope) is -2, which means that for every dollar increase in the ticket price, the demand drops by two passengers. This change in demand occurs for all price points, creating a constant negative slope.

While the linear demand function is simplistic and clear to understand, the fact remains that the demand schedule rarely has a perfect linear form. This makes intuitive sense as passenger demand does not drop off evenly with price increases, but instead drops off in steps. Frequently, there is a major inflection point where demand decreases dramatically. As demonstrated in Figure 3.1, for the JFK–LAX route, this point of inflection occurs around $350. Above this point, demand is much less responsive to changes in price. Below that point, demand increases rapidly. In other words, in our example, a $300 decrease in price from $1,800 to $1,500 will only increase demand by 1,010, whereas a $300 decrease in price from $600 to $300 will increase demand by 44,800.

Two other types of demand functions are the semi-log function and the log-linear function.[3] The general forms for the two functions are:

Semi-log: $LnQ_D = \beta_o + \beta_1 P$

Log-linear[4]: $LnQ_D = \beta_o + \beta_1 LnP$

Where:

- Q_D is the quantity demanded;
- P is the ticket price;
- β_o and β_1 are the coefficients for the constant term and the price variable respectively.

For the airline industry, it is usually assumed that the typical demand function takes the log-linear shape. The general shape of a log-linear function is contained in Figure 3.4.

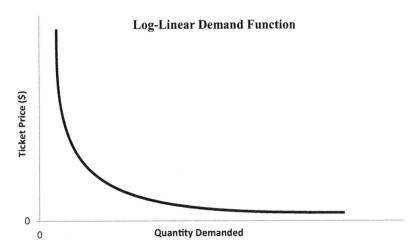

Log-Linear Demand Function

Ticket Price ($)

Quantity Demanded

Figure 3.4   Conceptual Log-linear Demand Curve

---

3   The log-linear function is also sometimes referred to as the log-log function.
4   Note that for log-linear demand functions, the elasticities of demand are the coefficient of the variables and are constant. For instance, the price elasticity of demand would be simply β_1.

Unlike a linear demand function where the slope of the function is constant throughout, the slope of a log-linear demand function changes. In Figure 3.4, the initial slope of the function is fairly steep, indicating that a unit drop in the ticket price does not generate a similar increase in the quantity demanded. This is partly a result of the fact that the ticket price is still considered expensive by the majority of potential customers. Eventually, as the price decreases, the quantity demanded for the product or service becomes greater and greater to the point where a small drop in the price generates a large increase in the quantity demanded. Once again, referring to Figure 3.1, the demand curve for the JFK–LAX route can be best approximated by a log-linear function, both before and after the demand inflection point.

Determinants of demand for air transportation

Some of the factors that affect this demand are:

- ticket price;
- competitor's ticket price;
- passenger income;
- state of the economy;
- availability of other modes of transportation;
- passenger loyalty;
- in-flight amenities;
- frequency of service;
- safety;
- aircraft type, location of airport;
- random factors, such as volcano ash cloud,[5] SARS, and 9/11 or threat of terrorism.

To illustrate the impact that these factors might have on the demand curve, consider the previously used demand schedule for the New York to Los Angeles flight. Table 3.3 provides an updated version of that demand schedule.

In order to highlight the effect of the ticket price on the demand curve, refer to Figure 3.5. As we said earlier, a change in the ticket price is always defined as a *ceteris paribus* movement along the demand curve. For example, if the current ticket price for the flight was set at $3,000 and then lowered to $2,000 (all other things remaining constant), the only effect this change would have would be on the quantity demanded. This is reflected in the demand schedule from Table 3.3 in column two where the quantity demanded moves from 315 to 465. This movement is displayed graphically in Figure 3.5, where the quantity demanded moves from point A to point B along the demand curve.

Ticket price is the only determinant of demand that causes a movement along the demand curve; changes in the other determinants of the demand cause a shift in the entire demand curve. Continuing with the New York to Los Angeles example, suppose that a competitive airline decreases air fares. This will cause demand for this airline's flight to increase as some consumers switch to this airline. This increase in demand is reflected in Table 3.3's third column, where the quantity demanded at each price level increases by

---

5   In 2010, the Icelandic volcano eruption forced most countries in northern Europe to shut their airspace, grounding more than 100,000 flights and an estimated 10 million travelers worldwide.

Table 3.3   **Demand Schedule for JFK–LAX**

Ticket Price JFK–LAX	Quantity Demanded Number of Passengers	New Quantity Demanded Number of Passengers
$0	765	975
$200	735	945
$350	713	923
$500	690	900
$1,000	615	825
$1,500	540	750
$2,000	465	675
$2,500	390	600
$3,000	315	525
$4,000	165	375
$5,000	15	225
$5,100	0	210

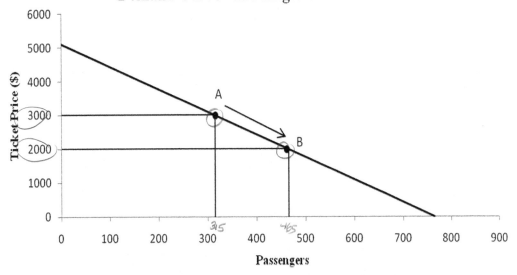

Demand Curve - Los Angeles to New York

Figure 3.5   **Change in Quantity Demanded for JFK–LAX Flight**

210 passengers from column two. The increase in the quantity demanded across all price levels creates a rightward shift in the demand curve from D_1 to D_2, as depicted in Figure 3.6. For every price level, the quantity demanded has increased by 210. For instance, at the price of $3,000, 525 passengers would be willing to travel from Los Angeles to New York instead of 315. This is represented by a shift in the demand curve from A to B. Conversely, a negative impact on demand would create a leftward shift in the demand curve. Finally, an increase in demand also causes an increase in the choke price of a market from D to C. In other words, the absolute maximum price beyond which a consumer would not be willing to purchase air travel will increase, since the quantity demanded has increased at every price level.

The demand for a single flight is affected by multiple factors but for many, especially price-sensitive leisure travelers, the price of the flight and the price of competing flights are probably the most important variables. With the advent of the Internet and numerous price aggregation websites like Expedia, Travelocity, and so on, price and competitors' prices have probably become even more important as airline ticket price information is readily available to potential customers. However, these price variables affect different segments of the population in different ways. For time-sensitive travelers, ticket price versus a competitor's ticket price may not be as important as for price-sensitive travelers.

For the competitor's price variable, the coefficient would be expected to be positive, since an increase in the competitor's price would make the competition's product less competitive, and ultimately increase demand for the company's product. Consider airlines A and B who both have flights on the same route and both have introductory fares of $400. If airline B were to increase its airfare to $450, this would make it less competitive and ultimately increase demand for airline A. Conversely, if airline B were to drop its airfare to $350, this would make it more competitive, which would decrease demand for airline A and increase demand for airline B. Regardless of the viewpoint, the price coefficient would be negative and the cross-price coefficient would be positive.

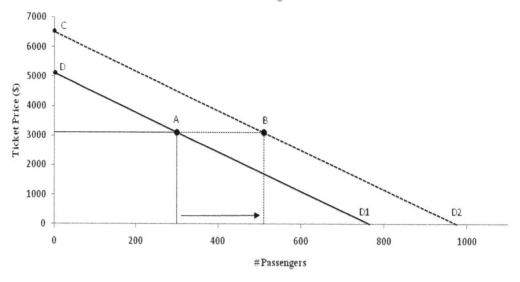

Figure 3.6   Change in Demand

Inverse demand function

In the airline industry, revenue managers are mostly interested in knowing their pricing power and the factors influencing passengers' behavior in response to variation in ticket prices. The inverse demand function is also called the price function, and it treats price as a function of quantity demanded. The inverse demand function is given by $P(Q) = a + b(Q)$, where a is the intercept and b is a number always less than zero. To compute the inverse demand function, simply solve for P in the demand function. For example, if the demand function is $Q = 350 - 0.50P$ then the inverse demand function would be $P = 700 - Q$.

Suppose the inverse demand function for an airline flying from London Heathrow (LHR) to Beijing China (PEK) is given by:

$$P = 15,000 - 75Q$$

According to the above equation, the maximum the airline can charge to sell 160 seats is $3,000 per seat.

Other demand functions

The general form of a log-linear demand function can be expanded to explicitly include the competition's price:

$$LnQ_D = \beta_o + \beta_1 LnP_X + \beta_2 LnP_Z$$

Where:

- β_o is a constant;
- P_X is the own price;
- P_Z is the competitor's price;
- β_1 and β_2 are the coefficients, β_o is the intercept.

Income is another important determinant of demand. Consumers with higher incomes are able to purchase more goods and services; therefore, an increase in disposable income will provide an increase in demand for air travel. Additionally, increased consumer income is usually correlated with increased business activity, indicating a higher demand for business travel. Because of this direct relationship between demand and income, the coefficient for consumer income is positive. The general log-linear demand function for air transportation can now be expanded to include income:

$$LnQ_D = \beta_o + \beta_1 LnP_X + \beta_2 LnP_Z + \beta_3 LnY$$

Where:

- β_o is the intercept;
- β_1, β_2, and β_3 are the coefficients.

While own price, substitute price, and income are the three major determinants of demand for air transportation, there are other factors that affect demand. An important one of these is the availability of substitutes. For air travel, this includes other modes of transportation. In the US, driving is a reasonable substitute for many short flights, while in Europe high-speed rail can greatly impact the demand for select air services. For example, the high-speed Brussels to Paris rail line has created a situation where there is little demand for Brussels to Paris air transportation. However, in situations where there is a lack of other modes of transportation, for example, air services to a remote Caribbean island, demand for air travel can be expected to increase. Based on this discussion, the expected coefficient for the availability of substitutes could be either positive or negative.

The final pricing variable that impacts the quantity demanded is the price of a complementary product or service. A complementary product or service is a product or service that is usually used jointly with the primary good. Examples of a complement for the airline industry are the hotel and rental car industries. Since many leisure and business travelers have to stay in hotels while on their trip, the price of the hotel will impact the demand for air travel. This is a result of demand for air transportation being derived (that is, being generated by something other than air travel). For example, if the average price of a night's stay in Cancun were to increase, less people would want to take a vacation in Cancun. Therefore, the demand for flights to Cancun will be reduced. Based on this typical example, one would expect the coefficient for the price of a complement to be negative.[6]

Other variables that might affect the demand for air travel include the quality of service offered, product differentiation, and so forth. These are typically known as 'quality' or 'hedonic' price determinants, and can play an important role in the demand for a good or product. Some of the hedonic indicators for airline demand are:

- flight frequency;
- connectivity between city-pairs;
- customer loyalty programs;
- service levels;
- on-time performance;
- customer service.

We will explore each in turn.

In a competitive market environment, frequency of an airline's service between two cities will also impact demand. Flight frequency is especially important for business travelers since they are generally more time-sensitive than leisure travelers. An airline with a large number of flights between two cities has a greater probability of meeting a traveler's schedule rather than an airline with only a few flights. Moreover, a robust flight schedule provides the traveler with greater flexibility in case of schedule changes. This is a primary reason why regional jets have become more popular; that is, they enable airlines to provide increased flight frequency while holding the total number of seats offered in the market steady. We would therefore expect a positive coefficient between flight frequency and demand.

Whether an airline has non-stop or connecting service between two cities will also affect demand. The availability of a non-stop flight will generally increase demand since passengers usually prefer non-stop flights over connecting flights. However, this assumption may not apply to all markets, especially ultra-long-haul markets where passengers may appreciate a stopover. Therefore, we would expect that a non-stop flight

---

6   While the coefficient for a competitor's price is positive.

variable would generally have a positive relationship with demand (with some possible exceptions on long-haul flights).

Customer loyalty is another key determinant of demand for air transportation. In this regard, one of the more successful marketing tools that the airline industry has implemented has been loyalty or frequent flier programs. By offering free flights and perks for loyalty, airlines have been successful in obtaining repeat business, especially among business travelers.

While customer loyalty is important, it is also important to attract new customers. Here airlines have stressed service. Service for airlines can include a host of factors including aircraft seat placement (more room), in-flight entertainment, food and beverages, airport amenities, baggage handling, and most importantly, friendly customer service. While service is ultimately an intangible variable, it does impact demand. Airlines that are perceived to provide a high level of service will have a greater demand for their flights. Airlines such as Virgin Atlantic, Emirates, and Singapore Airlines have been very successful at generating increased demand through a perceived level of strong service.

On-time performance is a metric gathered by the Bureau of Transportation Statistics (BTS) to rank airlines every quarter. This is often used as a proxy for customer service, and is an important determinant in public perception of an airline's quality. On-time performance influences the demand for an airline as poor on-time performance indicates an unpredictable and inconvenient service as compared to an airline with an excellent on-time performance record. By the same token, the public nature of the on-time performance metric sets up incentives for airlines to circumvent the process by practices like enplaning passengers and departing the gate on time, but shifting their expected delay to the ramp or taxiway area instead of delaying flight departure from the airport.

Finally, there are stochastic and random factors that can materially affect demand. For example, the terrorist attacks of September 11, 2001 were events that crippled demand for air travel for quite some time.

Characteristics of demand for air transportation

The demand for air travel also has many unique characteristics that present problems for the airline industry. While all of the following characteristics shape the demand for air transportation, only the first five will be examined in detail.

- fluctuations
- cyclicality
- seasonality and peaking
- directional flow
- perishability
- schedule wait time
- airport access time
- flight time
- hub connection time
- denied boarding time.

The first major characteristic of demand for air transportation is that, unlike the demand for many products, it is constantly fluctuating. Because of the numerous determinants highlighted in the preceding section, demand for individual flights is constantly changing. Moreover, no two routes exhibit the same properties of demand, making every route unique in its demand characteristics.

Cyclicality, a second characteristic of demand for air transport, refers to a long-term trend of peaks and troughs of economic activity. The national economy has long been known to experience such cyclicality. And, since the airline industry is highly correlated with the national and global economy, it is not surprising that it also experiences some cyclicality.

Figure 3.7 depicts the number of passengers carried by US airlines since 1970. While the chart shows overall steady growth, there remain three pronounced drops in US airline industry enplanements: the early 1980s, early 1990s, and post-9/11. The graph below indicates a cyclicality with general economic cycles, showing a dip in scheduled revenue passengers around the early 1980s then again in the early 1990s, coinciding with the two recessions in the last two decades. The slope coefficient of the linear trend line suggests that US enplanements have, on average, increased each quarter by 983,939.

Another major characteristic of air transport demand is peaking, or more commonly called seasonality. Unlike cyclicality, which is a long-term cycle, peaking is more of a short-term event where demand spikes. This is demonstrated by the quarterly peaks and troughs in demand in Figure 3.7 below, which represents within-year variations in demand that are nearly independent of economic cycles. The most common form of peaking is seasonality where demand increases during the summer months and then declines during the winter months. This trend is particularly true of leisure destinations, where the weather is more favorable and individuals have greater time off. For example, Mediterranean resort destinations are in high demand during the summer months, but demand declines during the winter months. This situation also applies to most domestic routes in the US. As a specific example, consider the Chicago to Seattle route. Figure 3.8 displays a 10 percent sample of the route's passenger enplanements per quarter since 2001. As Figure 3.8 shows quite clearly, the route experiences tremendous demand during

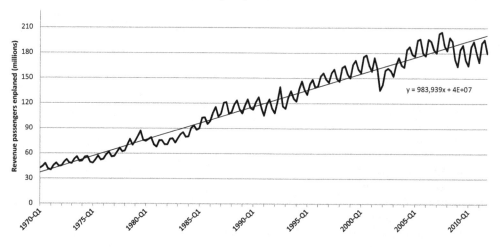

US Airline Industry Enplanements: 1970-2011

$y = 983,939x + 4E+07$

Figure 3.7   US Airline Industry Passenger Enplanements Since 1970

Source: Compiled by the authors using Back Aviation Form41 data

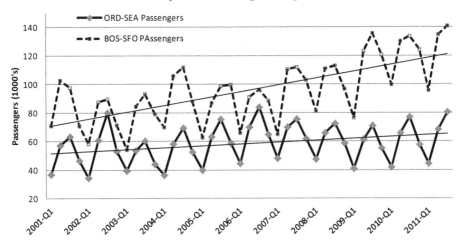

Figure 3.8   Seasonality of Chicago to Seattle and Boston to San Francisco Enplanements

Source: Compiled by the authors using Back Aviation O&D data

the third quarter of every year, while demand is quite low for the first quarter of every year. This is a fairly regular pattern partly attributed to Seattle's summer cruise ship industry and the Pacific-Northwest's more favorable summer weather. In addition to seasonality, other peaks in demand include Thanksgiving, Christmas holidays, and other public holidays.

Since peaking is fairly predictable, airlines can add capacity, if they choose, by either increasing the frequency of the flights on an existing route, or by introducing seasonal-only service. However, in order for airlines to be able to add seasonal capacity, they either have to take the aircraft off other routes or have excess capacity in reserve, which is very expensive. Ideally, the airlines would like to add aircraft during the summer months and retire them during the winter; however, short-term leases for aircraft are rare. Therefore, seasonality can present a sizeable financial, operational, and scheduling burden for airlines, as they want to be able to meet the seasonal demand, but must also bear the assets for the remainder of the year. This is one reason why North American carriers have robust aircraft maintenance schedules during the winter months when their schedule is not as busy. Some carriers, such as American Airlines and Air Canada, have been successful at moving capacity to Central and South America during the northern hemisphere winter season.

Another characteristic of airline demand that is similar to peaking is directional flow. Directional flow relates to the increased demand of passengers in one direction for a period of time. While cyclicality spans decades and peaking spans years, directional flow is usually assessed on a weekly basis and is fairly short term. An example of directional flow could include customers flocking to a city a few days before a major sporting event (such as the Super Bowl) and then immediately demanding to leave after the event has finished. The key to note is that directional flow is essentially one-way for a short period of time. This creates another unique scheduling problem for airlines, since, in order to accommodate the directional flow of passengers, some aircraft will be flown in a relatively empty condition in the opposite direction.[7]

The major problem with cyclicality, peaking, and directional flow is that demand for air transportation is perishable. The moment the plane leaves the gate, any empty seats are lost as revenue-generating products. This is, of course, not the same as a manufacturing industry where the company can keep its product in inventory for sale on another day. Therefore, the close matching of demand and supply is essential for success in the aviation industry. Because of situations such as peaking and directional flow, airlines are faced with the prospect that a good portion of their seats will go unsold, simply due to the nature of demand and the structure of their operation. Based on this, pricing is extremely important to help offset issues related to the structure of demand. Airline pricing policy and yield management will be covered in greater detail in Chapter 11.

BASICS OF SUPPLY

One of the key reasons why the airline industry has faced financial difficulties and has profit margins well below many other industries is that its demand fluctuates constantly but its supply is relatively fixed. Lack of flexibility in the supply function makes it very difficult to manage capacity effectively. The perishability of air transport services, the high fixed costs, and the predetermined capacity in the form of schedules that are published well in advance of the flight, make supply relatively unresponsive since an airline cannot shift its supply at short notice. Therefore, it is important to understand the factors impacting airline industry supply in some detail. The next section introduces these factors, while the following chapter will discuss airline production in greater detail.

Factors affecting supply of airline services

Following the previous definition of demand, supply refers to a firm's willingness and ability to provide a specific number of seats at a given price, time period, and market. In the airline industry, supply is the capacity of an airline to transport passengers, a function of offered routes and available aircraft. Supply is usually expressed in available seat miles (ASMs) or available ton miles (ATMs). An ASM is simply one seat carried through the air for one mile, regardless of whether it contains a passenger or not. The presence of a revenue passenger in the seat is the key difference between RPMs (demand) and ASMs (supply), as RPMs only measure seats that have a revenue passenger in them. Dividing RPMs by ASM gives us a key performance indicator for airlines, the *load factor*. This is defined as the percentage of capacity that has been matched with demand. Similarly, for cargo or freight transportation, the equivalent unit to RPMs is RTM, which may be matched up with ATMs to yield a cargo load factor. Load factors are critical to airline

---

7   While this previous example highlights a one-time directional flow event, rush hour traffic is an example of a continuous directional flow in demand. In most major cities, the roads are clogged with people attempting to head into the city center in the morning, while the other direction is usually fairly empty. For continuous directional flow in the aviation industry, Las Vegas is probably the best example. On Friday evenings there is demand for air travel to Las Vegas, while on Sunday evenings there is demand for travel out of Las Vegas as people want to spend a weekend in Las Vegas without skipping work. Directional flow of demand presents problems for airlines as they may want to capture the one-way demand, but will have problems filling their aircraft on the return flight for the same day.

performance, since they determine aircraft utilization, drive the profitability of a given route, and indicate the useful utilization of capacity.

> The Law of Supply states that at higher prices, producers are willing and able to produce more products. The quantity suppled increases as prices increase, and decreases as prices decrease, given everything else constant.

Some of the factors affecting supply are:

- ticket price;
- price of resource inputs:
  - fuel prices
  - labor costs
  - landing fees
  - aircraft costs
  - maintenance costs
- navigation charges;
- technology;
- availability of other mode of transportation;
- government regulation;
- stochastic factors:
  - weather condition
  - strikes.

The implicit supply function for the airline industry can be written as:

$$Q_S = f\{P, P_{RES}, Tech, Comp, Rand, GOV\}$$

Where:

- P is the ticket price;
- P_{RES} is the price of resources;
- Tech represents technological improvements;
- Comp is the behavior of the competition;
- Rand represents random factors;
- GOV is government regulation.

The major determinant of supply, just like demand, is the ticket price of the good or service. This relates to the law of supply in that the quantity of a good supplied in a given time period increases as its price increases, assuming all else is held constant. In the airline industry this simply means that airlines are willing to supply more seats as ticket prices increase. Based on the law of supply, the supply curve slopes upward up so that any change in price is simply a movement along the supply curve and this is referred to as an increase in the quantity supplied. Table 3.4 provides a hypothetical supply schedule for the New York to Los Angeles flight.

The next major determinant of supply is the price of resources. For the air transportation industry, production resources include, but are not limited to, aircraft, fuel, maintenance,

Table 3.4   Supply Schedule for JFK-LAX

Ticket Price JFK–LAX	Quantity Supplied Number of Passengers
200	15
500	315
1,000	465
1,500	540
2,000	615
2,500	653
3,000	690
3,500	701
4,000	713
4,500	724
5,000	735

labor, and landing fees. All these factors, and many more, impact supply because they affect the cost of production. If the cost of production increases, then the airline's total costs increase, causing a leftward shift in the supply curve. In the airline industry, an increase in the costs of production may force the airline to cut some flights that would no longer be profitable. This reduction in flights is a leftward shift of the supply curve (that is, fewer seats are offered at the same ticket price). Conversely, if the price of resources decreases, then the supply curve shifts to the right. (That is, more seats are offered at the same ticket price.)

Applications of the impact that the price of resources have on supply are numerous. For example, during bankruptcy protection, Delta Air Lines was able to significantly reduce its costs by receiving wage concessions and reducing aircraft leasing rates. As a result of this, Delta dropped many domestic destinations and redeployed resources to new international markets. Similarly, with skyrocketing fuel costs, airlines eliminated a number of flights that were previously viable, since revenues earned on those routes would not justify offering the flights.

Figure 3.9 exhibits a linear supply curve. As ticket price goes up, so does the number of seats offered. At a price of $3000 per ticket, an airline would be willing and able to supply 4,520 seats. The degree of responsiveness of quantity supplied to a change is price is called the price elasticity of supply.

The next major factor that determines supply for air transportation is technology. The impact of technology on the supply of air transportation has been vast. Technology has advanced civil aviation to the point where it is one of the safest modes of transportation. The introduction of the Boeing 747 in 1970 created a rightward shift in the supply curve as

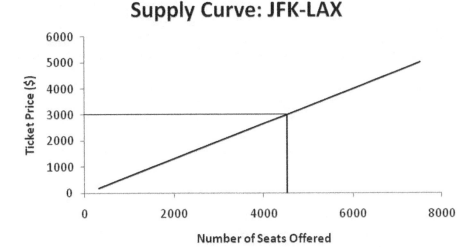

Figure 3.9   Supply Curve for JFK–LAX

the jumbo-jet was able to carry more passengers on a flight than any other aircraft (Boeing, 2007). Similarly, the 787-8 Dreamliner use 20 percent less fuel for comparable trips than today's similarly sized airplane (Boeing, 2010). The new Airbus 380 has also caused a shift in the supply curve of airlines that received the aircraft. However, perhaps one of the best examples of technology's impact on air transportation supply is the introduction of ultra-long-range aircraft. Prior to ultra-long-range aircraft such as the Boeing 777-200LR and the Airbus A340-500, non-stop routes such as Singapore to Los Angeles and New York to Dubai could not be operated. Finally, the introduction of Extended Twin-Engine Operations (ETOPS) enabled a greater supply of air transportation over the Atlantic and Pacific oceans through the use of smaller twin-engine aircraft.

Competitive factors are another important determinant of supply for air transportation. Since airlines have historically aggressively competed over market share, competition from other airlines has impacted supply considerably. Airlines regularly adjust capacity and supply in markets in response to competition and changing market forces; however, there have been cases where airlines have taken this to an extreme. For example, in 1999 American Airlines faced anti-trust lawsuits over its competitive actions against smaller rivals, particularly Vanguard Airlines out of Love Field (American cleared, 2003). The lawsuit alleged that American Airlines had "dumped" capacity on routes where Vanguard Airlines was competing with American (American cleared, 2003). Eventually American Airlines was cleared of the charge. Similarly in 2000, El Salvador-based Grupo TACA accused Continental Airlines of capacity dumping following a liberalization of the US–El Salvador bilateral air agreement (Knibb, 2000). These examples highlight many occurrences of how the supply curve will shift either to the left or right for an airline depending on the actions taken by other airlines. Since the financial downturn in the US aviation industry, airlines have begun to compete less on market share and focus more on profits. Just as in demand, random factors play a large role in affecting supply. While the tragic events of September 11, 2001 created a dramatic one-time shift in the demand curve, they also impacted the supply curve, albeit not to the same extent. Because of the terrorist attacks, all commercial flights were grounded for two days, then, when flights resumed, new security procedures were introduced that added significant costs. Additionally, for the two carriers involved in the terrorist incident, the lost aircraft reduced their fleet size and ability to transport passengers.[8] As another example, the airspace over Central and Eastern Europe was closed for several days following a volcano eruption in Iceland. The volcano caused widespread disruption at airports in Britain and other parts of northern Europe, grounding 1,000 flights and delaying hundreds of thousands of passengers.

Deregulation and liberalization of air transport have also significantly affected the supply curve. Since regulation generally prohibits market forces from determining supply, there is usually an artificial cap set on supply. Therefore, when air transport deregulation occurred, the artificial cap on supply was withdrawn and supply subsequently increased. The US–EU open skies agreement which took effect in 2008 eliminated air service restrictions between the US and Europe and allowed airlines to fly for the first time between any EU city and any US city. A more current example would be the US–China air transportation agreement. The current agreement limits the amount of supply between the two countries, thereby placing an artificial cap on supply. However, whenever additional rights are granted, the supply of air transportation between the two countries will shift to the right. Further discussion of international aviation is contained in Chapters 4 and 5.

The final factor impacting air transportation supply is government regulation. Despite the fact that aviation industry has gone through massive privatization and financial deregulation, almost all aspects of safety and operations remain highly regulated. In the US, the federal government completely controls all US airspace, and airlines are totally dependent on the Federal Aviation Administration (FAA) to operate that airspace in a safe, efficient, and affordable manner. On January 20, 2005, the FAA mandated that the vertical separation between aircraft above the US at altitudes from 29,000 to 41,000 feet be reduced from 2,000 feet to 1,000.[9] The Domestic Reduced Vertical Separation Minimum (DRVSM) rule was designed to increase airspace capacity. This action has allowed more available routes for airlines and has had the effect of shifting the supply function to the right (Figure 3.10).

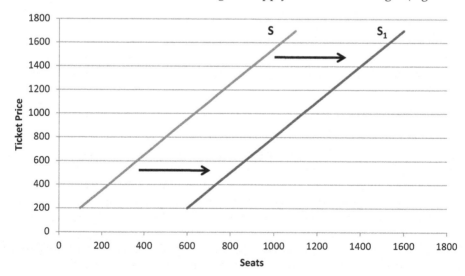

Figure 3.10   Impact of Deregulation on Supply

---

8   American Airlines and United Airlines.
9   FAA November 6, 2009.

Characteristics of supply for airline services

Two major characteristics of supply that help shape the air transportation industry are seasonality and rigidity. Both these characteristics make it difficult for airlines to match supply and demand.

One of the major characteristics of air transportation demand is that demand constantly fluctuates. Because of this, airlines must react to this fluctuation by adjusting the supply to match the passenger demand. In order to accommodate seasonality, airlines need to either pull capacity off existing routes, or have idle capacity available to accommodate additional flights. Both of these options have imbedded costs, and airlines typically use a mix of both options. While the increased costs of a seasonal schedule can be offset by increased revenues, airlines have greater difficulty adjusting supply on a short-term basis due to the second major characteristic of airline supply, rigidity.

An airline's supply is fairly rigid as it can be difficult for airlines to reduce and/or increase supply dramatically. Since an airline creates a schedule at least six months out, and accepts bookings up to a year out, the airline must adhere to the schedule or face re-accommodation fees. Fixed costs, such as investment in infrastructure at hub airports, aircraft leases, and labor contracts have to be paid regardless of the schedule, making it impractical for airlines to reduce capacity on short-term notice. This is particularly a problem for those major US carriers that operate in a hub-and-spoke network and is one reason why a non-hub carrier such as Southwest Airlines has greater flexibility with supply. Ultimately, this rigidity in supply limits the airlines' ability to match supply and demand effectively

MARKET EQUILIBRIUM

So far, we have analyzed supply and demand separately, considering the characteristics of both. Matching supply and demand occurs when both parties agree on a price that determines the allocation of the resource. At any given price, there will be a quantity demanded, and a quantity supplied. At one price, however, quantity demanded will exactly equal quantity supplied, and this point is known as equilibrium. In other words, market equilibrium is the setting of a price such that the quantity demanded and the quantity supplied of a good are exactly equal. It is the point where both supply and demand curves intersect, and the price is known as the market clearing price.

Figure 3.11 presents a supply and a demand curve on the same set of axes, with price on the y-axis and quantity demanded on the x-axis. We find equilibrium price at P* where supply and demand intersect. At P* buyers want to buy exactly the same amount that sellers want to sell. Although we tend to think of businesses as being in control of prices, it is fair to say that consumers "set" prices just as well as producers. Airlines always want high ticket prices and passenger consumers always desire low prices; P* is a compromise that forces both sides to take into account the other's needs. Producers must receive a high enough price to cover costs and motivate production; consumers must receive a price consistent with their budget.

P* is achieved through a process of trial and error. The firm estimates demand and plans a level of output and charges a price based on that estimate. Suppose an airline underestimates demand and therefore charges too low a ticket price, P_1 and offers too little output, Q^s_1, as in Figure 3.11. At this low price the amount demanded by passengers,

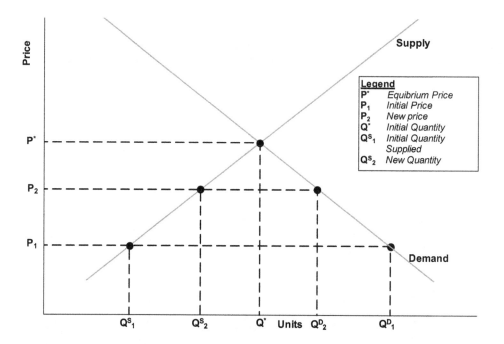

Figure 3.11   Price Movements: Average Fare is Below Equilibrium

Q^d_1, is much greater than the output supplied. Airlines, or any other business in this situation, will see that tickets are sold at unusually high rates; that is, the aircraft will begin to fill up its seats much faster than normal. In fact, if airlines don't respond there will be a shortage; they will soon be sold out and turn away numerous customers. However, airlines typically respond fairly quickly. Airlines will raise the price and, since the higher price can cover the higher associated per unit costs, will increase output. Suppose that airlines raise output to Q^s_2 and increase average fare to P_2 so that quantity demanded falls to Q^d_2. This is a step in the right direction but there is still excess quantity demanded; the price is still too low so that seats continue to fill overly fast. So, price and output will be raised again and will not settle into equilibrium until P^* is reached, at which point Q^s and Q^d are equal, at least approximately, at Q^*. Once the average fare reaches P^* seat inventories behave normally and quantity supplied is brought into balance with quantity demanded as well as is feasible under the given conditions.

Airlines differ from most businesses in that inventories perish—that is, empty seats are worthless once the plane takes off. We will address this complication in some detail in the chapter on revenue management (Chapter 11). For now, let us simply acknowledge that selling air travel is more complex than selling, for instance, canned vegetables. The producer of canned vegetables can maintain inventories for some time, whatever is unsold today can be sold in the future; thus, it is possible to more closely match Q^s to Q^d. Since airlines inventories are perishable, it generally isn't feasible to have 10 percent load factors, or to fill every seat on every flight. So, for airlines, Qs and Q^d are only approximately equal.

There is a dynamic process for reaching P^*, equilibrium price, if airlines initially overestimate demand. In that case, the average fare is too high and very few bookings are received. If this persists there will be a surplus of available seats and aircraft will depart with many more empty seats than normal. Of course, airlines want to avoid this and will therefore bring price down, and reduce capacity until equilibrium is reached at P*.

Once equilibrium is achieved then price and quantity will remain at P* and Q* as long as both supply and demand remain constant. In practice, supply and demand for air travel tend to shift often so we see almost constant changes in price and quantities (Mankiw, 2007).

The same concept is presented in a mathematical fashion below.

Suppose that the demand and supply functions for DirectJet's flights from New York to Seattle are the following:

$$Q^D = 500 - 5P_X + 2P_Z + 0.01Y$$

$$Q^S = 800 + 3P_X - 2P_{RES}$$

Where:

- Q^D is the quantity demanded;
- P_X is the price of DirectJet's tickets;
- P_Z is the price of a competitor airline's tickets;
- Y is consumer income;
- P_{RES} is the price of resources

Assuming that the competitor's ticket price is $300, the annual average income is $50,000 and the cost of resources for the flight is $100, both the demand and supply functions can be rewritten solely in terms of the ticket price.

$$Q^D = 500 - 5P_X + 2(300) + 0.01(50,000)$$

$$Q^D = 500 - 5P_X + 600 + 500$$

$$Q^D = 1,600 - 5P_X$$

$$Q^S = 800 + 3P_X - 2(100)$$

$$Q^S = 800 + 3P_X - 200$$

$$Q^S = 600 + 3P_X$$

In order to find the market equilibrium price for the flight, the demand and supply functions need to be set equal to each other.

$$Q^D = Q^S$$

$$1,600 - 5P_X = 600 + 3P_X$$

$$1,000 = 8P_X$$

$$P_X = \$125$$

Based on this calculation, the market equilibrium price for DirectJet's flight between New York and Seattle is $125. At this price point, the quantity demanded equals the quantity supplied. Since the supply and demand curves are equal to each other at the equilibrium price, either the demand or supply function can be used to determine the market equilibrium quantity. Both functions are used below to illustrate this point.

$$Q^D = 1,600 - 5P_X$$

$$Q^D = 1,600 - 5(125)$$

$$Q^D = 1,600 - 625$$

$$Q^D = 975$$

$$Q^S = 600 + 3P_X$$

$$Q^S = 600 + 3(125)$$

$$Q^S = 600 + 375$$

$$Q^S = 975$$

As depicted in Figure 3.12 below, at a uniform price of $125, 975 consumers will demand DirectJet's flight between New York and Seattle and DirectJet will supply 975 seats. At this point, supply and demand are perfectly matched. However, if the price of the flight were to decrease, the quantity demanded would exceed the quantity supplied. This excess demand is depicted by brace "A-B" in Figure 3.12 below and is sometimes called spillage. If the price of the flight were to increase from the equilibrium level, the quantity supplied

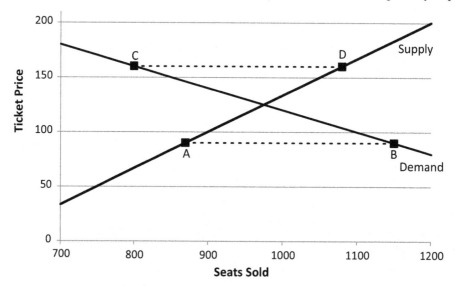

Figure 3.12   Spoilage and Spillage for DirectJet's Flights from New York to Seattle

would exceed the quantity demanded. This situation would be represented in the form of empty seats on the aircraft. Since demand is perishable, the case of supply exceeding demand is sometimes called spoilage and is depicted by brace "C-D" in Figure 3.12.[10]

The market is an allocating mechanism because at the equilibrium price of $125, exactly 975 consumers whose willingness to pay is at or above $125 get airline seats. If a consumer's willingness to pay is below $125, he or she will opt out of the market and not choose to buy a seat.

Changes in equilibrium

Analyzing changes in equilibrium is straightforward as long as one proceeds by first deducing which curve is shifting in which direction. Suppose, for instance, that the air travel market is in equilibrium initially and then changed when a new event causes passengers to more fully realize how safe commercial flying is compared to other modes of travel. Though the airlines are very pleased with this development this does not change production costs; therefore we know that the supply curve doesn't move. It stands to reason that a more realistic assessment of airline safety will result in a greater general willingness to fly—for a given price people want to fly more than before. This causes an *increase in demand*; the demand curve shifts right as we see in Figure 3.13. Now, we simply read the graph, based on where the new demand curve, *demand_{new}*, intersects supply to see that equilibrium price rises and equilibrium quantity also rises. Supply doesn't change since the supply curve is not shifting but *quantity supplied increases* as we move along the existing supply curve.

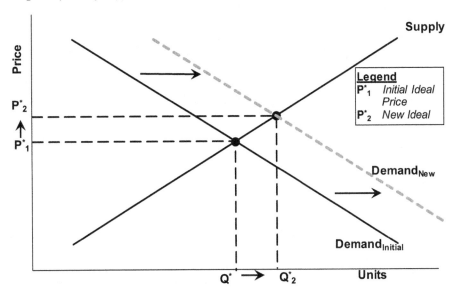

Figure 3.13   Shift of Demand Curve

---

10   However, in the airline industry the supply is not a smooth upward sloping curve. An aircraft can accommodate only a fixed number of passengers. This would create the situation where the supply curve for a single flight would move up step-wise according to the number of aircraft in the airline's fleet.

Next, suppose we are initially in equilibrium once again when wages for airline employees increase. This will not shift demand because consumers do not *directly* care much about the details of airline employee compensation. Since higher wage costs do increase production costs we will shift the supply curve to the left as presented in Figure 3.14. We see from the intersection of demand and the new supply that equilibrium price rises from P^*_1 to P^*_2. Since consumers do, of course, care about the price of air travel (thus they *indirectly* care about how expensive pilots are, which is embodied in the price of air travel) *quantity demanded* falls with equilibrium quantity decreasing from Q^*_1 to Q^*_2.

The basic effects of any given shift in supply or demand can be deduced by following these same procedures. As another example, consider the market for oil. The initial supply and demand curves would be at position Supply$_{initial}$. When the suppliers decide to collaborate and supply less oil for every price, this causes a backwards shift in the supply curve, to Supply$_{New}$.

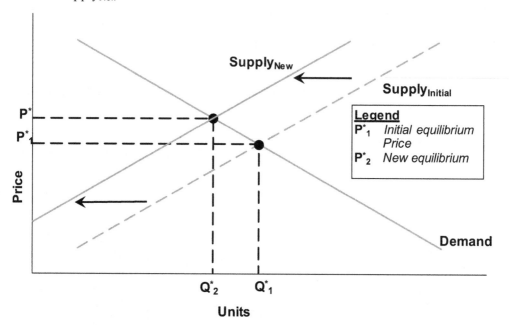

Figure 3.14   Shift of Supply Curve

Equilibrium price maximizes consumer well-being

The equilibrium price is generally the best possible price for consumers given the reality of producer costs. Consumers would love for air travel to be provided for free but most have enough sense to realize that a law requiring airlines to give their product away would simply result in airlines shutting down. Price must be high enough to motivate sellers to provide their product at appropriate quality. Sometimes price may seem higher than necessary to motivate needed production—but things are not always as they seem. It is generally optimal for consumers to let price move wherever supply and demand may send it.

It is useful to initially illustrate this point outside of the air travel market; then we can apply a similar logic to a more complex aviation example. Let us first analyze the widely misunderstood case of pricing crucial consumer goods in the aftermath of a natural disaster. Suppose a severe hurricane knocks out electrical power in an area with drastic consequences for, say, the ice market. Most of the ice in the area will have melted while simultaneously the demand for ice surges way above normal level. The drastic reduction in supply combined with huge demand will raise equilibrium price far above the norm, as depicted in Figure 3.15 where we see equilibrium price at $20 for a bag of ice that would normally sell for less than $2.

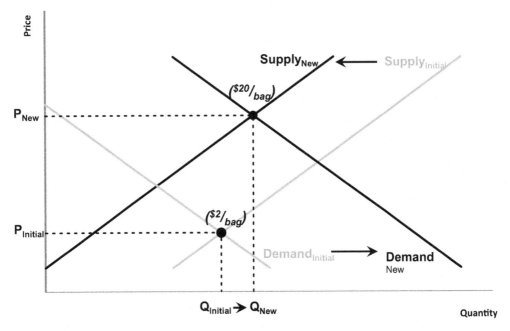

Figure 3.15   Movement of Both Supply and Demand Curves

The natural, emotional reaction to this is to feel that sellers are engaged in "outrageous price gouging" but economic logic leads to a very different conclusion. In this case the high price actually helps consumers to better deal with the emergency — the problem here is the hurricane; the high price helps people cope far better than a "normal, fair price" would.

The harsh reality is that the hurricane leaves the city with an ocean of demand for ice but only a few drops of ice available. It is impossible to get the product to all who want it; so it is important to guide this crucial resource to those that have the most urgent needs. Everyone is thirsty for something cold to drink but a few people have life-saving medicines that will spoil unless they are preserved with ice. The $20 price will convince most of those who are merely thirsty to leave the ice alone while those who face a literal life or death need for ice will not hesitate to pay the exorbitant price. Thus, the high price *rations* the good to those who need it most urgently. Of course, to truly minimize human suffering we need more than just a high price, we need some charity as well to buy the ice for those who have urgent health needs but are too poor to afford to buy the ice themselves. The high price maximizes their chance to find and purchase ice in time to save lives. In fact, any philanthropist rushing in with ice would do more good to sell it for the $20 than they would by giving it away randomly—ice selling at $20 is mainly going to those with urgent needs while ice randomly given away is mainly going to thirsty people. Naturally, it would be ideal to give the ice only to those with urgent needs but it is difficult to quickly identify those with greatest need under such chaotic conditions.

This seemingly unfair, outrageous price also motivates extreme measures to bring in more supplies. In such situations, young pilots have rented aircraft and flown in ice by helicopter and sea plane, a mode of transport normally unaffordable for most young pilots. But with a bag of ice going for $20, pilots of modest means could afford to rent the aircraft and bring in the life-saving supplies.

It would be nice if private charities or, perhaps, government officials were able to miraculously bring in enough ice to solve all problems. But this is inherently not the case— no consumer would pay $20 for something they could readily receive as a free handout from the Salvation Army or a Government Relief agency. The high price is conclusive evidence that charities and governments are overwhelmed and an urgent response is needed by anyone capable of bringing ice in quickly.

It may be that some people bringing in the ice do so out of purely selfish motivation, unconcerned about saving lives and minimizing human misery, they are simply rushing in to "make a fast buck." We might fret about the soul of such a person, but if the ice they bring saves the life of a sick child by preserving her antibiotics then that child is no less alive because she was saved by a selfish money grubber rather than a generous philanthropist. When generous people can't do enough it's nice to have a high price to get everyone else motivated to help as well.

This whole scenario is an example of what economics' founding father, Adam Smith, termed the *Invisible Hand.* Voluntary trade in free enterprise often motivates behavior that helps society even as individuals are mainly trying to help themselves and their own families. A pilot with some spare time looking only to enrich himself is guided, in Smith's phrasing, "as if by an invisible hand" to fly in ice that saves lives. Similarly, even a greedy consumer giving no thought to the crucial health needs of others will tend to leave the ice for more needful neighbors simply because he refuses to pay such a high price. Working through the price system, the invisible hand defeats the greed of the consumer and redirects the greedy impulse of the pilot into highly productive service to others. This invisible hand principle is the underlying foundation that makes individual freedom feasible and a free enterprise system so productive.

Invisible hand solutions also tend to be directly proportional to the problems posed. In the immediate aftermath of the hurricane, price is at its highest as the need for careful rationing and new supply efforts are most crucial. As power begins to be restored demand will decrease while supply of ice increases. Both effects will reduce price so that people with lower priority needs will begin to buy ice. In a few weeks if power is completely restored, price falls to a normal level, perishable medicines are back in refrigerators and the typical ice use is once again to chill drinks at parties. The high price is with us only as long as we need it.

Price controls

It has been argued that the primary objective of price control is to prevent extreme, runaway inflation and all the evils that go with it, but some other negative consequences may happen as well. For example, in the early 1970s, the US President Richard Nixon imposed price controls on August 15, 1971 to contain inflation. A controlled price will allocate resources, but not in accordance with supply and demand. Suppose government decrees that the price of ice must be reduced from $20 to $2. The siren call of low prices is appealing to consumers anxious to get a bargain but the result is tragic. As Figure 3.16 illustrates, the low price will drastically reduce quantity supplied, from Q^* to Q^s_g. Renting helicopters or driving refrigerated trucks from far away is no longer so affordable or appealing with price an artificially low $2. So the flow of ice slows to a trickle, suffering and even death become more likely because there is so little ice. At the same time quantity demanded surges now to Q^d_g because price is depressed—most anyone is thirsty enough to pay $2 to ice down some drinks. So what little ice is available is now mostly snapped up for casual use.

Perhaps the greatest irony is that the government price control results in consumers typically "paying" more for ice then the market rate of $20. This follows from the fact that there is more to life than cash; time, in a manner of speaking, is money, too. Figure 3.16 shows that the amount of ice available, Q^d_g, could be sold for P', which is obviously well above the $20 market rate. If P' is, say, $32, then we know consumers would pay that price to buy up all the ice. Normally consumers compete for scarce products through the price but in this case errant government regulation precludes that. So, consumers instead compete by getting to limited supplies ahead of the crowd; that is, they arrive early and wait in line. Since the cash price is artificially limited to $2, consumers are willing to use up an additional $30 worth in time. If the average ice consumer values their time at $10 per hour than the average wait in line will be three hours. Of course, the chaotic uncertainty of the situation will result in some people waiting much longer or giving up altogether, others may luckily stumble into an unexpected delivery and face little waiting. But the average price paid will be $30 worth in time plus $2 cash. One way or another, massive demand in the face of miniscule supplies will bid up price. In the absence of regulation, people pay $20 for ice; with the price control there is much less ice and people pay a higher "price" for it—only now the greatest cost is time with a much smaller cash cost. But it's that total cost that counts and renders the price control a sort of "fool's gold."

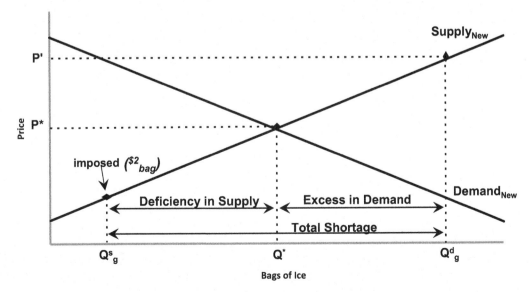

Figure 3.16   Supply Shortage

CONSUMER AND PRODUCER SURPLUS

Consumer surplus is the benefit accrued by consumers whose willingness to pay exceeds the market equilibrium price. The lower the equilibrium price, the higher the consumer surplus. Referring to Figure 3.17 below, region C marks the area of consumer surplus— these are the consumers that "benefit" from the market equilibrium price of $125. Consumer surplus is defined as the difference between the total amount that consumers are willing and able to pay for a good or service and the total amount that they actually do pay.

Similarly, producer surplus is defined as the benefit that accrues to producers whose willingness-to-supply lies below the market equilibrium price. In other words, if airlines were willing to supply the market with seats at prices below the $125 per seat market equilibrium shown in Figure 3.17, then these airlines are able to sell their seats at the higher price and therefore reap the benefits of the higher price. This is depicted as section D in Figure 3.17 below.

Now, consumer surplus results only because some consumers in the market have a willingness to pay that lies above the equilibrium price. Supposing an airline could identify these customers and charge them each what they would be willing to pay—it could derive the benefit of charging some consumers higher prices, while leaving them no worse off because they are still paying the price at or below their expressed willingness to pay. Further, with a multi-tiered pricing strategy (known as *price discrimination*), the airline could maximize its producer surplus, since total producer surplus would now equal areas C + D. This forms the basis of revenue management, a topic we will address at some length in Chapter 11.

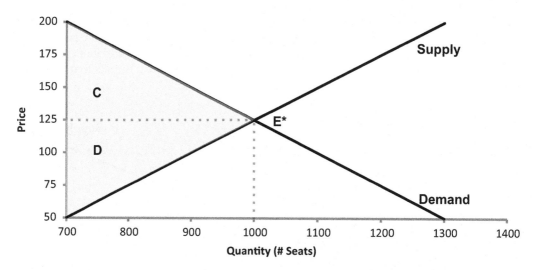

Figure 3.17   Consumer and Producer Surplus at Equilibrium

Airport landing fees and airport congestion

Following the logic of the preceding discussion, excessive and persistent airport congestion may be a function of a price that is not in equilibrium; that is, the landing fee is kept too low by government. (Most of the world outside the US is moving toward privatizing airports but major private airports, such as Heathrow, typically have landing fees mandated by government.) Other factors, of course, may contribute to airport congestion—stringent environmental regulation that prevents airport expansion or dated technology that forces aircraft to maintain wider separation, and so on. But regardless of other factors, the correct price can eliminate excessive congestion.

The situation is illustrated in Figure 3.18. The supply curve is represented as vertical, because in the short run (and usually in the long run as well) airport supply is fixed. Adding runways or ramp capacity takes time and investment, and is not a decision that can be undertaken at short notice. Sometimes expansion may be impossible due to nearby development. Therefore, airport supply is fixed no matter what the price for its services. Airport demand, on the other hand, is downward sloping as airlines can choose to fly into an airport or stop flying into an airport depending on the prices charged for use of airport resources. Market equilibrium landing fee for airline traffic is P*. This fee eliminates excessive congestion at the airport, since the price is set such that the number of planes landing at the airport closely matches its capacity. However, most airports face government regulation on the fees that they can charge landing traffic. Indeed, in the US, most general aviation traffic at medium or small airports cannot be charged a landing fee at all, effectively subsidizing the use of airport facilities. However, assume the landing fee for the airport is set at P', below P*. This implies that L' takeoffs and landings will take place at the airport, which is higher than the optimal market quantity L*. The difference between L' and L* is airport congestion— airport operations in excess of capacity capabilities. Of course, if the government were to set the landing fee at the optimal quantity of P*, it would naturally eliminate most congestion. However, P* is usually unobserved, and reached by a process of trial and error by buyer and seller. It is usually extremely challenging to estimate P* in absence of market data and mispricing is generally the outcome of price controls, leading to congestion.

Thus, the theoretical optimal landing fee for the airport is P*, determined by the supply and demand for the airport resources.

Of course, it must be noted that L* does not imply zero congestion, merely optimal congestion. It may be optimal for an airport to have some degree of congestion, if

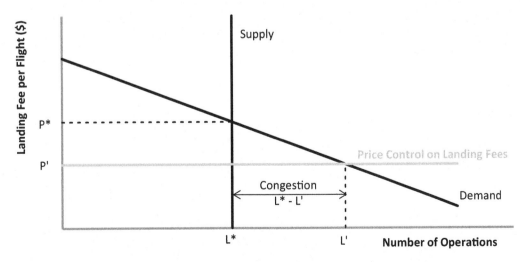

Figure 3.18   Price Controls on Landing Fees in Airports

only because that represents the efficient use of its resources. For instance, while having many fewer landings per day at Atlanta might indeed mean zero congestion, it also represents a suboptimal use of the resources at the airport. Both airlines and passengers are willing to accept some degree of congestion for the benefit of using that particular airport. An airline might conclude that having a takeoff clearance wait time of ten minutes might be acceptable given the convenience of using that airport. Therefore, L* merely represents the optimal level of congestion at the airport. In the absence of government regulation, if the airport initially set its landing fees too low, and congestion exceeded L*, airlines would eventually stop using that airport due to the unacceptably high costs of delays, which would return the level of congestion to L*.

Price floors

A price floor is a minimum price, generally above the equilibrium price, set by the government on a product or service. Since the price is set above the equilibrium price, a surplus develops because more is being produced than consumers are willing to purchase at that price. When the market price hits the floor, it can fall no lower and the market price equals the floor price. There is no effect on the price or quantity if the price floor is below the equilibrium. A government may impose a price floor to protect a favored industry and are particularly common in agricultural price supports. A good example of a price floor in the US is the minimum wage. This is the lowest wage that an employer is allowed to pay a worker.

Suppose the market demand and demand functions given by:

$Q_d = 150 - 30P$

$Q_s = 30 + 30P$

Equating demand and supply yields:

$150 - 30P = 30 + 30P$

$60P = 120$

$P = \$2$

To determine the equilibrium quantity, we simply plug this price into either the supply or the demand junction, we find that:

$Q = 30 + 30(2)$

$Q = 90$ units.

Look what would happen if the government imposes a price floor of $3. Since the price is above the equilibrium price, a surplus would develop. At the P = $3:

$Q_d = 150 - 30(3) = 60$ units

$Q_s = 30 + 30(3) = 120$ units

Thus, there is a surplus of $120 - 60 = 60$ units.

Price ceiling

Opposite to a price floor is a price ceiling where a maximum price is set by the government for particular goods and services. The government may perceive the current price is too high and potentially harm consumers and the general economy. In order to be effective, price ceilings must be set below the natural equilibrium price. If a price ceiling were imposed above the equilibrium price, it would have no impact on the market. Generally, price ceilings lead to shortages. A possible consequence of the shortage created is the generation of a black market where the good is sold illegally above the price ceiling. An example of a price ceiling could be in the 1970s the government controlled the prices of gasoline. In 1971, President Nixon imposed wage and price controls. The wage and price controls were mostly abandoned by April, 1974. Rent control is another example of a price ceiling and exists in many countries around the world. In the US, New York State has had the longest history of rent controls, since 1943 (Baar, 1983).

Now, suppose that, the government imposes price ceiling of $1 as shown in Figure 3.19.

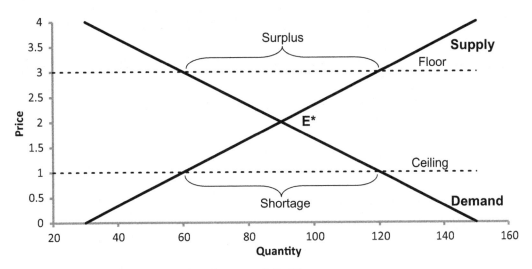

Figure 3.19   Price Controls: Floor and Ceiling

When the P = \$1 quantity supplied and demanded are:

Q_d = 150 – 30(1) = 120 units

Q_s =30 + 30(1) = 600 units

Thus, there is a shortage of 60 – 120 = -60 units.

DISEQUILIBRIUM

While markets tend toward equilibrium the process is not always smooth or automatic. Indeed, there can exist extended periods of disequilibrium, where price does not allocate the quantity demanded and the quantity supplied, even in the absence of government control. Of course, when one adds regulatory pressures to the analysis, the effects of disequilibrium can potentially be extended far longer than they would in a free market.

Disequilibria can occur as a result of both microeconomic and macroeconomic shocks. An example of a macroeconomic shock was the terrorist attacks of 9/11 in 2001. Immediately following the attacks, all aircraft were grounded in the Continental US. This had the effect of reducing the supply of air transportation to zero within the space of a few hours. There was no equilibrium price, because the market lacked any means of supply. The market was in disequilibrium with no means of clearing it until supply was allowed to resume. We are all familiar with the effects of such a disruption, with passengers and aircraft stranded at various locations all over the country, and passengers who were forced to seek alternative modes of transportation since air traffic was suspended until further notice. Recovery from this disequilibrium was a long process, with passengers requiring reimbursement, airlines receiving aid from the Federal Government as compensation for service disruptions, and so forth.

An example of a microeconomic shock may be bad weather at airports. This has the effect of shifting the supply curve to the left. In normal days, traffic can be sequenced out of airports under Visual Flight Rules (VFR). This helps to reduce the burden of maintaining aircraft separation for the controllers, since aircraft are essentially asked to watch out for each other. A less stringent set of air traffic regulations apply, most importantly in terms of wait times between takeoffs and landings, and when an aircraft is deemed "clear" of the airspace. Air traffic controllers also face less extensive vectoring responsibilities. In contrast, when an airport is encountering bad weather, it restricts the supply of the airport. Bad weather may not always imply full airport closure. More often than not, Instrument Flight Rules apply and aircraft may still leave the airport, but under a much more stringent set of rules, specifically involving longer separation times between aircraft, close and extensive vectoring by air traffic control, and reduced airspace capacity around the airport. Therefore, the supply curve shifts to the left during periods of bad weather, a phenomenon which might be resolved by an increase in the landing price or a change in technology. Demand outstrips supply, and congestion results. Figure 3.20 illustrates this.

When the airport encounters bad weather, the number of aircraft that can be allowed to conduct operations in the airspace falls. This is represented by shifting the airport's supply curve to the left from S to S1. If the weather was to worsen, the supply curve would shift all the way to S2, causing the market to go into disequilibrium until the weather clears.

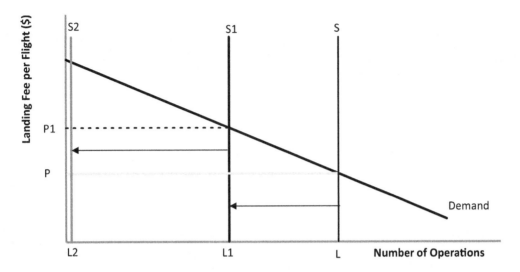

Figure 3.20   Airport under Bad Weather

As a broader point, it must be noted that the state of disequilibrium may last a few hours, as in the case of bad weather, or a few days, or a few years, as with airport landing fees and price controls.

ELASTICITY

An important economic principle that can aid airline or airport managers in their economic decision-making process is elasticity. The formal definition of elasticity is the percentage change in the dependent variable (quantity demanded) resulting from a one percent change in an independent variable (factor of demand). Informally, elasticity measures the responsiveness of one variable to changes in another.

The basic formula for elasticity is:

$$Elasticity = \frac{\%\Delta Q_D}{\%\Delta P}$$

Where:

- $\%\Delta Q_D$ = the percentage change in quantity demanded; and
- $\%\Delta P$ = the percentage change in price.

In measuring elasticity, there are two types of variables: endogenous and exogenous. Endogenous variables are variables that the airline can directly control while exogenous variables are variables that are out of the airline's control. In the airline industry, both price and service would be endogenous variables, while factors such as consumer income, competitor's price, and price of complementary goods are exogenous variables. It is useful to know the effects of these exogenous variables on demand, since this information allows the airline to manage capacity and demand more efficiently. The three major elasticities that will be explored in greater detail are:

- price elasticity;
- cross-price elasticity;
- income elasticity.

Price elasticity

While there are numerous types of elasticity, the price elasticity of demand is probably one of the most useful elasticities for airline managers. Using the general definition of elasticity, price elasticity is the percentage change in the quantity demanded resulting from a one percent change in price. Therefore, price elasticity enables managers to perform "what-if" scenarios to see the effects on the quantity demanded that a change in price would have.

Since elasticity is not generally constant throughout the entire demand curve, there are two ways to measure elasticity.

Point elasticity measures the elasticity of the function at a specific value, while *arc* elasticity measures the elasticity of the function over a range of values. Thus, arc elasticity is an average of the elasticities over a specified range of values, while point elasticity is the exact level of responsiveness at the specific price. The basic formulae for point price elasticity and arc price elasticity are:

Arc Price Elasticity:

$$E_p = \frac{\dfrac{\Delta Q}{Average Q}}{\dfrac{\Delta P}{Average P}}$$

$$E_p = \frac{\Delta Q}{\Delta P} \times \frac{P_2 + P_1}{Q_2 + Q_1}$$

Point Price Elasticity:

$$E_p = \frac{\% \Delta Q}{\% \Delta P}$$

$$E_p = \frac{\Delta Q}{\Delta P} \times \frac{P_A}{Q_A}$$

To illustrate the difference between the two, consider Figure 3.21, which reproduces the demand curve for the route JFK–LAX.

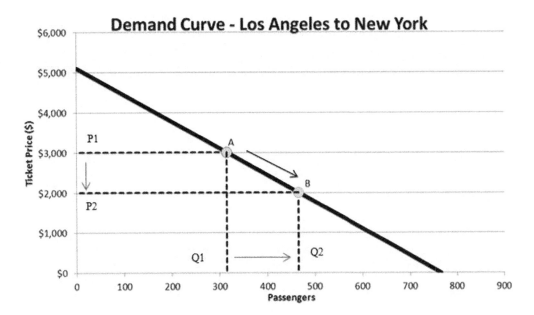

Figure 3.21   Price Elasticity of the JFK–LAX Route

The arc price elasticity of demand in this case is calculated as follows:

$$E_D = \frac{\%\Delta Q_D}{\%\Delta P}$$

$$E_D = \frac{Q_2 - Q_1}{P_2 - P_1} \times \frac{P_2 + P_1}{Q_2 + Q_1}$$

Where:

- $Q_2 = 465$
- $Q_1 = 315$
- $P_1 = 3000$
- $P_2 = 2000$

Therefore, Price Elasticity:

$$E_D = \frac{150}{780} \times \frac{-5,000}{1,000}$$

$$E_D = -0.96$$

Note the negative sign, which implies that the price of the product is negatively related to the quantity demanded, which is given by the law of demand. In this case, if the airline decides to raise the ticket price by 1 percent it will find that quantity drops by 0.96 percent.

Elasticity can range from zero to infinity. An elasticity = 0 would indicate no change in demand no matter what the change in price. An infinite elasticity means any price increase will drop demand from an infinitely large amount to zero.

In order to calculate price elasticity, the derivative of the demand function needs to be calculated.[11] As a numerical example, consider the following explicit short-run demand function for a flight, where we assume that the competitor's ticket price is $120 and that the average annual income in the market is $40,000.

$$Q_D = 800 - 2P_X + 1.5P_Z + 0.0005Y$$

The first step is to re-compute the demand function based on the assumptions concerning the external market. Substituting those values into the demand function yields:

$$Q_D = 800 - 2P_X + 1.5(120) + 0.0005(40,000)$$

$$Q_D = 800 - 2P_X + 180 + 20$$

$$Q_D = 1,000 - 2P_X$$

Based on this new demand function that is only related to price, price elasticity can now be calculated. The first step is to take the first-order derivative of the demand function which results in a value of -2 (see footnote 11). The next step is to find the quantity demanded based on a single price for point elasticity, or for multiple prices for arc elasticity. With a ticket price of $100, the quantity demanded would be 800 seats. Using the point price elasticity formula, the point price elasticity is:

$$\epsilon_d = \frac{\partial Q}{\partial P} \times \frac{P}{Q} = -2 \times \left(\frac{100}{800} \right) = -0.25$$

A point price elasticity value of -0.25 means that for every one percent increase in the ticket price, from the $100 level, the quantity demanded would decrease by 0.25 percent.

The arc price elasticity for ticket prices ranging from $100 to $200 can be calculated as follows:

$$E_p = \frac{\Delta Q}{\Delta P} \times \frac{P_2 + P_1}{Q_2 + Q_1}$$

$$E_p = \frac{-200}{100} \times \frac{200 + 100}{600 + 800}$$

$$E_p = -2 \times \frac{300}{1400}$$

$$E_p = -0.43$$

---

11   For those readers without calculus, the derivative is simply the change in the dependent variable for a one unit change in the independent variable. In this case that would be dQ/dP or -2. That is, a one unit change in P will produce a minus two unit change in Q.

In this case, if the ticket price goes up by 10 percent, the volume of traffic will go down by 4.3 percent. Price elasticity is usually categorized into one of three groups based on its numerical value and its impact on demand:

$|E| > 1$ Elastic

$|E| < 1$ Inelastic

$|E| = 1$ Unitary Elastic

Price elasticity with an absolute value of less than one is termed inelastic. Inelastic demand occurs when a one percent increase in price results in a less than a one percent decrease in demand. (In the above example, since the absolute value of elasticity was calculated as less than one, the price elasticity would be inelastic). In these situations, the consumers have a strong desire to purchase the good or service and therefore price is not a central concern. In price-inelastic situations, firms can increase the price to increase total revenue as the price effect dominates the quantity effect. However, since point elasticity is not constant throughout a linear demand curve (in fact, as will be discussed later, every linear demand curve has both an elastic and inelastic region) this practice can only continue up to a certain price when the demand becomes less inelastic (or more elastic). A good example of a price-inelastic product includes the demand for air travel in the short run. In the few days leading up to the day of departure, the majority of travelers who need to take the flight are willing to pay for an expensive ticket, since they are likely to have an important reason to travel and the convenience of the flight is more important than price to them.[12]

At the opposite end of the spectrum is elastic demand. Elastic demand occurs when the coefficient of elasticity has absolute values greater than one. With elastic demand, consumers are more sensitive to changes in price, so that a 1 percent decrease in price will be offset by a greater than 1 percent increase in the quantity demanded, and total revenue increases with price cuts. However, as above, the benefits of price cuts will be exhausted at some point as the firm will eventually reach a portion of the demand curve where demand becomes inelastic. Longer-term demand for air travel is more price-elastic than short-term demand as many passengers (especially leisure travelers) will choose to take a flight based solely on price.

The final category of elasticity is unitary elastic demand, which has an absolute value equal to one. Under unitary elastic demand, the quantity and price effects are equal, creating a situation where a 1 percent increase in price is directly offset by a 1 percent decrease in the quantity demanded.

And, as the preceding discussion indicates, the point at which the company achieves the optimal price level is the point of unitary elasticity. Therefore, the managerial rules of thumb are simple; if the demand for the product is inelastic, then the company should raise prices; if the demand for the product is elastic, then the company should lower prices; if the demand for the product is unitary, then the company should retain the present price. These pricing decisions are displayed graphically in Figure 3.22. For simplicity, the graph focuses only on revenue maximization; the true goal, profit maximization will be discussed more fully in subsequent chapters.

---

12   That is part of reason why last-minute tickets are expensive and the airlines use revenue management techniques to save seats for last-minute time-sensitive passengers.

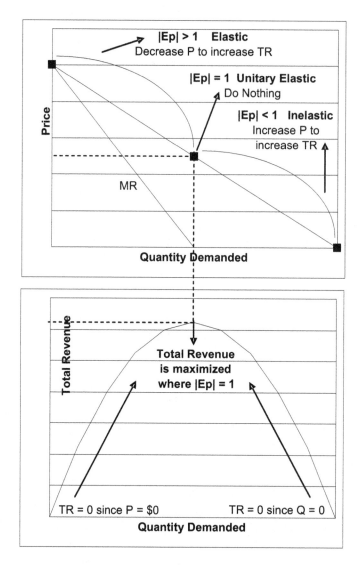

Figure 3.22   Pricing Decision Based on Elasticity

Since a clear relationship exists between price and total revenue for different states of elasticity, Table 3.5 summarizes the impact that a change in price has on total revenue. As mentioned previously, an increase in price will increase total revenue for goods with inelastic demand, while decreasing total revenue for products which are price elastic. The direct opposite is true for a decrease in price. For unitary elastic products, a change in price will not affect total revenue since total revenue is maximized at the point where unitary elasticity is achieved.

While elasticity varies along a linear demand curve, certain products and services can be categorized based on their normal own price elasticity. As mentioned earlier, air travel exhibits tendencies of both inelastic and elastic demand depending on the time frame. Table 3.6 provides a list of various services and products with their corresponding estimated price elasticity.

Table 3.5   Relation between Price Changes, Total Revenue and Elasticity

Relation Between Price Changes, Total Revenue and Elasticity			
	Elastic	Unitary Elastic	Inelastic
Price Increase	↓ in TR	No Impact	↑ in TR
Price Decrease	↑ in TR	No Impact	↓ in TR

Table 3.6   Estimated Price Elasticities of Demand for Various Goods and Services

Estimated Price Elasticities for Goods and Services	
Goods	Elasticity of Demand
Inelastic	
Salt	0.10
Toothpicks	0.10
Airline travel, short-run	0.10
Gasoline, short-run	0.20
Gasoline, long-run	0.70
Coffee	0.25
Tobacco products, short-run	0.45
Physician services	0.60
Automobiles, long-run	0.20
Approximately Unitary Elastic	
Movies	0.90
Housing, long-run	1.20
Private eductaion	1.10
Tires, short-run	0.90
Tires, long-run	1.20
Elastic	
Restaurant meals	2.30
Foreign travel, long-run	4.00
Airline travel, long-run	2.40
Automobiles, short-run	1.20 - 1.50
Fresh tomatoes	4.60

Source: Compiled by authors from Anderson, McLellan, Overton, & Wolfram (1997)

An important application of price elasticity in airline management is revenue management. Although we will cover this topic in depth in Chapter 11, revenue management involves segmenting the passenger base of an airline into varying price elasticities, and pricing each segment differently. For example, last-minute business travelers will tend to be relatively price insensitive. Their priority is the travel, and their price elasticity will be closer to 0 than to 1. On the other hand, leisure travelers who book well in advance will tend to be more price elastic—a drop or rise in ticket prices will likely make them change their travel plans. Although the traveler's characteristics are unobservable, the airline can make a fairly reasonable guess regarding the characteristics of the traveler based on observable variables like advance booking, booking over the weekend versus the work week, and the number of children, the number of people in the party, and so forth. Based on these variables, firms will prefer to price the price-elastic consumer lower than the price-inelastic consumer. In other words, airlines practice systematic price discrimination based on consumer elasticities in order to maximize their revenues, and maximize producer surplus. For example, a ticket booked to Chicago on a Monday a few days in advance will probably belong to a price-inelastic customer, who may be charged a higher price than someone who books a ticket two months in advance for a weekend in Orlando for a party of four. As demonstrated in Figure 3.22, firms can maximize revenue by pricing high and low elasticity customers differently, and this forms the core of revenue management.

Cross-price elasticity

Another type of elasticity is cross-price elasticity of demand; this elasticity measures the responsiveness of demand for one product or a service following a change in the price of another product or service. Cross-price elasticity of demand helps determine if the related firm is either a substitute (competitor) or a complement. The basic formulae for both point and arc cross-price elasticity is:

Point Cross-Price Elasticity:

$$\epsilon_{xy} = \frac{\%\Delta Q}{\%\Delta P}$$

$$\epsilon_{xy} = \frac{\partial Q_X}{\partial P_Y} \times \frac{P_Y}{Q_X}$$

As the following formula indicates, the arc elasticity computes the percentage change between two points in relation to the average of the two prices and the average of the two quantities. This provides the average elasticity between the two points.

Arc Cross-Price Elasticity:

$$E_{xy} = \frac{\dfrac{\Delta Q}{Average Q}}{\dfrac{\Delta P_y}{Average P_y}}$$

$$E_{xy} = \frac{\Delta Q}{\Delta P_y} \times \frac{P_{y2} + P_{y1}}{Q_2 + Q_1}$$

Using the above formulae and the preceding example, the point cross-price elasticity of demand can be found assuming that the ticket price of the firm's flight is $100, the competitor's ticket price is $120, and the annual income is $40,000.

$$Q^D = 800 - 2P_X + 1.5P_Y + 0.0005M$$

$$Q^D = 800 - 2(100) + 1.5P_y + 0.0005(40,000)$$

$$Q^D = 580 + 1.5P_y$$

$$\epsilon_{XY} = \frac{\partial Q}{\partial P_Y} \times \frac{P_Y}{Q}$$

$$\epsilon_{XY} = 1.5 \times \frac{P_Y}{Q}$$

$$\epsilon_{XY} = 1.5 \times \frac{120}{760}$$

$$\epsilon_{XY} = 0.24$$

From this example, the cross-price elasticity of 0.24 indicates that the related competitor can be considered a mild substitute to the firm, since the cross-price elasticity of demand is greater than zero. In this situation, a 1 percent increase in the competitor's ticket price will cause a 0.24 percent increase in the airlines' ticket sales. In the airline industry, where price competition is fierce, the cross-price elasticity of demand is undoubtedly highly positive. From the perspective of revenue management, fare transparency has forced airlines to redouble efforts to stay competitive. With information from the Internet, passengers are now able to view almost all the pricing options for their flights, and therefore some passengers can be extremely price conscious (especially since there is little difference between the product; that is, an airline seat). As a result, the cross-elasticity of demand for the airline industry is generally very high. Therefore, because of liberalization and deregulation the competition in the airline market is very vigorous. When several airlines compete with each other on the same route, they must consider how one airline might react to its competitor's price change. Will many passengers switch? Will the other airlines match a price rise? Will it follow a price reduction?

If the cross-price elasticity of demand is found to be less than zero, then the related firm's product is determined to be a complementary good. A complementary good is a good which increases the demand for the firm's good. Examples of complementary goods to the airline industry are both hotels and rental cars, as the price of accommodation and transportation directly relate to the demand for air transportation. As an example, if the cross-price elasticity of demand was found to be -0.55, then a 1 percent increase in the complementary good's price would create a 0.55 percent decrease in the quantity demanded. As in the case of own-price elasticity, cross-price elasticity can be categorized into one of three groups based on its numerical value and its impact of a related firm's price on demand:

$E_{X,Y} > 0$ Substitute

$E_{X,Y} < 0$ Complement

$E_{X,Y} = 0$ Independent

For example, the cross-price elasticity of demand between Ryan Air and easyJet is expected to be strongly positive. Likewise when there is a strong complementary relationship between two products (hotels and airlines), the cross-price elasticity will be highly negative. Independent or unrelated products have a zero cross elasticity.

Income elasticity

A third type of elasticity is income elasticity. Income elasticity determines the sensitivity that changes in the annual income of consumers have on the quantity demanded for a product. Disposable income, personal income, and Gross Domestic Product (GDP) are all good measures for this variable. Disposable income is income available to spend on leisure travel, while business income is a part of the equation since increased business activity will likely spur an increased need for business travel. Since the income variable is comprised of two parts for air transportation, GDP is the best proxy variable for income, since it takes into consideration both household disposable income and business activity. The formulas for income elasticity are similar to both the price and cross-price elasticity formulas:

Point Income Elasticity:

$$\epsilon_{xy} = \frac{\%\Delta Q}{\%\Delta Y}$$

$$\epsilon_{xy} = \frac{\partial Q_X}{\partial P_Y} \times \frac{Y}{Q}$$

Arc Income Elasticity:

$$E_Y = \frac{\dfrac{\Delta Q}{AverageQ}}{\dfrac{\Delta Y}{AverageY}}$$

$$E_Y = \frac{\Delta Q}{\Delta Y} \times \frac{Y_2 + Y_1}{Q_2 + Q_1}$$

Using the same example, assuming that the ticket price is $100 and the competitor's ticket price is $120, the arc income elasticity between $40,000 and $50,000 is:

$$Q^D = 800 - 2P_X + 1.5P_Y + 0.0005Y$$

$$Q^D = 800 - 2(100) + 1.5(120) + 0.0005Y$$

$$Q^D = 780 + 0.0005Y$$

$$Q^D_1 = 780 + 0.0005(40,000) = 800$$

$$Q^D_2 = 780 + 0.0005(50,000) = 805$$

$$E_Y = \frac{\Delta Q}{\Delta Y} \times \frac{Y_2 + Y_1}{Q_2 + Q_1}$$

$$E_Y = \frac{5}{10,000} \times \frac{50,000 + 40,000}{805 + 800}$$

$$E_Y = 0.0005 \times \frac{90,000}{1,605}$$

$$E_Y = 0.028$$

As with own price elasticity and cross-price elasticity, the good or service can also be classified according to its income elasticity:

$E_Y > 1$ Superior good

$E_Y > 0$ Normal good

$E_Y < 0$ Inferior good

If the product's income elasticity is greater than zero, then the good is categorized as a normal good; that is, the quantity demanded of a normal good increases with any increase in the consumer's income. As we might expect, the vast majority of goods and services can be classified as normal goods, as in the above example. A sub-category of normal goods are superior goods, whose income elasticity is greater than one. Superior goods have a proportional increase in the quantity demanded that is greater than the increase in consumer income. Superior goods usually encompass high-end luxury products such as fancy sport cars and business jet travel.

The other goods categorized according to income elasticity are inferior goods. Inferior goods have income elasticity values less than zero, indicating that for any increase in income, the quantity demanded decreases. This peculiar situation occurs when products have a price advantage over competitors but are generally not perceived as quality goods. Therefore, when consumers' income increases, they are more willing to purchase the perceived better product. Examples of inferior goods might include generic products versus brand names or, in some markets, coach travel against first-class seats.

The concept of elasticity is critical to understanding the pricing policies of any industry, especially the air transportation industry. And, as the earlier discussion has shown, elasticity can be used to determine the optimum price level where total revenue is maximized. Ultimately, revenue management has its foundation in this concept as elasticity can be used to help manage both pricing and capacity.[13,14]

---

13   The topic of yield management will be discussed in full in Chapter 11.

14   The principles of elasticity enable managers to see the impact that changes in competitors' prices, advertising campaigns, and economic booms and recessions will have on the airline's operations. From a foundation based on elasticity, a competitive plan and strategy can be created for the airline. Therefore, the concept of elasticity is invaluable to decision-makers in all industries, especially highly volatile industries such as air transportation.

Pricing and elasticity application

Since the ultimate goal of any pricing policy is to maximize total revenue, and total revenue is known to be maximized where price elasticity is unitary elastic, formulas can be derived to determine the optimum price. In order to understand the formula derivation, consider the inverse demand function with the general form of:

$$P = a - bQ$$

The basic, general formula for point price elasticity can be slightly modified to reflect an inverse demand function. The derivative of the demand function is simply the inverse of the derivative for the inverse demand function. Since the derivative, or slope, of the inverse demand function is (b), the price point elasticity formula can be rewritten as:

$$\varepsilon_p = \frac{dQ}{dP} \times \frac{P}{Q}$$

$$\varepsilon_p = \frac{1}{slope} \times \frac{P}{Q}$$

$$\varepsilon_p = \frac{1}{-b} \times \frac{P}{Q}$$

Additionally, the price variable (P) can be replaced by the general form of the inverse demand function. Therefore, the elasticity formula would be:

$$\varepsilon_p = \frac{1}{-b} \times \frac{a - bQ}{Q}$$

Since the goal is to determine the optimum pricing point where total revenue is maximized, the elasticity formula needs to be set equal to -1, or where unitary elasticity is achieved. Once the equation is set equal to negative one, the optimum quantity where total revenue is maximized can be determined. The optimum quantity is:

$$\frac{1}{-b} \times \frac{a - bQ}{Q} = -1$$

$$\frac{a - bQ}{Q} = b$$

$$Q = \frac{a}{2b}$$

While this formula provides the optimum quantity demanded for where total revenue is maximized, it needs to be placed back into the inverse demand function to obtain the optimum price level. Total revenue is maximized where:

$$P = a - b(\frac{a}{2b})$$

$$P = \frac{a}{2}$$

Based on the derivations of the inverse demand function and the price point of elasticity formula, the total revenue maximizing price point can be found. Using the previous example, the revenue maximizing price and quantity would be:

$$Q_D = 1000 - 2Px$$

$$P = 500 - 0.5Q$$

Therefore, the revenue maximizing price is $250, which creates a demand of 500 seats. At this point, the point price elasticity of demand is negative one, or unitary elastic. These formulas are the building block for yield management, which is covered in more detail in Chapter 11.

SUMMARY

In this chapter, we introduce the reader to supply and demand analysis in the context of aviation. We commence by defining demand as the willingness and ability to pay for a good or service, then outline the law of demand which postulates an inverse relationship between quantity demanded and price. We go on to examine the determinants of demand, analyzing the impact of own price, the price of substitutes, the price of complements and other influences on the demand for a product. We go on to classify movements in the demand curve as movements along a curve caused by price shifts, and shifts in the curve caused by external factors. Peaking and cyclicality are covered as some of the unique characteristics of airline demand.

Supply is then defined as the willingness and ability of a producer to supply a good to the market at a given price. The unique characteristics of airline supply are examined, including the perishability of the product, a lack of inventories and a general lack of flexibility in longer-term provision of the product. We analyze supply movements, and some of the determinants of supply such as own price, resource prices, technology and competitive factors. We then discuss market equilibrium as an interaction between supply and demand, and the market clearing price as the price at which the quantity demanded matches the quantity supplied. We point out the effects of government price controls in the case of airport congestion pricing, and market disequilibrium is discussed with appropriate examples. Finally, we discuss price elasticity of demand, income elasticity of demand, and the cross elasticity of demand. We present the formulas and interpretation of elasticities, and analyze revenue management as an application of price elasticity of demand. Supply and demand analysis is the cornerstone of economic thought, and the concepts presented in this chapter will be used as analytical tools throughout the book.

DISCUSSION QUESTIONS

1. What is the price elasticity of demand? Please provide some examples of products that have elastic, and those with inelastic demands, and explain your reasoning.
2. Explain the difference between shortage and surplus.
3. Define the law of demand and supply. Explain the difference between demand and quantity demanded. Give an example for the airline industry and show it graphically.
4. Is the price elasticity of demand for airline industry in short-haul markets more or less than long-haul markets? Why?
5. What are the factors that influence the elasticity of demand for pilots?
6. Suppose the income elasticity of demand for a good is -4. Is this good a normal good or an inferior good? If it is a normal good, then is it a luxury or a necessity? Why?
7. What are some of the determinants of demand for narrow-body commercial aircraft such as Boeing B-737 and Airbus A-320?
8. With changes in fuel prices what kind of effects should we see in the market for air travel?
9. Explain direct demand and derived demand. Which of these two types of demand is the airline industry most characterized by?
10. Suppose JetBlue raises the ticket price by 6 percent and the ASM decreases by 3 percent. Is the price elasticity of supply elastic, unit elastic, or inelastic? Why?
11. Suppose an airline increases its average ticket price from JFK to CDG from $435 to $515, and consequently the number of seats sold drops from 800 to 735:
    a   What is the price elasticity of demand for this market?
    b   Is the demand elastic, unitary elastic, or inelastic in this price range?
    c   What is the interpretation of that price elasticity of demand–what does it mean?
    d   Suppose the price elasticity of demand calculated in the first bullet (price elasticity of demand) above is the exact number representing passengers' responsiveness to a price change for this airline for this market. If there is a 10 percent decrease in the ticket price, what would the percentage change in the quantity demanded be equal to? If the ticket price was to rise by 15 percent, what would the percentage change in the quantity demanded be equal to?
    e   What happens to total revenue of the airline when the price rises from $435 to $515? How is this related to the price elasticity of demand for this market?
    f   What could cause JFK–CDG to have this elasticity of demand calculated in the first bullet (price elasticity of demand) above?
    g   Suppose the cross-price elasticity of demand between these two markets is equal to 2. Would these markets be substitutes or complements? Why?
12. With the recent increase in decrease in jet fuel prices, what kind of effects should we see in the air transport market?
13. Assume your demand and cost functions are the following:

    $Q = 100 - 0.50P$ and

    $TC(Q) = 104 - 14Q + Q^2$

    a   Find the inverse demand function for your firm's product.
    b   What price you should charge if you are planning to sell five units?
    c   Calculate your total cost.

14. The JetGo Company is a major fixed-base operator (FBO) at a regional airport. Management estimates that the demand for the jet fuel is given by the equation:

$$Q_{JF} = 10,000 - 2,000P_{JF} + 0.2Y + 200\,P_C$$

Where:

- Q_{JF} is demand for jet fuel in thousands of gallons per year;
- P_{JF} is the price of jet fuel in dollars per gallon;
- Y is GDP per capita (thousands of dollars);
- P_C is the price of jet fuel in dollars per gallon on the adjacent airport. Initially, the price of jet fuel is set at $2 per gallon, income per capita is $40,000 and the price of competition is $1.60 per gallon.

a   How many gallons of jet fuel will be demanded at the initial prices and income?
b   What is the point income elasticity at the initial values?
c   What is the point cross elasticity demand? Are these two product substitutes or complements?

REFERENCES

American cleared again on predatory pricing issue. (2003). *Airline Business.* Retrieved on February 1, 2007 from Air Transport Intelligence.

Anderson, L., McLellan, R., Overton, P., and Wolfram, L. (1997). Price Elasticity of Demand. *Mackinac Center for Public Policy.* Retrieved on January 29, 2007 from http://www.mackinac.org/article.aspx?ID=1247.

Baar, K. (1983). Guidelines for Drafting Rent Control Laws: Lessons of a Decade. *Rutgers Law Review.* Vol. 35, Vol. 4, 721–885.

Boeing. (2007). *The Boeing 747 Family.* Retrieved on February 1, 2007 from http://www.boeing.com/commercial/747family/background.html.

Boeing (2010) *The 787 Dreamliner.* Retrieved on August 16, 2012 from http://www.boeing.com/commercial/787family/background.html.

Knibb, D. (2000). Play by the rules. *Airline Business.* Retrieved on February 1, 2007 from Air Transport Intelligence.

Mankiw, N.G. (2007). *Principles of Microeconomics* (4th edn). Mason, OH: South-Western.

4

Cost and Production Analysis: The General Concepts

Certainly, airlines like JetBlue airlines and Southwest, they've got a different cost structure, they've got a different rationale, they've got a different model.

David Field, *Airline Business Magazine*

Any financial analysis of the profitability of the airline industry suggests that the industry should probably focus on its cost structure and should strive to achieve higher levels of productivity. The airline industry is highly capital intensive, and consequently its profitability is strongly affected by fuel efficiency and aircraft utilization in addition to high labor costs. The theory of production and cost is one of the principal theories of economics and is the foundation upon which many business strategies and policies are based. In the airline industry, a proliferation of low-cost models and cost competitiveness helps low-cost airlines to outperform their competitors in the market and remain profitable. Low-cost airlines are generally more cost competitive than the legacy airlines, mostly due to their substantially higher input productivity and/or lower labor costs. In recent years, several legacy airlines have reduced their costs and returned to profitability. This section adopts a modern approach to the principles and practice of airline cost and production analysis. It introduces prospective airline managers to the technical cost accounting knowledge that can be applied across a whole range of organizations. And, although the chapter has been prepared primarily for students, it may also be useful to industry and government agencies. Cost management and cost restructuring are important to the survival of the airline industry. Cost restructuring involves more than just lowering labor costs; it also involves optimizing the entire operational and financial strategies. As one important example of this process of cost reduction, airlines may have to restructure their fleets to newer, more fuel-efficient aircraft. This process will entail transitioning to more cost-efficient aircraft like the Airbus A320 and Boeing 737 MAX and retiring older, more expensive airframes. More specifically, Alaska Airlines finished a fleet transition in 2008 that included retiring MD-80s and 737-200s. Delta Air Lines originally retired its DC-9 fleet in 1993 and is in the process of retiring 39 DC-9-50s that it acquired from its merger with Northwest Airlines. The full phase-out of Delta's DC-9s, along with the reduction of its fleet of older, less efficient aircraft like the Saab 340 and CRJ 100/200s is expected to be completed in 2012.[1]

---

1   *Air Transport World*, February 4, 2011.

In this chapter, we discuss a number of topics related to production and cost theory. We are specifically interested in applying this theory to the airline and aviation industries. This chapter provides a body of knowledge that will allow readers to understand how the costs of production are minimized through either the mix of inputs or the reallocation of resources across multiple plants. The chapter also illustrates which costs are relevant for specific airline decisions. And finally, the chapter discusses returns to scale, scope, and density. The specific topics are the following:

- Cost Division
  - Cost classification
    - Historical costs
    - Current costs
    - Sunk costs
  - Components of cost
    - Total costs, fixed costs, and variable costs
    - Marginal costs
  - Cost treatments
    - Opportunity costs
    - Accounting versus economic costs
- Cost Functions
  - Linear cost functions
  - Cubic cost functions
  - Using cost functions for managerial decisions
- Economies of Scale, Scope and Density
  - Economies of scale
  - Diseconomies of scale
  - Economies of scope
  - Economies of density
- Airline Industry Cost Structure
  - Fuel costs
  - Flight/cabin crew expenses
  - Maintenance costs
  - Other operating costs
  - Indirect operating costs
- Airline Economies of Scale, Scope, and Density
- Airline Breakeven Analysis
- Operating Leverage
- Airline Operating Leverage
- Summary
- Discussion Questions
- Appendix: Airline Cost Classification and Division

COST DIVISION

Profitability in any industry can be improved by increasing revenues or decreasing costs. The amount of profitability is the difference (contribution margin) between unit revenues and unit costs. As a concrete example of these concepts, Table 4.1 displays the unit revenue (passenger revenue per available seat mile: RASM), the unit costs (cost per available seat mile: CASM) and the contribution margin for major US airlines in 2011. Continental Airlines (CO) had the widest spread (RASM-CASM) among US airlines as a result of having the lowest operating CASM. United Airlines (UA), which completed its merger with CO in 2011, enjoyed the second highest contribution margin, despite having higher operating CASMs than Alaska Airlines and JetBlue.

> "Today, the situation is exacerbated with costs exceeding revenues at four times the pre-September 11 rate. Today, we are literally hemorrhaging money. Clearly this bleeding has to be stopped—and soon—or United will perish sometime next year."
>
> James Goodwin, Chairman and CEO of United, 2001

Table 4.1   Average Cost and Revenue for US Airlines, 2011

	AA	AS	B6	CO	DL	UA	US	WN
Passenger RASM	11.54	11.44	10.93	11.87	11.62	11.96	11.51	12.07
Operating CASM	8.32	7.30	7.30	6.97	7.80	7.72	8.12	7.85
RASM−CASM	3.22	4.14	3.63	4.90	3.81	4.24	3.38	4.21

Source: Compiled by the authors from OAG Form41 data

Table 4.2 presents the RASM and CASM for 2010. From this table we see that all but one carrier (CO) recorded a decrease in their RASM-CASM spread despite increases in the passenger RASM. Southwest remained among the top airlines, with a slight marginal decrease between the two years owing to a 0.90 cent increase in CASM and a smaller 0.75 cent increase in passenger RASM. However, it is important to note that consistency in profitability is the long-run goal of any airline and Southwest has consistently achieved this goal.

Table 4.2   Average Cost and Revenue for US Airlines, 2010

	AA	AS	B6	CO	DL	UA	US	WN
Passenger RASM	10.94	10.67	9.83	10.67	10.68	11.04	10.46	11.32
Operating CASM	7.27	6.39	5.90	6.08	6.54	6.75	6.87	6.96
RASM−CASM	3.67	4.28	3.92	4.59	4.14	4.29	3.59	4.36

Source: Compiled by the authors from OAG Form41 data

Recently, US commercial airlines have been more successful at reducing costs, and increasing revenues by opening up new routes, dropping unprofitable ones, selecting the most efficient aircraft type, and implementing advanced revenue management practices. Up until 2001, airlines largely lacked a detailed focus on cost reductions, but the downturn in demand has shown that reductions in costs are quite important. In an attempt to cut costs, the major airlines have progressively eliminated numerous services and reduced many cabin amenities. Through reduced in-flight amenities, labor outsourcing, automation, and aircraft retirement, legacy carriers have made efforts to lower their cost and optimize their cost structure. Low-cost carriers (LCCs) have also been successful due to their ongoing focus on managing costs. The lesson from the past few years is clear; it is vitally important to understand the cost structure and the variables that impact profitability. The remainder of this chapter is devoted to these issues.

Cost classifications

The economic definition of a cost is the foregone alternative use of resources in the production, transformation, use, and delivery of services. For example, for a manufacturer, the costs could be the alternative uses of raw materials, labor, buildings, and general overhead supplies. In the airline industry costs incurred include labor (of various types), fuel, maintenance, aircraft, catering, and airport landing/usage fees. Typically, fuel, labor, maintenance, and aircraft ownership costs are the four largest costs for any airline. Costs can be separated into different categories and classifications: Historical, Current, and Sunk.

Historical costs   Costs can be classified and grouped according to time. Historical costs are the costs actually incurred when acquiring an asset, whereas current costs are the costs incurred under prevailing market conditions. For example, when measuring aircraft costs, one must decide whether to use historical costs, current costs, or some other measure. Assume DirectJet bought an aircraft in 2011 for $25,000,000, yet today the aircraft is worth $20,000,000 in the open market. In this example, the historical cost for the aircraft would be $25,000,000 while the current cost would be $20,000,000. Under Generally Accepted Accounting Principles (GAAP), the book value of a company's assets is based on its historical costs. The book value of an asset is determined by its historical cost, less capital consumption allowance (depreciation). Most airlines' depreciable lives for their aircraft hover around 30 years with an estimated residual value of generally 5 percent of the historical.

Current costs   Historical costs are used in accounting to measure the book value of assets; they provide information on depreciation, depletion and asset impairment value. However, current costs provide the best representation of the present situation. And, quite obviously, future costs are the expected costs that may be incurred sometime in the future; these will be affected by such macroeconomic variables as the (uncertain) rate of inflation and the rate of interest. Replacement costs are the costs required to duplicate the productive capabilities using current technology. For example, in the airline industry, an older generation 737 could be replaced by a modern 737 MAX for the cost of the 737 MAX.

Sunk costs   Finally and perhaps most importantly, are sunk costs. Sunk costs are the investments made in the current planes and airport facilities that can't be recovered; these expenses have been incurred in the past and are not recoverable. Since these costs have already occurred and they are not recoverable, they should never influence business decisions. Unfortunately, they commonly do. An example of such a misguided decision-making occurs in the following: suppose an airline has already purchased eight additional lavatory units and is considering installing these units in new aircraft. However, additional detailed analysis shows that these lavatory units would increase fuel costs, decrease passenger revenue (by removing seats), and increase maintenance costs. Suppose further that there is no secondary market for these lavatories. Therefore, installing the aircraft lavatories would be costly to the airline, but managers may want to install the lavatories anyway since, "we already have them." In this example, the pre-purchase of the lavatories has now become a sunk cost, and therefore should not affect the final decision as to install them. Another example of a sunk cost is the cost of a seat on an airplane to the airline. The cost of an airline seat is primarily fixed and sunk with a small portion attributed to variable costs (VC) if one considers the fuel, airport charges, and cabin services.[2] Therefore sunk costs should not be included in a capital budgeting analysis or any financial decision-making process; rather, only opportunity costs should be included in the process.

Components of cost

Proper application of cost requires a strong understanding of the relationship between cost and the level of production. Cost functions specify the technical relationship between the level of production and production cost. Production costs can be subdivided as total, fixed, and VCs.

Total costs, fixed costs, and variable costs   While costs categorized by time are important in accounting, in economics the most common and practical method of classifying costs is by their relation to output. In the short run, total costs (TC) consist of two categories of cost: total fixed costs (FC) and total VCs. Costs that remain fixed in the short run, regardless of the level of output, are termed FCs. Costs that directly vary with changes in production are termed VCs. When FCs and VCs are added together, the result is TCs. In simple terms the formula below provides the equation for determining TC:

TC = FC + VC

As pointed out above, the timeframe is important when categorizing costs based on their relation to output. In the short run some costs are fixed since they have already been incurred and they are difficult or impossible to change. In the long run however, all costs are assumed to be variable since over time a company is fully able to change FCs (by selling and/or altering the FC asset). An example of this could be long-term aircraft leasing contracts. Since airlines are legally and contractually obligated to pay aircraft lease payments to the lessee, in the short run the airline would be unable to avoid these payments no matter how it adjusts output. Therefore, lease payments are a FC in the short run, but in the long run they are variable since eventually the contractual obligations are subject to termination. At this point the airline can cut the aircraft from the fleet in order to adjust capacity. There is usually no set timeframe when FCs will turn into VCs; however, the specific situation will dictate when all costs become variable.

Firms and industries have different ratios of fixed to VCs. The airline industry tends to have high FCs, which increases barriers to entry of the industry. The ratio of fixed to VCs is called operating leverage, and this topic will be covered later on in the chapter. Figure 4.1 displays fixed, variable, and TC functions.

If we divide the TC function by output we obtain another important category of costs, called the average costs (Boyes, 2004). By definition then, average TC is the total

---

2   Assuming that the airline would fly these routes regardless of whether the seats are sold or not.

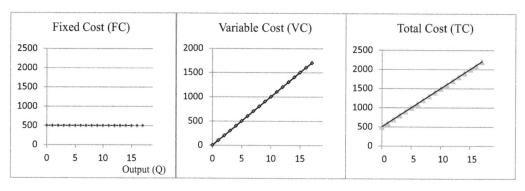

Figure 4.1   Disaggregated Cost Functions

amount of costs (both fixed and variable) per unit of output. Average fixed cost (AFC) is the FC per unit of output and average variable cost (AVC) is the total VC per unit of output.

$$AFC = \frac{FC}{Q}$$

Because FCs remain constant regardless of output, AFCs will decrease with increases in output, since the FCs are spread out over a greater range of outputs. This provides an incentive for a firm to increase output if they have high levels of FCs. Airlines which fly long-haul markets generally enjoy lower CASM. Longer stage lengths allow the FCs of each flight to be distributed over more available seat miles (ASMs). Conversely, an increase in unit cost may be attributed to a shorter stage length.

The AVC function measures the VCs per unit of output. In some cases, AVCs tend to decrease over the first units of production (since so few units are initially produced) and then increase as production increases.

$$AVC = \frac{TVC}{Q}$$

Finally, average total cost (ATC) is the amount calculated by dividing the TC by the units of production. ATCs will behave like AVCs since it is made up of fixed and VCs. The formula below describes the relationship between ATCs, AVCs, and AFCs.

$$ATC = \frac{TFC}{Q} + \frac{TVC}{Q}$$
$$ATC = AFC + AVC$$

In the airline industry, the average cost is generally represented by CASM.[3] The output for an airline is passenger miles. CASM is the primary measure unit of cost in

---

3   Or, Cost per Available Seat-Kilometer (CASK).

the airline industry and is obtained by dividing the operating costs of an airline by ASM.[4]

$$CASM_{Total} = \frac{Total\ Cost}{ASM}$$

$$CASM_{Operating} = \frac{Operating\ Cost}{ASM}$$

Another type of cost related to changes in outputs is mixed costs. Mixed costs exhibit characteristics of both FCs and VCs. That is, the costs are fixed for a certain range of outputs and then increase for a different range of outputs. While FCs are fixed in the short run for all outputs, mixed costs have smaller bands of FCs and fluctuate to a greater degree. For example, labor can be a mixed cost since it may be difficult to adjust staffing levels to rapidly changing levels of output, but eventually labor costs will have to be adjusted upward or downward to accommodate higher or lower levels of production. Therefore, a mixed cost appears like a step function with various levels.

Marginal cost   A final and extremely important type of cost relating to output is marginal cost (MC). "Marginal" refers to the change in the dependent variable caused by a one unit change in an independent variable. Assume that Ryanair offers five daily flights from Liverpool to Dublin, and adding an additional flight increases the cost by £13,000 (or £100 per available seat), hence the MC of the last flight is £100 per seat. Algebraically, the MC is the change in TCs resulting from an increase in one additional unit of output (Maurice and Thomas, 2005).

$$MC = \frac{dTC}{dQ} = \frac{\Delta TC}{\Delta Q}$$

MCs are simply calculated by the change in TCs divided by the change in output. For example, it might cost a company $1,000 to produce 20 units and $1,100 to produce 21 units. In this example the MC would simply be $100. Additionally, it might cost $1,250 to produce the 22nd unit so that the MC of this unit would be $150 and so forth. It is important to note that MCs generally do not remain constant throughout; therefore, every MC value is unique to that particular change in output.

The airline industry generally has very high FCs and low MCs. There is a small increase in cost for each additional passenger since, regardless of the number of passengers, airlines have to pay the high FCs associated with aircraft ownership, terminal expenses, and maintenance facilities. Consequently, airline managers cannot cut costs quickly if this becomes necessary.

Cost treatments

Another way by which we can categorize costs is by explicit and implicit costs. Explicit costs are the costs represented by actual out-of-pocket expenditures, while implicit costs are generally non-cash expenditures. For example, an implicit cost would be the time and

---

4   In other countries (except US and Canada) available seat kilometer (ASK) is used as a measure of supply.

effort that an owner devotes to his company rather than working for another firm. Implicit costs are best thought of in terms of opportunity costs.

Opportunity costs   As mentioned above, since the pursuit of any economic activity represents a choice between two options, the opportunity cost is the value of the next best option. Therefore, the chosen economic activity must provide a better rate of return than the next best alternative; otherwise, the company would have been better off pursuing the other alternative. For example, according to Airlines for America (A4A)[5] estimates, the booking of Federal Air Marshals into first-class seats and displacing passengers is producing opportunity costs running as high as $180 million annually (Stewart, 2006).

In order to better illustrate opportunity costs, consider the situation an airline faces when it is looking to purchase either a Boeing or Airbus aircraft. Assume that both aircraft have the exact same seating capacity and have similar performance characteristics. The Boeing aircraft costs $70 million while the Airbus aircraft costs $72 million. The fuel costs for each aircraft type have been estimated to be $1,350 per block hour for the Boeing and $1,370 per block hour for the Airbus. Based on these specifications the airline purchases one Boeing aircraft. Here the explicit costs would be the purchase price of $70 million and the hourly fuel consumption rate of $1,350. The opportunity costs in this example would be $2 million of savings in the aircraft purchase price and $20 additional fuel costs per block hour, since the Airbus aircraft was the next best alternative for the airline.

Accounting versus economic costs   The final major method to classify costs is either accounting or economic costs. Accounting costs generally recognize only explicit costs, while economic costs include both explicit and implicit costs. Therefore, accounting costs do not take into consideration opportunity costs. The two costs also deal with depreciation differently. Accounting costs calculate depreciation based upon a predetermined historical usage rate applied against the cost incurred to acquire the asset. On the other hand, economic cost is defined as the value of any resource used to produce a good in its best alternative use. Both accounting and economic costs are useful in their respective context, with accounting costs used primarily for financial accounting purposes and economic costs for managerial decision-making process.

Economic Cost = Accounting Cost + Opportunity Cost[6]

Economic Cost = Explicit Cost + Implicit Cost[7]

Suppose you have started a small fixed-based operation (FBO) at a regional airport. Your explicit costs are:

Labor	$1,000,000
Materials and supply	$750,000
Finance charge, insurance and others	$250,000
Total explicit costs (accounting)	$2,000,000

You are not receiving a payment for your services, because the business is new and does not have enough income. Also assume you have rejected a position, in which you could

5   Formerly Air Transport Association (ATA).
6   Accountants are generally concerned with explicit costs r for financial reporting purposes.
7   Economists are generally interested with economic costs for decision-making purposes.

have earned $200,000 a year. So in this case: Total Economic Costs = $2,000,000 + $200,000 = $2,200,000 rather than the $2,000,000 of explicit accounting costs.

COST FUNCTIONS

A cost function is a mathematical relationship between TC and units of quantity produced. A cost function is an analytical tool which allows airline managers to forecast and estimate cost of operations at different level of services. While every company has a unique cost function, there are two more general forms of cost functions: linear and cubic cost functions.

Linear cost functions

The linear cost function can be represented by the formula:

TC = FC + VC

TC = a + bQ

Where a and b are both constants.

In a linear cost function, the constant "a" would represent the FC, the constant "b" would represent the average cost[8] component, and Q would represent the total units produced. The underlying assumption of a linear cost function is that there are constant returns to scale.[9] That is, successive units of output can be produced for the same cost (in this case the constant b) so that the average and MC are equal. While linear cost functions may not be attainable for all levels of production in the long run, in the short run they may be approximated within a given production range. A linear cost function generally applies best to a mechanized and automated production line where unit costs are approximately the same. For example, if the production capacity of an assembly line is 300 cars per day, a linear cost function can approximate this process up to that point. For units above 300 a day, a new assembly line would have to be built and the cost function would shift accordingly.

Consider a linear cost function of:

TC = 50 + 5Q

Assume that the capacity of the production line is 15 units. Table 4.3 breaks down the cost function into the various classifications.

Figure 4.2 displays the FCs for the cost function. Figure 4.3 displays the AFC function, and Figure 4.4 provides a vertical summation of the average fixed and AVC functions (ATC). Finally, Figure 4.5 displays the average and MC functions for the linear example, with MC being equal to the AVCs and declining average TCs and FCs for increases in unit output.

To summarize, linear costs occur when the company experiences constant MCs of input charges. While the linear cost functions are not typical of most industries, they may be useful approximations over a given range of outputs.

---

8   For this simple cost function the average variable (bQ/Q) and marginal costs are equal.
9   Average cost remains unchanged as output varies.

Table 4.3   Cost Classifications

Units	TC	FC	VC	AFC	AVC	ATC	MC
1	55	50	5	50.0	5	55.0	
2	60	50	10	25.0	5	30.0	5
3	65	50	15	16.7	5	21.7	5
4	70	50	20	12.5	5	17.5	5
5	75	50	25	10.0	5	15.0	5
6	80	50	30	8.3	5	13.3	5
7	85	50	35	7.1	5	12.1	5
8	90	50	40	6.3	5	11.3	5
9	95	50	45	5.6	5	10.6	5
10	100	50	50	5.0	5	10.0	5
11	105	50	55	4.5	5	9.5	5
12	110	50	60	4.2	5	9.2	5
13	115	50	65	3.8	5	8.8	5
14	120	50	70	3.6	5	8.6	5
15	125	50	75	3.3	5	8.3	5

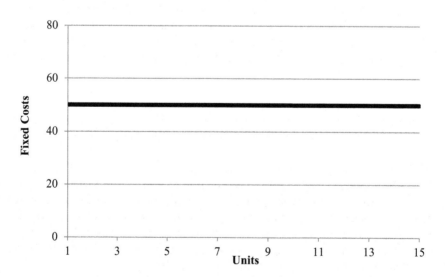

Figure 4.2   Fixed Cost Function

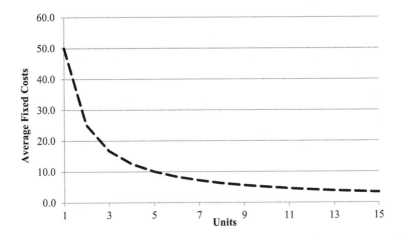

Figure 4.3   **Average Fixed Cost Function**

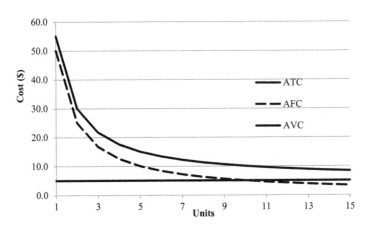

Figure 4.4   **Average Variable, Fixed, and Total Cost Functions**

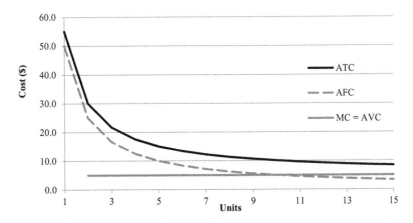

Figure 4.5   **Average and Marginal Cost Functions**

Cubic cost functions

Another major cost function is the cubic cost function. The general form of a cubic cost function takes the following form (Maurice and Thomas, 2005):

$$TC = a + bQ + cQ^2 + dQ^3$$

A cubic cost function represents the normal theoretical cost function. This is a common cost function for many industries since it first exhibits increasing marginal returns and then diminishing returns to scale as demonstrated by its "U-shaped" average cost function. Consider a company whose TC function is:

$$TC = 100 + 40Q - 8Q^2 + (2/3)Q^3$$

By plugging this formula into an Excel spreadsheet, Table 4.4 provides the numerical values of the various cost classifications. Also, from this equation we can apply the cost definitions covered earlier to review the various categories of cost (Maurice and Thomas, 2005).

FC = constant "a" of the general form (in this case 100)

VC = $40Q - 8Q^2 + (2/3)Q^3$ (costs that change with output)

Table 4.4   Cost Classifications

Units	TC	FC	VC	AFC	AVC	ATC	MC
1	133	100	32.7	100.0	32.7	132.7	
2	153	100	53.3	50.0	26.7	76.7	20.7
3	166	100	66.0	33.3	22.0	55.3	12.7
4	175	100	74.7	25.0	18.7	43.7	8.7
5	183	100	83.3	20.0	16.7	36.7	8.7
6	196	100	96.0	16.7	16.0	32.7	12.7
7	217	100	116.7	14.3	16.7	31.0	20.7
8	249	100	149.3	12.5	18.7	31.2	32.7
9	298	100	198.0	11.1	22.0	33.1	48.7
10	367	100	266.7	10.0	26.7	36.7	68.7
11	459	100	359.3	9.1	32.7	41.8	92.7
12	580	100	480.0	8.3	40.0	48.3	120.7

$$AFC = \frac{100}{Q} \quad \text{(FCs divided by output)}$$

AVC = 40 − 8Q+ (2/3)Q² (VCs divided by output)

$$MC = \frac{\Delta TC}{\Delta Q} = \frac{TC_n - TC_{n-1}}{Q_n - Q_{n-1}} \quad \text{(the change in TCs for a one unit change in output)}$$

MC = 40 − 16Q + 2Q²

In this scenario, the company has FCs of $100, which remain constant regardless of the level of output. Since the FCs remain constant throughout, the AFC declines for every unit increase in output. This cost function also contains a VC function that may include costs such as labor and raw materials. Figure 4.6 displays the FC function for this example, while Figure 4.7 displays the typical VC function curve, and Figure 4.8 provides vertical summation of the cost curves, displaying the TC function and the FC line.

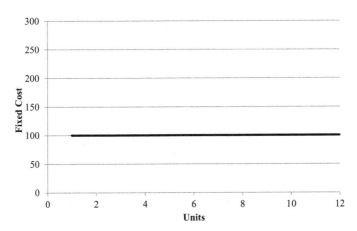

Figure 4.6   Fixed Cost Function

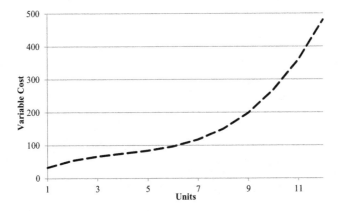

Figure 4.7   Variable Cost Function

Figures 4.8 display the total, fixed, and VC functions for the production facility. Since FCs are the same for all levels of production, the FC line remains horizontal throughout. The VC curve changes slightly for different rates of output with costs escalating dramatically for higher levels of output. Finally, the TC curve is simply the VC curve shifted upward above the FC line.

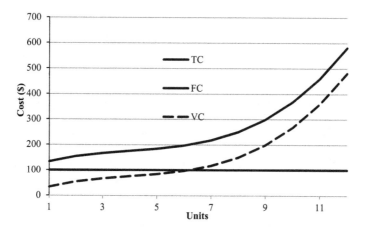

Figure 4.8   Total Cost Function

Figure 4.9 displays the MC curve and the average cost curves for all three cost classifications. From Figure 4.9 we see that the AFC curve will always slope downward at a declining rate where it will eventually be asymptotic with the x-axis. The average TC curve is u-shaped which is similar to the AVC since they differ by only the AFC. The MC curve crosses both the AVC and average TC curves at their minimum point and continues above them as output rises. The intuitive reason for this is that at first we are adding unit costs that are less than the average—causing the average costs to fall. Then, as these additional unit costs increase, they eventually are equal to the average (at the minimum of the average). Above this point, we are adding unit costs that are above the average, hence the average is rising. A cubic cost function represents the normal theoretical cost function, which exhibits both decreasing marginal and average costs and increasing marginal and average costs.

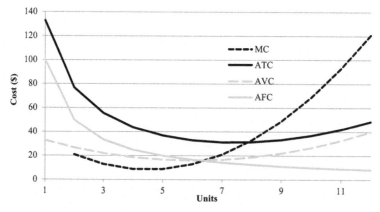

Figure 4.9   Average and Marginal Cost Functions

Using cost functions for managerial decisions

While it is instructive to review the cost categories, it is even more important to understand how they affect managerial decision-making. In this regard the cost function provides the following information for managerial decision-making:

- MC tells us where (how much) to produce;
- Average TC tells us whether we should produce;
- AVC tells us when we should cease production.

Since MC is the cost for extra units of output, then it is obvious that as long as the revenue from the extra unit exceeds its cost then we will want to produce more. Conversely, if the cost for that unit exceeds the revenue, then we will not want to produce that unit. Hence, we will continue production up to the point where MC and marginal revenue (MR) are equal but not beyond. So we continue to produce as long as MR≥MC.

$$ATC = \frac{TC}{Q}$$

$$ATC = \frac{TFC + TVC}{Q}$$

$$TC = ATC \times Q$$

Average TC can be multiplied by output to give TCs. As long as total revenues exceed TCs we will of course continue production. In the event that TC falls below total revenue but is still above VCs then we would still continue production (in the short run) since we are still making something to help pay our FCs.[10] Recall that FC must be paid in the short run whether we are producing or not. Hence, average TC tells us whether we should continue production.

Finally, we have AVC which can be translated into total VC by multiplying by output. In this case, if our total revenues do not cover our total VCs, then we should cease production since we will be losing more money on every unit we produce.

Shut-Down	
Short Run	Long-Run
TC>TR Loss ATC>P>AVC operate in Short Run ATC>AVC>P Shut Down	TR>TC Operate TR<TC Shut Down

A third type of cost function typically prevalent in the industry is the quadratic function which has a general form of:

$$TC = a + bQ + cQ^2$$

---

10   The difference between price and the variable costs (P-VC) is called contribution margin.

This can exhibit decreasing returns to scale for the range of production possibilities since the curve is upward sloping at an increasing rate. The TC curve then becomes asymptotic at the capacity level of the production range.

The TC function can be derived to determine both the average cost function and the MC function for all three types of cost functions. In order to determine the average cost function, simply divide the TC function by Q. The MC function is found by taking the first order derivative of the TC function.

Consider the linear cost function:

$$TC = 50 + 5Q$$

By dividing the function by Q, the average cost function for a linear cost structure is:

$$ATC = \frac{50}{Q} + 5$$

Taking the first-order derivative of a linear function, the MC = 5, which is also the constant b and the VC of the linear function.

For a cubic cost function, both the AVC and MC functions are found to be quadratic. From the equation:

$$TC = 100 + 40Q - 8Q^2 + \frac{2}{3}Q^3$$

The average cost function is:

$$AVC = 40 - 8Q + \frac{2}{3}Q^2$$

The MC function for the cubic cost function is:

$$MC = 40 - 16Q + 2Q^2$$

This MC curve is quadratic in nature (the MC curve is U-shaped), indicating that at the beginning of the production, MC drops but later for every increase in output, MCs start to increase at a greater rate. This is the basic definition of decreasing returns to scale.

ECONOMIES OF SCALE, SCOPE, AND DENSITY

Having covered the cost curves in some detail we can now use these concepts to define economies of scale, scope, and density. These factors can play a crucial role in management decision-making and the cost structure of not only the company, but the industry as a whole.

Economies of scale

The concept of economies of scale refers to the advantages gained when long-run average costs decrease with an increase in the quantity being produced. Economies of scale are common in highly capital intensive industries with very high FC such as aircraft manufacturing, airline industry, railroads, and steel industry.[11] For example, in the airline industry, network airlines may have an inherent cost advantage over smaller airlines because of economies of scale.

The cost of adding one more flight to the network is relatively low, while the financial benefits are much higher. The merger between United and Continental Airlines was expected to bring cost savings and at the same to increase revenues by providing more routes and more effective competition against rivals. The merger and consolidation could provide significant cost advantages for the joint operation of the airlines.

The following are some of the sources of economies of scale:

- lower cost and higher productivity due to division of labor;
- lower average cost labor due to specialization;
- increase in labor productivity due to division of labor and concentration on a fewer number of tasks and more experience on the job (learning curve);
- lower average cost due to higher seat density;
- lower average cost due to higher aircraft utilization;
- lower average cost due to using uncongested secondary airports;
- single aircraft type.

Diseconomies of scale

Diseconomies of scale occur when average unit costs increase with an increase in production quantity. Some of the causes for diseconomies of scale follow:

- inability to efficiently monitor and coordinate material flows and manage employees' performance due to larger facilities;
- slow decision-making ladder;
- workers and management becoming more segregated and communication becoming less effective;
- inflexibility;
- capacity limitations on entrepreneurial skills (hiring qualified employees).

In the airline industry, economies of scale are claimed as a justification for mergers and acquisitions amongst major airlines. Economies of scale have the potential to provide larger airlines with a cost advantage over smaller airlines. The Delta–Northwest merger was completed in October 2008 and, because of the consolidation, the accounting, revenue management, and other administrative offices of two merged airlines can be smaller than the sum of these departments in the two separate airlines. Depending on the firm's cost function, economies of scale usually do not exist for every level of production. Companies will have levels of quantity where economies of scale are present and levels where diseconomies of scale exist. This is exactly what occurs for cubic cost functions. Referring back to Figure 4.9, the average TC function is u-shaped, indicating economies of scale for production quantities of one through seven and diseconomies of scale for production quantities eight and above.

---

11   Because of very high fixed cost only a few companies can stay in market. As a result, there are only two aircraft manufacturing companies in the world (Boeing and Airbus) that manufacture large commercial aircraft.

Economies of scope

Economies of scope refer to the situation where the company can reduce its unit costs by leveraging efficiencies through sharing of resources for multiple projects or production lines. Put more simply, multiple projects/processes can be more cost-efficient when they are done together rather than when they are done individually. As an example, on April 30, 2012 Delta Air Lines announced that it will pay $150 million for a refinery near Philadelphia aiming to cut jet fuel cost by $300 million a year.[12] Delta is planning to reduce fuel costs by eliminating speculators and marketing by intermediaries.

The presence of economies of scope provides benefit for a company to house activities together and concurrently. Possible synergies achieved through economies of scope could include shared labor, shared knowledge, and shared capital equipment. For example, Boeing houses four production lines (747, 767, 777, and 787) at its large Everett production facility (Boeing, 2010). By operating three production lines in the same building, Boeing is able to share resources such as labor and equipment between all four lines to maximize resource efficiency. If all four production lines were individually located throughout the country, Boeing would not be as cost-effective in manufacturing aircraft since resources could not be shared among all three production lines. Additionally, Boeing is able to leverage capital knowledge by reducing research and development (R&D) expenses through utilizing technology developed in other projects. These savings can be considerable when developing new aircraft such as the Boeing 787.

Another example of economies of scope would be Aer Lingus, which in 1970 began to seek new sources of revenue by offering engineer training, maintenance services, computer consulting, and data-processing services to other airlines (Rivkin, J., 2005). Airlines globally are depending on ancillary revenue by charging for things once included in the ticket price to selling all-new products and services that travelers may want. Traditionally, airlines have earned ancillary revenue from catering, co-branded credit cards, duty-free shopping, ground handling, maintenance, and other activities. To increase on-board revenue, Ryan Air offers the cabin crew 10 percent commission on all on-board sales. Charging for luggage is becoming more and more common among all airlines in addition to charging for overweight bags. However, not all airlines charge their passengers for certain types of luggage.[13]

Economies of density

Economies of density are achieved through the consolidation of operations. The airline industry is a good example of this as it has developed the so-called hub-and-spoke system for air travel. That is, airlines have found it more cost-effective to consolidate operations at a single airport rather than operate a point-to-point service. Under a hub-and-spoke network, aircraft are more likely to fly at full capacity, and can often fly routes more than once a day. For example, consider five airports that could all be connected together either by using one airport as a hub (Figure 4.10), or by flying between each city (Figure 4.11). Using a hub, all the airports can be connected to each other with a minimum of four flights, while the point-to-point service would require ten flights. The difference in the

---

12   *The Florida News Journal*, May 1, 2012.
13   Virgin America is now responding to its bag fee-induced cabin crowding by allowing anyone without carry-on luggage to board first.

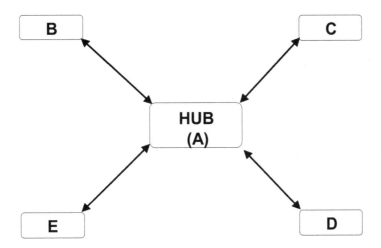

Figure 4.10   Hub-and-Spoke Route Network

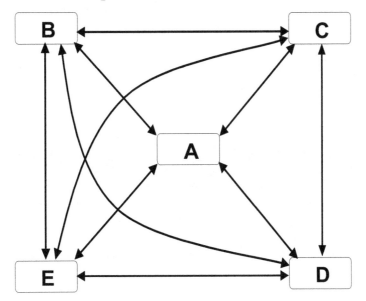

Figure 4.11   Point-to-Point Route Network

number of required flights is a cost saving as the airline can provide service to all the cities with less resources. Additionally, the advent of the regional jet has enabled airlines to further strengthen their hub networks by increasing flight frequency as well as connecting markets of smaller capacity that would be unprofitable with larger aircraft. Of course, there are other benefits to flying point-to-point services such as being able to offer non-stop flights for the passengers. However, the hub-and-spoke network is still the dominant flying structure in the US and will probably continue to be so for the foreseeable future.

The economies of scope and density of a hub-and-spoke network can also be stated mathematically. Figure 4.10 displays a typical hub-and-spoke diagram for a network utilizing five airports, while Figure 4.11 displays a point-to-point route network also for the same five airports. Clearly, by simply looking at the diagrams, the hub-and-spoke network has significantly less flights. The number of flights in a hub-and-spoke network can be calculated by using the formula:

Number of Flights = $n-1$

Where n is the number of airports.

Using this formula, the number of flights required to connect five airports in a hub-and-spoke network is 4 (5-1). These four flights are displayed in Figure 4.10.
    The mathematical number of flights required in a point-to-point system can also be determined by using the formula:

Number of Flights = $\dfrac{n \times (n-1)}{2}$

Using the above formula, the number of flights required in a point-to-point network for five airports is ten. Or similarly, in a network with ten airports, the hub-and-spoke system requires only nine routes to connect all destinations, while a true point-to-point system would require 45 flights.

Number of Flights = $\dfrac{10 \times (10-1)}{2} = 45$

This reduction in flights is the major reason why the hub-and-spoke network has been adopted by almost all carriers in the US in the post-deregulation period. Southwest Airlines is the sole major US carrier that has decided to adopt the point-to-point route network, yet even the majority of Southwest passengers make connections between flights. One negative aspect of a hub-and-spoke system is the delays at the hub that can result in delays throughout the network. Similarly, the delays at a spoke can also affect the entire network.

AIRLINE INDUSTRY COST STRUCTURE

First, we have to lower our costs to levels that are more competitive. This will prevent the lower-cost airlines from pushing us out of the markets we want to serve. We've made great progress on this front, but we need to keep pushing.

Gerard Arpey, CEO American Airlines

Since every airline's operation is unique, it can be difficult to compare airline operating costs from airline to airline. The most common metric used to standardize airline costs are CASMs. An available seat mile (ASM) is one aircraft seat, flown one mile, regardless of whether it is carrying a revenue passenger. Costs per ASM, or CASM, are the cost of flying one aircraft seat for one mile. CASMs can be created for a variety of costs, such as operating costs, total operating costs, or simply crew costs. Table 4.5 provides a breakdown of various CASMs for eight major US airlines in 2011 with the TC per ASM representing

Table 4.5   US Airline Cost per Available Seat Mile Breakdown, 2011

	AA	AS	B6	CO	DL	UA	US	WN
Fuel Cost per ASM	4.64	4.05	4.35	4.08	4.93	4.42	4.57	4.44
Maintenance Cost per ASM	1.50	0.94	0.91	1.05	1.09	1.57	1.43	1.11
Crew Costs per ASM	1.17	1.19	1.09	0.92	1.11	0.81	0.91	1.43
Other Operating Costs per ASM	1.00	1.12	0.96	0.92	0.67	0.91	1.21	0.88
Total Operating Costs per ASM	**8.32**	**7.30**	**7.30**	**6.97**	**7.80**	**7.72**	**8.12**	**7.85**
Non-Operating Costs per ASM	7.39	7.05	3.92	8.08	8.48	8.90	9.62	4.56
Total Costs per ASM	**15.71**	**14.35**	**11.22**	**15.05**	**16.28**	**16.62**	**17.75**	**12.41**

Source: Compiled by the authors from OAG Form 41 data, 2011

the direct operating costs (DOC) of fuel, labor, maintenance, and other operating and non-operating costs (typically called overhead costs). The four DOCs can be considered VCs while the non-operating costs can be considered FCs. Figure 4.12 also displays this information graphically.

Figure 4.12 displays the composition of the airlines' various costs. For all the airlines, the non-operating costs comprise the largest share of the airline's TCs. In many cases, this share is over 50 percent of the airline's TCs (except for the LCCs Southwest and JetBlue where non-operating costs were 37 percent and 35 percent of TCs respectively). Since non-operating costs are fixed with output, this chart confirms that airlines have large FCs. Acquisitions such as aircraft and investment in airport infrastructure are large capital expenditures; these in turn create an industry with high barriers to entry. Recently, however, these capital requirements have been lowered with attractive aircraft leasing options, thereby enabling easier entry into the market.

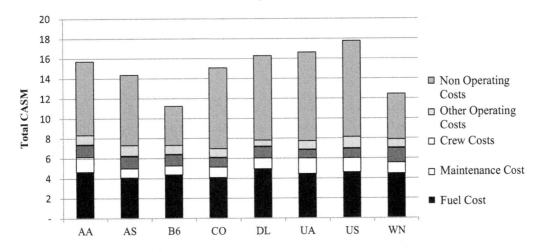

Figure 4.12   US Airline Cost per Available Seat Mile Breakdown, 2011

Source: Compiled by the authors from OAG Form 41 data, 2011

DOCs are those costs that are directly attributable to the airline's operations, and are incurred each time when the aircraft is flying. The major categories of DOCs are listed below, and they will be discussed in more detail in the following paragraphs.

- fuel costs;
- flight deck and cabin crew expenses;
- direct maintenance expenditures;
- other operating costs, including landing fees and capital equipment charges;
- en route (Air Traffic Control expenses);
- aircraft rentals.

Fuel costs

Fuel costs in 2011 represented the greatest share of an airline's DOCs. This was due to sustained high prices of oil that were experienced in 2011. The general formula for airline fuel costs is:

$$\text{Fuel Costs} = \text{ASM} \times \cfrac{\text{Fuel Price per Gallon}}{\cfrac{\text{ASM/Block Hour}}{\text{Gallons/Block Hour}}}$$

An airline's fuel cost per ASM is a result of two factors: the price of fuel and fuel efficiency. While the price of fuel is generally out of the airline's control, airlines can lessen the impact of this cost by using more complex investment strategies; airlines purchase options for the price of oil and not for the price of aviation fuel. By purchasing aviation fuel at the market-bearing rate and offsetting this with investment gains from oil options, the costs of increases in the price of fuel are lessened. However, these gains are recorded as investment gains in the airline's consolidated financial statements and not as fuel cost benefits.

> "Fuel makes up a significant portion of an airline's total costs, and fuel efficiency plays a significant role in airline profitability. When fuel prices remain high, you can expect ticket prices to increase and profitability to decline."
>
> Airlines for America (A4A), 2012

The second way airlines can adjust their fuel costs is by being more fuel efficient. The simplest way to do this is to operate new fuel-efficient aircraft over older, less fuel-efficient aircraft (as pointed out in the introduction to the chapter). While the capital expenditures required to purchase new aircraft are considerable, they may outweigh the fuel costs associated with operating older aircraft. Of all the airlines listed in Table 4.5, Delta had the highest fuel cost per ASM at 4.93 cents. This is largely due to the fleet of aging aircraft which Delta acquired after merging with Northwest Airlines. The reverse is true of airlines such as Southwest and JetBlue that operate mostly newer generation aircraft that are more fuel efficient.

Other fuel efficiency methods center around technological advances; for example, the installation of blended winglets. Aviation Partners Boeing (2006), the joint-venture company that manufacturers blended winglets, estimates that the winglets lower fuel burn by 3.5

percent–4.0 percent on flights greater than 1,000 nautical miles for Boeing 737NGs. The winglet technology does not provide substantial savings on short flights as the fuel burn advantage is offset by the increased weight. While winglets were originally offered on just 737NG aircraft, their success has led to their installation on a number of different aircraft types.[14] Airbus's A320neo aircraft will come standard with 2.5 meter tall winglets called "sharklets" which are expected to cut fuel burn up to 3.5 percent by reducing aerodynamic drag. These sharklets will replace the smaller A320 wingtip fences currently used.

While the methods highlighted above usually require substantial capital investments to reduce fuel costs, more subtle fuel management strategies by airlines can also increase fuel efficiency. One strategy commonly employed by airlines is to use only one engine during normal taxiing procedures, thereby reducing the fuel costs associated with operating an engine during taxi. The more congested the airport is, the greater the amount of taxiing, and the greater amount of savings such a program can provide the airline. Additionally, the airline can selectively shutdown an engine(s) during ground delays when the aircraft will be sitting idle.

Flight planning plays a significant role in fuel efficiency as optimized flight planning can help plan flights for minimum fuel-burn routes and altitudes. Flight planning can also be enhanced by measuring on-board weight more accurately in order to avoid carrying extra fuel.

Altering the location where fuel is purchased allows airlines to take advantage of lower fuel prices in certain regions. Employing this strategy involves a cost-benefit analysis as the fuel cost savings from lower prices needs to be compared with the additional fuel burn generated from the additional weight involved in "tankering" the extra fuel to reach the desired region. Airlines can also pool resources when purchasing fuel in order to achieve bulk discounts. This is a strategy a few Star Alliance carriers have explored. It was estimated that joint fuel purchasing could provide around $50 million in savings for alliance members over three years (Ionides, 2007).[15]

At airports, airlines can employ self-imposed ground delays to reduce airborne holding and the airlines can also redesign hubs and schedules to reduce congestion. All these factors can and do contribute to airlines being more fuel efficient. Figure 4.13 provides a comparison of fuel efficiency for major US airlines in 2011, in terms of domestic ASMs per gallon. In Figure 4.13, the metric is ASMs per gallons of fuel used; therefore, a higher value is more desirable since airlines want more ASMs to be flown with one gallon of fuel.

Stage length is important since longer flights burn less fuel per ASM (Figure 4.14). The reason for this is the fact that the takeoff and landing phases of flight use the most fuel per ASM. Therefore, the longer the flight, the more fuel-efficient ASMs there are to dilute the less efficient takeoff and landing phase. Figure 4.15 shows the correlation between fuel efficiency and average stage length.[16] An anomaly to this correlation was Southwest Airlines which has relatively high fuel efficiency, but the lowest average stage length of all the airlines. Southwest has managed to overcome its short stage length with operating procedures that conserve fuel; these include using only one engine when taxiing and opting to fly out of less congested airports.

---

14   Winglets have being installed by airlines on 727s, 757s, and 737Classics. Airlines such as Continental and Southwest heavily promoted the use of winglets, which resulted in decreased fuel costs per ASM.
15   This initiative was being employed at four major airports worldwide: Los Angeles, London Heathrow, Paris Charles de Gaulle, and San Francisco.
16   Fuel efficiency is also affected by an airline's schedule as longer flights burn less fuel per ASM, where the takeoff and landing phases of flight burn the most fuel. On longer flights, these phases are spread out over greater distances, thereby making fuel costs per ASMs smaller for longer flights.

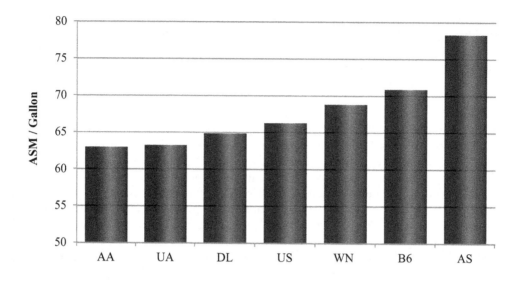

Figure 4.13   US Airlines Fuel Efficiency, 2011

Source: Compiled by the author from OAG Form 41 data

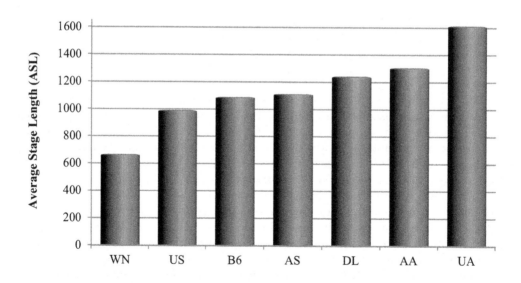

Figure 4.14   US Airlines Average Stage Length, 2011

Source: Compiled by the author from OAG Form 41 data

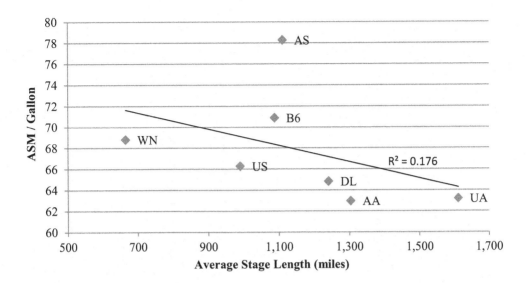

Figure 4.15   Correlation between Fuel Efficiency and Average Stage Length, 2011

Source: Compiled by the author from OAG Form 41 data

Flight/cabin crew expenses

The next greatest DOCs for US airlines are crew expenses. Prior to the recent downturn in the health of the aviation industry, crew costs constituted a much greater proportion of an airline's DOCs; however, mainly through bankruptcies, airlines have recently been able to dramatically reduce their labor costs. Since most airlines deal with a heavily unionized labor force, it can be difficult for airlines to adjust labor input to output, causing the crew costs to resemble mixed costs. That is, contractual agreements with labor groups make it difficult for the airline to furlough employees, causing long lag times for airlines to respond to decreasing travel demand and output. Productivity gains may also be limited by unions and government regulations concerning work rules. The general formula for flight personnel costs is:

$$\text{Flight Personnel Costs} = \text{ASM} \times \frac{\text{Labor Rate/Block Hour}}{\text{ASM/Block Hour}}$$

Figure 4.16 provides a comparison of crew costs (both flight deck and cabin crew) for 12 major US airlines. The data in Figure 4.16 closely mimics the crew costs per ASM in Table 4.5, but block hours are the most common measurement metric of crew costs in the airline industry.

The general assumption that LCCs always pay the least is somewhat inaccurate since Southwest Airlines is in the middle for crew costs. Southwest is effective through a more efficient use of its employees. That is, employees are expected to perform many different tasks in addition to their primary duties. These productivity gains help Southwest offset its higher pay rate. JetBlue benefits from being a relatively young company; therefore, they have a relatively younger workforce with a lower pay rate. Airlines such as American, Delta, and Continental have been around long enough that many crewmembers are quite

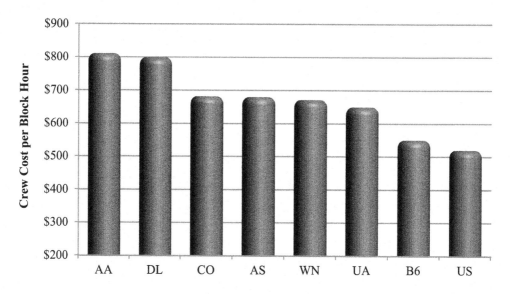

Figure 4.16   Crew Costs per Block Hour, US Airlines 2011

Source: Compiled by the author from OAG Form 41 data

senior and command a higher pay rate than junior crewmembers. This factor is enhanced when pension and Medicare issues are included in the crew cost calculations.

In 2011, American had the highest crew costs in the industry; however, Chapter 11 bankruptcy proceedings may very well enable them to reduce their crew costs per block hour.[17] Another example of the benefits of bankruptcy protection is United Airlines which at one time had the highest crew costs in the industry, but now has crew costs per block hour below that of Southwest Airlines. The recent trend in the aviation industry has been toward significant reductions in crew costs as airlines in the US have recently posted record losses. However, Southwest's crew costs per block hour have actually increased in the last three years and it remains to be seen if these costs are sustainable.

Maintenance costs

Another considerable cost for airlines is maintenance cost. However, since safety is the number one priority for every airline, maintenance costs are usually not under as much cost-saving scrutiny. That being said, airlines must still cost-effectively manage their maintenance operations and staffing levels, while also being safety conscious. In order to accomplish this, a major innovation in the maintenance area has been the outsourcing of maintenance activities to third-party vendors, especially for aircraft heavy checks. Maintenance costs can be calculated by using:

$$\text{Maintenance Costs} = \text{ASM} \times \frac{\text{Maintenance Labor and Materials/Block Hour}}{\text{ASM/Block Hour}}$$

---

17   Delta Air Lines emerged from bankruptcy, in April 2007, following a 19-month reorganization and a fight against a hostile takeover.

Each airline is unique in the amount of outsourcing that they do. American Airlines, for example, does the majority of its maintenance internally, and also does contract maintenance work for other airlines. On the other hand, Continental Airlines does all its wide-body heavy maintenance externally while it has internalized some maintenance operations with the opening of a 757 heavy check line. On the opposite end of the spectrum, JetBlue externally sources almost all its heavy maintenance and just performs line maintenance internally.

Figure 4.17 displays the maintenance costs for US airlines, including outside labor costs; these costs are standardized per flight hour since flight hours are the primary driver of an aircraft's maintenance cycle. Based on data from Figure 4.17, Continental and United had the highest maintenance costs while the LCCs, Southwest, JetBlue, and Frontier had the lowest maintenance costs in 2011. One possible explanation for the great differences in maintenance costs is that the LCC fleets are relatively new. Airlines receive a maintenance honeymoon on new aircraft since their costly heavy checks are delayed for a few years, but older aircraft have more frequent and costlier heavy checks. Therefore, there is some relation between the airline's average aircraft age and maintenance costs.

Another factor related to maintenance costs are maintenance checks. These are more expensive for larger aircraft than for smaller aircraft so that airlines such as Frontier benefit, maintenance wise, from operating only narrow-body aircraft, while airlines such as Delta, United, and American operate extensive wide-body fleets. Finally, aircraft commonality plays an important role in the maintenance costs for airlines. Airlines with diverse aircraft fleets usually must have spare parts on hand for each aircraft type thereby requiring a large inventory of parts. One way around this problem is to segment aircraft markets. This is the primary reason why American Airlines decided to operate only certain aircraft types out of certain hubs. For example, MD-80s do not operate out of the Miami hub, 737s do not operate out of Chicago, and the A300s do not fly to either Dallas or Chicago. This rationalization enables American Airlines not to maintain a spare parts inventory for every aircraft type at every major hub.

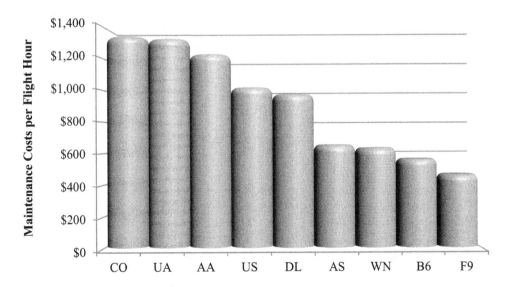

Figure 4.17   US Airline Maintenance Costs per Flight Hour for 2011

Source: Compiled by the author from OAG Form 41 data

Other operating costs

The final category of operating costs is other operating expenditures. These costs can include a variety of things such as airport-related expenditures (that is, landing fees, gate agents, and baggage handlers) and in-flight catering costs. These areas have seen dramatic reductions, particularly catering, since they were easy areas for immediate cost cutting. The majority of airlines now use third parties to supply their airport and catering services. Continental Airlines had the highest other operating costs per ASM which can possibly be explained by the fact that Continental still owns its own catering kitchen. If airlines once again begin to offer more elaborate services in the future, other operating expenses should be expected to increase.

Indirect operating costs (IOC)

The distinction between direct and indirect operating costs (IOCs) derives from the fact that some costs might change directly with the level of operation, while other cost may not. IOCs are those costs which an airline incurs whether the aircraft flies or not.

> Spreading out peak travel times at hub airports by several more hours each day, a concept known as rolling hubs, may be one way to make operations more efficient
>
> *Business Week*, October 23, 2002

There are many examples of indirect costs that occur in both small and large airlines and they can also be grouped as the following:

- distribution cost (sales and promotion);
- station cost;
- depreciation and maintenance;
- ground expenses;
- passenger services;
- general overhead and administrative expenses.

AIRLINE ECONOMIES OF SCALE, SCOPE, AND DENSITY

The airline industry is affected by the economies of scale, scope, and density discussed earlier in the chapter. Economies of scale play a significant role in the industry since, as we have seen, FCs are extremely high. These high FCs and marketing requirements encourage expansion in the industry. This is a major reason why there are very few small airlines in the industry and why there is continual consolidation.

Economies of scale refer to the reduction in average cost resulting from increased production, and they are generally achieved through operational efficiencies. From an operational standpoint, economies of scale also play a significant role. Due to pilot training and maintenance spare parts for aircraft, it is less costly to simplify aircraft fleets and to focus on just a few aircraft types. Therefore, airlines generally prefer to operate a minimum number of aircraft types where this is possible. The optimum number of aircraft required to achieve economies of scale is unknown, but JetBlue believes that it achieved all economies of scale from its 80-strong Airbus 320 fleet. In 2006, the airline decided to order an additional fleet of 100 Embraer 190 aircraft (JetBlue.com, 2006). Thus, while the airline has decided to use two types of aircraft, it believes that both fleets are of sufficient size to maximize economies of scale.

Hub airports also contribute significantly to economies of scale, in addition to the economies of density mentioned earlier. Hubs are extremely costly operations and the costs that they generate, such as multiple labor shifts, terminal leases, and ground equipment are FCs in the short term. Therefore, in order to spread the costs over more units of output (air seat miles), airlines have a strong incentive to use these assets as intensively as possible. While most airlines operate banked hubs to provide shorter connection times for their passengers, airlines such as American and Delta have experimented with rolling hubs in order to better utilize hub assets. With banked hubs, assets sometimes remain unused for extended periods of time between banks. On the other hand, rolling hubs use hub assets throughout the day, making the airline's use of assets more efficient and thereby achieving greater economies of scale. In operations, economies of scale can be achieved by:

- quantity discounts;
- higher efficiency through labor specialization;
- spreading FCs over more output units.

Economies of scope play an important role in the aviation industry, and are defined as the process of reducing the average cost of resources by spreading the use of productive resources over two or more products. As mentioned earlier, aircraft manufacturers such as Boeing capitalize on economies of scope when producing aircraft. Airlines achieve economies of scope by operating various ancillary programs/services such as frequent flier programs (FFPs), maintenance activities, catering, and ground handling. Ultimately, the amount of outsourcing an airline does depends on how many synergies exist between the organizations that are creating economies of scope. For example, FFPs are usually more effectively run by the airline itself, since a FFP's main cost is inventorying reward seats, and economies of scope are achieved when this is done internally (because the airline already has a staff to schedule seats). One FFP, Air Canada's Aeroplan, reversed this trend when it became its own publically traded loyalty program in 2005.[18] Airlines have also experimented with economies of scope by being involved in other related industries (such as cruise line and hotels), but history has shown that most of these external activities have failed (Sabena hotels Carnival Airlines) and the economies of scope that were imagined either did not exist or were not achieved.

Economies of density exist in the airline industry through the use of hubs and the consequent reduction of flights. Similarly, aircraft size exhibits both economies of scale and density. Economies of scale are realized when airlines can put more seats into the aircraft to reduce unit costs, but economies of density are also achieved by using larger aircraft. For example, suppose an airline could use a 100-seat aircraft or two 50-seat aircraft

---

18   A similar situation would be catering companies, but with airlines largely ignoring food services on flights, this no longer becomes an area where airlines can achieve significant economies of scope.

to service a route. On shorter domestic flights airlines may opt for a higher frequency of flights, while on longer domestic and international flights the airline will usually select the larger aircraft in order to capture the economies of density (that is, the extra costs of pilots, gate agents, landing fees, baggage handling for the high-frequency decision). While airlines typically have moved toward smaller aircraft to provide increased frequency on a route, if the aircraft depart relatively close to one another, a single larger aircraft will always be cheaper to operate than the two smaller aircraft—this is due to the various FCs required to operate a flight (that is, pilots, gate agents, landing fees, baggage handling). This tradeoff benefit is another example of economies of density and is common in the airline industry, especially on very long-range flights, where (due to time zones), most flights depart at around the same time.

AIRLINE BREAKEVEN ANALYSIS

An important measurement of any company's cost structure is its breakeven analysis. Breakeven analysis is the number of units or revenue required in order for the firm's costs to be recovered. By breaking even, an airline is not losing any money, but with the same token is not making any money.

Profit = Total Revenue - Total Cost

At Breakeven point:

Fixed Costs + Variable Costs = Revenue

$$FC + VC \times Q = P \times Q$$

$$FC = (P - VC) \times Q$$

$$Q_{B-E} = \frac{FC}{P - VC}$$

As the above equation shows, the breakeven quantity, Q_{B-E}, depends directly on FC, and inversely by per-unit VC and per-unit revenue over the whole business.

In manufacturing, the breakeven point is represented in product units. For example, the revised breakeven forecast for the Airbus A380 program is 420 aircraft (Shannon, 2007).[19] Airbus will have to sell 420 aircraft to simply recoup the FCs related directly to the A380 program. Assume that the average list price of an A380 is about US\$ 350 million, its VC of each aircraft produced is \$314 million, and the total development cost is about \$15 billion. Then we can calculate the Q_{B-E}.[20]

We can now calculate the breakeven level (Q_{B-E}) of sales for the Airbus Co by applying the following breakeven formula:

$$Q_{B-E} = \frac{\text{Fixed Costs}}{(\text{Price - Variable Costs})}$$

---

19   The breakeven for the A380 was initially estimated at 270 units.
20   Airbus Aircraft 2010 list prices.

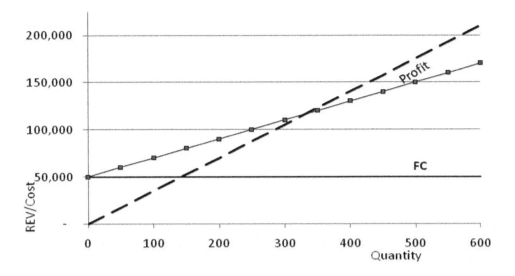

Figure 4.18   Manufacturing Breakeven Point

In the previous equation, the difference between the price and VCs of a good is called the contribution margin:

$$Q_{B-E} = \frac{FC}{P-V} = \frac{FC}{Contribution\ Margin}$$

$$Q_{B-E} = \frac{\$15,000}{\$350-\$314} = 416\ Aircraft$$

Breakeven in the airline industry is usually expressed as a percentage of total ASMs. This provides a breakeven load factor (BLF), or a load factor which the airline must meet to recover all FCs. BLF is the percentage of seats that must be sold on an average flight at current average fares for the airline's passenger revenue to breakeven with the airline's operating expenses. The general formula for airline breakeven is the following:

Profit = OP - OC

OP = RPM × RRPM

OC = ASM × CASM

Profit = O

RPM × RRPM − ASM × CASM=0

Where:

- RPM = Revenue Passenger Miles
- RRPM = Revenue per Revenue Passenger Mile
- ASM = Available Seat Miles
- CASM = Cost per Available Seat Mile
- OP = Operating Profit
- OC = Operating Costs

From this basic formula, two passenger load factors can be found—actual load factor and BLF. The formula derivations to achieve the two ratios are:

$$RPM \times RRPM = ASM \times CASM$$

$$\frac{RPM}{ASM} = \text{Load Factor}$$

$$\text{Load Factor}_{Breakeven} = \frac{CASM}{RRPM}$$

$$LF_{B-E} = \frac{CASM}{RRPM}$$

Obviously, if the actual load factor is greater than the BLF, then the airline is making enough money to cover FCs. However, if the actual load factor is less than the BLF, then the airline is losing money. Figure 4.19 shows the actual load factor and the BLF for nine major US airlines in 2011. Alaska Airlines, Continental, United, JetBlue, Delta, US Airways, and Southwest all have actual load factors that exceed their BLF. Therefore, these airlines are presently earning enough revenue to cover their FCs.

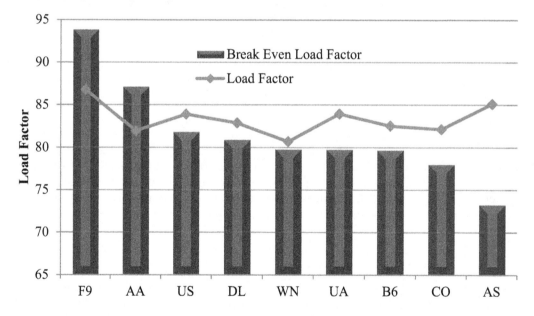

Figure 4.19   Comparison of Actual and Breakeven Load Factors for US airlines, 2011

OPERATING LEVERAGE

Operating leverage is a powerful metric that highlights the ratio between operating profit (OP) growth and sales growth. More directly, the degree of operating leverage (DOL) is an elasticity of the overall company's financial health with respect to sales growth. Operating leverage can also provide an indication of the company's cost structure, especially with respect to FCs. The general formula for the DOL is:

$$DOL = \frac{\%\Delta \ in \ EBIT}{\%\Delta \ in \ Sales}$$

The explicit formulas are as follows:

$$DOL = \frac{Q(P-V)}{Q(P-V)-FC}$$

$$DOL = \frac{S-VC}{S-VC-FC}$$

Where S is initial sales in dollars.

The DOL can vary considerably among companies in the same industry. Companies with a high DOL are much more reactive to changes in output. This is partly a result of the company having sizeable FCs, which can either be leveraged effectively during times of increasing sales (because of decreasing average costs), or become a burden to the company during decreasing sales. As a concrete numerical example, suppose the DOL is 2, then this means that a 1 percent increase in ticket sales will result in a 2 percent increase in the airline's OP. In this case, there would be increasing returns to scale. Unfortunately, the reverse is also true during downswings in the economy. Using the same DOL of 2 as an example, a 1 percent decrease in ticket sales would result in a 2 percent decrease in OP. Therefore, companies with a high DOL will experience greater volatility of OP than companies with smaller degrees of operating leverage.

Operating leverage can also be a negative value which indicates that the company is experiencing annual FCs greater than annual OPs. This scenario indicates decreasing returns to scale and that the company may have grown too quickly and is not effectively leveraging its FCs. Therefore, in order to improve its DOL, the company should reduce its FCs while keeping the same level of output, or increase its output with the same FC infrastructure.

Additionally, changes to the company's contribution margin will also dramatically affect the company's DOL. An airline may choose between a high level of fixed assets and a lower level of fixed assets. For instance, an airline may substitute self-service check-in machines at different airports for check-in agents. If labor is not replaced with check-in machines, FCs are held lower, and VCs are higher. With a lower level of operating leverage, the airline shows less growth in profits as sales rise, but faces less risk of loss as sales decline.

This overall understanding of a firm's cost structure is critical in the strategic planning phase of a company. It provides a comparison for the company between sales and OP, and depending on the value, can help guide strategic direction. It should be noted that operating leverage should not be confused with financial leverage, which deals with how much debt a company is using to finance its activities.

AIRLINE OPERATING LEVERAGE

As has been explained, operating leverage is the ratio between OP growth and sales growth. Table 4.6 provides the DOL for US airlines in 2011. The contribution margin was found using operating revenues and expenses, while FCs were assumed to be contractual obligations for the airline in the next year. These contractual obligations include firm aircraft purchases, regional capacity purchase agreements, and long-term debt. A breakdown of these contractual obligations is a mandatory requirement in SEC 10K forms. Finally, the number of enplaned passenger is used as a reference for the scope of the airline's operation.

As Table 4.6 depicts, the DOL for US airlines varies considerably among carriers. A positive DOL indicates that carriers' profits will increase at a greater rate than an increase in sales. Using Alaska Airlines (AS) as an example, a DOL of 1.203 means that for every 1 percent increase/decrease in sales, the airlines OPs will increase/decrease by 1.203 percent. A Carrier with an operating leverage between 0 and 1 (AA in this case) is an anomaly to the DOL equation because it can only exist when the firm is experiencing an operating loss and suggests that it will be better off by selling less tickets. As mentioned earlier, a negative DOL indicates a carrier would be experiencing an OP, but would also have incurred annual FCs that are greater than the OP, and thus incurring an annual net loss. This problem should be addressed by reducing FCs or increasing sales.

Table 4.6   US Airline Operating Leverage, 2011

	AA	AS	B6	DL	UA	US	WN
Operating Revenues ($ millions)	23,979	3,832	3,779	31,755	37,110	11,908	12,104
Operating Expenses ($ millions)	25,033	3,361	3,446	29,538	35,288	11,127	11,116
Annual Contractual Obligations* ($ millions)	1,518	208	198	1,944	1,315	436	644
Passengers (millions)	79	17	24	105	96	49	88
Degree of Operating Leverage	**0.410**	**1.788**	**2.467**	**8.121**	**3.594**	**2.264**	**2.872**

Source: Compiled by the author from SEC 10K filings and OAG Form 41 data
* Current Maturities of Long-term Debt and Capital Leases

SUMMARY

In recent years, many airlines across the world were forced to reduce their costs and return to profitability. The consistent rise in fuel costs has spurred investment in new fuel-efficient aircraft. To reduce fuel costs, airlines are placing orders for more cost-efficient aircraft like the Airbus A320neo, Boeing 737 MAX, and the Bombardier CSeries and retiring older, more expensive airframes. There are quite a few classifications of cost, and this chapter discussed the following categorizations: Marginal Cost, Total Cost, Fixed Cost, Total Variable Cost, Average Total Cost, Average Fixed Cost, and Average Variable Cost. The use of costs for managerial decision-making is then explained and the intuitive rationale behind these decision rules is explained. Various costs are compared across the spectrum of airlines. Finally, economies of scale, scope, and density are explained and discussed in detail.

DISCUSSION QUESTIONS

1. Identify the reasons why airlines would want to take over other airlines.
2. Please provide one particular industry as an example to illustrate that MC is not U-shaped.
3. Please provide one particular industry as an example to illustrate that MC is U-shaped.
4. What is the difference between economies of scale and economies of scope?
5. For many airlines in the short run, a major portion of the cost of production such as aircraft and terminal space are fixed. Should these very large FCs be ignored when the revenue managers are making output and pricing decisions? Why?
6. What is a sunk cost? Provide an example of a sunk cost in an airline. Should such costs be irrelevant in making future decisions?
7. Define and compare historical, replacement, and sunk costs.
8. Why it is important that an airline manager consider opportunity costs when making economic decisions regarding the airline?
9. Calculate BLF for the following airlines:

US Airlines, 2010				
	Airlines	Yield	CASM	BLF
1	American	14.09	12.84	?
2	Continental	14.96	11.48	?
3	Delta	13.07	11.84	?
4	United Airlines	15.15	12.03	?
5	US Airways	11.68	11.73	?
6	Southwest Airlines	10.66	11.27	?
7	JetBlue	11.75	9.77	?
8	AirTran	9.71	10.35	?
9	Frontier	10.85	10.07	?
10	Virgin America	12.21	9.44	?
11	Alaska	12.38	10.93	?
12	Hawaiian	9.72	11.88	?
13	Allegiant	7.51	9.08	?

10. Suppose the cost function is given by:

$$C = 50 + 4Q + 2Q^2$$

- What is the AFC of producing five units of output?
  a   What is the AVC of producing five units of output?
  b   What is the ATC of producing 5 units of output?
  c   What is the MC of producing five units of output?
  d   What is the AFC function?
  e   What is the AVC function?
11. At a certain manufacturer, the MC is:

$$MC = 3(Q - 4)^2 \text{ dollars per unit when the level of production is Q units.}$$

a   Express the total production cost in terms of the FCs and VCs.
b   What is the TC of producing 14 units if the FC is $400?

APPENDIX: AIRLINE COST CLASSIFICATION AND DIVISION

Airline Fixed and Variable Costs

FIXED COSTS
 	
• Fixed salaries, benefits and training costs for flight crews, which do not vary according to aircraft usage.
• Maintenance costs
  – Maintenance labor for maintenance scheduled on an annual basis
  – Maintenance contracts for maintenance scheduled on an annual basis
• Lease costs based on a length of time
• Depreciation
• Operations overhead
• Administrative overhead
• Self-insurance costs
 	
VARIABLE COSTS
 	
• Crew costs (travel expenses, overtime charges, wages of crew hired on an hourly or part-time basis)
• Maintenance costs scheduled on the basis of flying time or flight cycles
• Maintenance labor (includes all labor salaries, wages, benefits, travel, training)
• Maintenance parts (cost of materials and parts consumed in aircraft maintenance and inspections)
• Maintenance contracts (all contracted costs for unscheduled maintenance, maintenance scheduled on flight-hour basis, on flight-cycle basis or on condition)
• Engine overhaul, aircraft refurbishment, major component repairs
• Modifications
• Fuel and other fluids
• Lease costs (leasing costs based on flight hours)
• Landing fees, airport and en-route charges

Source: Foundations of Airline Finance, Methodology and Practice

Airline Operating and Non-operating Costs

Operating Cost (OC)

Direct Operating Cost (DOC)

- Flight crew
- Aircraft fuel and oil
- Airport fees (landing fees: cost per aircraft ton landed).
- Navigation charges
- Direct maintenance: labor and materials
- Depreciation/rentals/insurance: flight equipment

Indirect Operating Cost (IOC)

- Marketing costs
- Ground property and equipment
- Depreciation, insurance and maintenance
- Administration and sales
  - Servicing administration
  - Reservations and sales
  - Advertising and publicity
  - General
  - Servicing
    - Passenger services
    - Aircraft services
    - Traffic services

Non-operating Cost (NOC)

- Depreciation
- Interest
- Insurance
- Losses from the retirement of property
- Losses from affiliated companies
- Other loss items, such as those from foreign exchange transactions and sales of shares.

Source: Foundations of Airline Finance, Methodology and Practice

REFERENCES

Aviation Partners Boeing. (2006). *Fuel Savings*. Retrieved on November 9, 2006 from http://www.aviationpartnersboeing.com/.

Boeing. (2010). *Commercial Sites*. Retrieved on May 12, 2012 from http://www.boeing.com/commercial/overview/overview5.html.

Boyes, W. (2004). *The New Managerial Economics (1st ed.)*. Boston, MA: Houghton Mifflin.

Ionides, N. (2007). Star Sees Benefits from Joint Fuel Purchasing. *Air Transport Intelligence News*, April 1.

JetBlue Airways (2006). *Annual Report*. Retrieved on April 3, 2012 from www.jetblue.com.

Maurice, S. and Thomas, C. (2005). *Managerial Economics (8th ed.)*. New York: McGraw-Hill.

Rivkin, J. (2005). *Dogfight over Europe*. Harvard Business School. 9-106-033.

Shannon, D. (2007). Airbus Increases A380 Breakeven Level 55% to 420 Aircraft. *Air Transport Intelligence News*, March 5.

Stewart M., (2008) A risk and cost-benefit assessment of United States aviation security measures, *Journal of Transportation Security*, No. 1, 143–159.

5

Aviation Infrastructure: Operations and Ownership

So here we are. One of the worst summers on record for delays is headed for aviation's history books. Total delays are up 19 percent from where they were last summer... Yet despite the progress we're making, our air traffic system is still not even close to what it needs to be—what the flying public demands it should be. The system's too old and it's not nimble enough for today's activity. I'll underscore that this is why we need to move to a system for the next generation, NextGen. This is the modernization step we need to take.

Marion Blakey, FAA Administrator (2002–2007)

One of the most unique features of the aviation industry is the unprecedented amount of regulatory and operational control that the industry is subject to. These controls are manifest in the form of the many government agencies that regulate and control everything from the direct ownership of airports, the control of aircraft in the air, and the certification of aircraft production on the ground. While virtually all industries are regulated in varying degrees, very few have their day-to-day operations under the direct control of a government agency. Such control presents many unique challenges to the aviation industry.

Using the theoretical constructs of supply and demand that were introduced in the previous chapters, this chapter analyzes the situation and discusses alternative arrangements that have been suggested and, in some foreign countries, have actually been implemented. The chapter specifically focuses on the operational infrastructure of the industry, namely, air traffic control (ATC) and airports. We begin with a brief history of ATC, followed by an economic analysis of the existing system, and finally, the prospects for reform. The last part of the chapter is devoted to an economic analysis of airport ownership and the likely outcomes when public or private ownership is considered.

This chapter will have the following outline:

- Air Traffic Control System
- Institutional Problems in US Air Traffic Control
- Air Traffic Control in a Government Corporation
- Political Obstacles to Reform
- Solutions to Air Traffic Control Problems
  - Regulation
  - Air traffic control charges

- Airport Ownership and Management
  - Trends in airport ownership
  - Reasons for privitization
  - Opposition to privatization
  - Types of privatization
  - Privatization in the US airport industry
- Summary
- Discussion Questions
- Appendix

AIR TRAFFIC CONTROL SYSTEM

Early control of air traffic began in the 1920s and was mainly concerned with navigation rather than control per se. This rudimentary system involved the use of flags, lights, and bonfires to locate and identify airports and runways as a means to communicate with pilots. In the early 1930s, this system was replaced by a more formal set up that consisted of a series of light towers. The system was called the Transcontinental Lighted Airway and at one time consisted of over 1,500 beacons and 18,000 miles of airways. Thus, the early technology forced aircraft to rely on point-to-point navigation over predetermined routes rather than the more direct routing that the aircraft were capable of. Gradually, as radios became more technically advanced, they replaced the earlier systems. As traffic between major metropolitan areas began to increase it was apparent that a more centralized system was needed to provide separation as well as facilitating navigation.

Accordingly, in the mid-1930s an airline consortium established the first three centers to pool information on specific flights so as to provide better flight separation. These centers were taken over by the Bureau of Air Commerce within the Department of Commerce when it assumed responsibility for air traffic in the US. Separation was accomplished mainly through flight scheduling over the already established prescribed routes. Again, the existing technology was forcing the aircraft to fly from ground-based fix to ground-based fix over the predetermined routes. At the end of the 1930s, Congress passed the Civil Aeronautics Act which transferred civil aviation responsibilities from the Department of Commerce to a newly created agency called the Civil Aeronautics Authority. In 1940, President Roosevelt separated the Authority into two agencies, the Civil Aeronautics Administration (CAA) and the Civil Aeronautics Board (CAB). The newly created CAA was tasked with certification, safety, airway development, and ATC (Kent, 1980). The CAB was split off from the CAA and charged with the economic regulation of the transport industry.[1] The Federal Aviation Agency was then created by the Federal Aviation Act of 1958 and tasked with the responsibilities of the old CAA and the new responsibility of safety rulemaking. Finally, in 1967 the Agency's name was changed to the Federal Aviation Administration and it was placed under the DOT.

Throughout the 1970s the Federal Aviation Administration (FAA) installed new radars, computers, and radio communications to upgrade and enhance the ATC system. The problem with this development was the fact that it was created under the same premise as the older systems. That is, aircraft were expected to travel from ground-based navigation point to ground-based navigation point along the predetermined and pre-existing airway route structure. This had the effect of lining up the traffic in a linear fashion along the

---

1   The effect of this regulation is covered in greater detail in later chapters.

various routes. Controllers would then use various techniques to maintain predetermined separation standards. As one would expect, the flow of traffic would be metered by the slowest aircraft and/or the largest separation distances. This of course is exactly analogous to a ground-based highway system, and it was often called the highway in the sky system. The trouble with this is the fact that it ignores the capability of an aircraft to travel in three dimensions and directly through the airspace. Nonetheless, the system worked reasonably and efficiently with the then existing volume of traffic and level of technology.

However, as traffic increased and technology advanced, it became increasingly evident that the system was becoming outdated and there was an urgent need to modernize and update.[2] Unfortunately, bureaucratic tendencies and various political considerations now took over and meaningful reform and modernization became increasingly difficult if not impossible.[3] Some expensive efforts were undertaken to improve the system, but none of these were very effective. It should be pointed out that one of the principal reasons that the system could function with the older equipment and procedures was the undeniable fact that it could operate reasonably well under good weather conditions. Under these conditions pilots could see and be seen so that separation standards could be reduced. Since good weather conditions are generally the norm, the problems with the system and the procedures were generally not evident. But, by the end of the 1990s and with the advent of the twenty-first century, the volume of traffic was increasingly overwhelming the system. The tragic events of 9/11 slowed this growth in volume, but traffic returned to and exceeded earlier levels by the year 2006. Hence, we have the comments by Marion Blakey at the beginning of the chapter.

One of the principal advantages of air travel is the speed with which an individual can arrive at his or her destination. Therefore, factors that contribute to delay in the system certainly reduce the attractiveness of air travel. As these factors increase in magnitude they increasingly reduce the quantity demanded of air traffic. The economic effects of delay can be analyzed using the supply and demand models introduced in the earlier chapters. Consider Figure 5.1 overleaf.

As in the earlier chapters we see that there is an equilibrium price and quantity supplied of air travel. This equilibrium is represented by P_E and Q_E and is the market equilibrium based on supply and demand. Recall from the earlier chapters that any price above the equilibrium price will result in a surplus of air travel supplied; that is, too many empty seats. In this situation, competition between airlines will lower the price to fill the seats and return to the equilibrium position. Any price below the equilibrium will result in a shortage and consumers will bid up the price to obtain the seating. Now suppose that the ATC system imposes repeated and prolonged delays in the form of ground holds or airborne holding. This can result from the system not being capable of handling the volume of traffic either because of separation standards or the inability of the human controllers to handle the volume of traffic. This situation can be thought of as an externally imposed cost to both the consumers and the producers of air travel. To the consumers it is an unanticipated delay that can be monetized as the cost of the consumer's time, or in the case of the business travelers, as forgone opportunities. To the producer the cost is also real and can be measured, among other costs, in terms of higher crew wages, more fuel burn, and the loss of the ability to utilize the airplane for extra flights. Therefore, the delay that is imposed on the system can be thought of as a tax on both the consumers and suppliers of air traffic.

---

2   Principal among these developments were precise methods of navigation and the ability to accurately locate the position of any aircraft in the sky.

3   Among these considerations were the location of facilities and the question of union job loss.

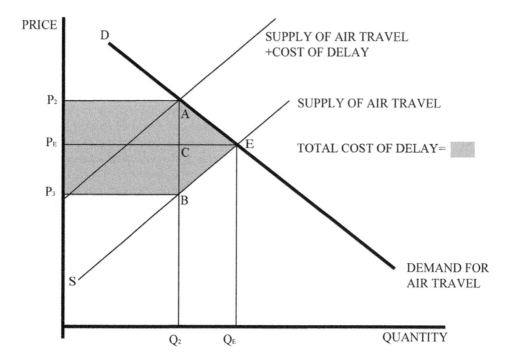

Figure 5.1   The Short-term Economic Effects of Air Traffic Delay

In Figure 5.1 these costs are shown as the straight line AB that joins the demand and supply curves. We can think of the costs as a parallel shift in the supply curve so that the new supply curve intersects the demand curve at point A. However, since these are extra costs, the new equilibrium quantity and price will be defined by the intersection of the new supply curve with the old demand curve at P_2 and Q_2. On the other hand, the actual price that the producers receive is determined from the old supply curve net of the cost introduced by the externally imposed delay at P_3. We can also see that the price that the consumers must now pay is the new equilibrium price P_2, which is clearly higher than the original equilibrium price. Therefore, the cost of the delay is shared between the consumers and the producers. The consumers bear part of the cost in the form of a new and higher ticket price, while the producers bear part of the cost by receiving less from the actual new ticket price. The total amount of the cost is equal to the rectangle ABP_3P_2 plus the triangle ABE. And, as we can see from the diagram, the total cost is borne by both the consumers and the producers. For consumers, the costs are equal to the rectangle P_2ACP_E plus the triangle ACE. For producers, the costs are equal to rectangle P_ECBP_3 plus the triangle CEB.

The cost of delay is analogous to the impact of a conventional tax on air travel, though congestion costs are worse. Suppose that a tax, equal to congestion cost, had been imposed. In conventional demand and supply tax analysis, the entire rectangle ABP_3P_2 is the amount that the taxing authority receives from the tax, while the triangle ABE is the dead weight loss that results from the imposition of the tax. This dead weight loss can best be thought of as the transactions between buyers and sellers that do not take place because of the imposition of the tax. That is, the consumers who would have purchased tickets absent the presence of the tax and likewise the producers who would have sold them the tickets absent the tax. In the case of a delay that is imposed by the ATC system on both consumers and producers, the situation is in reality worse than a tax, because the entire area of rectangle ABP_2P_3 plus the triangle ABE is a dead weight loss. In other words, consumers and producers suffer a loss of wealth—wasted time, wasted fuel, higher maintenance costs, and so on—but there are no tax proceeds being transferred to government. Instead, all of the costs are borne by the consumers and producers in the form of a dead weight loss.

The question of who bears a larger amount of the costs is more complex but can still be addressed using supply and demand analysis. As may be clear from Figure 5.1, the question of who bears a greater amount of the cost depends on the slopes of the demand and supply curves. As the supply curve becomes more and more inelastic, that is, as the slope gets higher and higher, it becomes more and more difficult for the producers to shift the burden of the cost to the consumers. Figure 5.2 below shows the situation in more detail. We can imagine a limiting situation where supply is perfectly inelastic at some given time. In this case, the supply is represented by a perfectly vertical line and is fixed regardless of price. Since supply is fixed by definition, it cannot shift when the extra cost is introduced. Instead, we can show the cost of delay as a downward shift in the demand curve. In this case, the ticket price remains the same for the consumers, and the entire cost of the delay is borne by the producers and is equal to the rectangle P_EEAP_2. We have presented the limiting case where the producers bear all of the costs since it is highly likely that in the aviation industry supply is relatively inelastic, at least in the short to intermediate term, as compared to demand. Therefore, we can expect that the greater part of the cost of delay in the industry will be borne by the producers. And, as we shall see in the latter part of the chapter, that this appears to be true when we observe the preferences of the market participants.

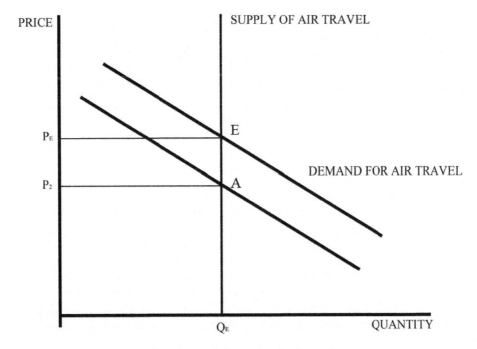

Figure 5.2   The Cost of Delay with Inelastic Supply

But all of this is not the end of the story. Unfortunately, there are other longer-term supply and demand effects that must be considered. On the supply side, if the cost of delay is persistent and longer lasting, as it appears to be for the foreseeable future, then the value of the specialized resources presently in use in the aviation industry will be diminished accordingly. These resources include, among other inputs, the production of aircraft, the supply of spare parts, the manufacture of avionics, airport-related concessions, air travel-related accommodations, automobile rental concessions, income for pilots, flight attendants, mechanics, and a host of other related factors. As these specialized resources wear out, they will not be replaced at the same rate as previously, and this will further reduce output in the industry. This process is illustrated in Figure 5.3 where the initial supply decrease occurs and then a further decrease takes place as these specialized resources wear out or exit the industry. As we can see from the figure, the long-term effect further lowers quantity and raises price.

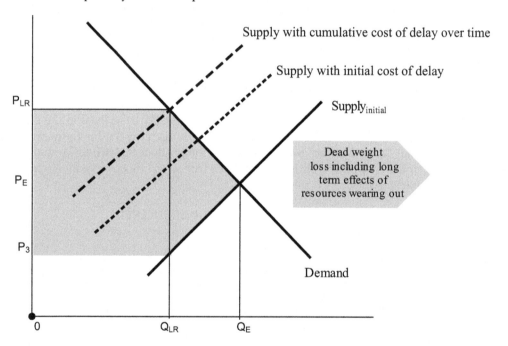

Figure 5.3   Costs of Delay over Time

On the demand side, we can expect a shift in demand away from air travel to alternate modes of travel where this is feasible. Although air travel is clearly the fastest mode of travel when considering relative speed, it is the total trip time that is of primary interest to the traveler. As things stand now, with delays, long security lines at some airports at critical times of the day, and persistent air traffic system delays, the inherent speed advantage of air travel is severely compromised. If this situation continues, more and more consumers will opt for surface travel which will most likely be by private automobile. Since automobile traffic is inherently more risky than air travel, there will be a concurrent rise in accidents and fatalities.

This process is illustrated in Figure 5.4 with both an initial demand decrease then a longer-term addition decrease. That is, demand decreases immediately with delay since the quality and speed of air travel is reduced. However, this initial reaction is multiplied over time, reducing demand further, as air travelers have more time to adjust their behavior, work out other travel arrangements, utilize modern communications to cut down on number of trips, and so on.[4] The demand decreases reinforce the decline in total air travel caused by the supply decreases; the industry is substantially smaller and less efficient than it would be without these delays. However, the impact on price becomes theoretically ambiguous since falling demand tends to reduce price while supply decreases tend to raise it.

The preceding analysis used the theoretical tools of supply and demand to illuminate a very real and pressing problem in the aviation industry. The fact is that the final product of the industry is under the direct control of an outside government agency (the Federal Aviation Adminstration or FAA) that does not have the same incentives or goals as the industry, especially those concerning profitability. This has led to large external costs that are increasingly imposed on the industry. Moreover, that agency itself has come under increasing criticism as the appropriate agency for operational ATC. The next sections discuss these criticisms in more detail.

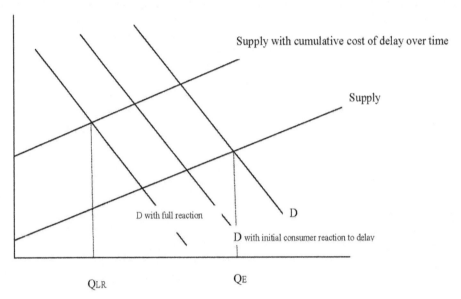

Figure 5.4   Market Reaction to Changes in Supply

INSTITUTIONAL PROBLEMS IN US AIR TRAFFIC CONTROL

A number of studies have concluded that US ATC management is inherently flawed and in need of major reform, including:

- the Aviation Safety Commission in 1988;
- the National Commission to Ensure a Strong Competitive Airline Industry in 1993;

---

4   In essence, this is merely a way of expressing the standard principle that demand (and supply) in the long run is more elastic than in the short run. For simplicity, we have omitted the long-run curves and focused on the additional shifts in the short-run curves.

- the National Performance Review in 1993;
- the Secretary of Transportation's Executive Oversight Group in 1994; and
- the National Civil Aviation Review Commission (Mineta Commission) in 1997.

The fact that funding flows from an unpredictable revenue stream subject to the federal budget process is a commonly raised issue. At times, there also seems to be an inability to attract and retain needed managers and engineers who are skilled at implementing complex technology projects, stemming in part from limitations in the civil service system.

The lack of mission focus and clear accountability is at the root of much of the criticism because authority is shared by Congress and FAA management in a confusing bureaucracy. The future implementation of NextGen, for example, will be complicated by the fact that numerous outdated FAA operations around the US will need to be closed. However, Congressional Representatives are reluctant to vote in favor of shutting down any operation that "provides jobs in their district."

A possible solution to these sorts of problems, according to some economists, might be to move to a private, non-profit corporation, as Canada did in 1996 with its NavCanada Corporation. However, it is commonly believed that this is not politically feasible. There is broad support for establishing a government corporation to address the concerns that have stymied previous attempts at major ATC reform.

AIR TRAFFIC CONTROL IN A GOVERNMENT CORPORATION

A key feature of a government corporation is non-political funding as user fees replace taxes and Congressional budgeting. The existence of an independent revenue stream allows access to private capital markets to fund modernization. In turn, the elimination of tax funding creates an exemption from government procurement rules that have previously tended to impede the acquisition of new technology. Likewise, independent funding allows exemption from Civil Service regulations that can make it difficult to attract, manage, and maintain the appropriately skilled work force.

New Zealand converted its ATC operation from a government division to a self-supporting government corporation in 1987. As of 2007 over 40 countries implemented similar commercialization reforms, including Australia, France, Germany, Switzerland, the UK, the Benelux countries, and Scandinavia. Only a few of these are privatized in the sense of being outside of government; most are government corporations. All of these commercialized, self-supporting Air Navigation Service Providers (ANSPs) belong to the Civil Air Navigation Services Organization (CANSO), which has become a key participant in international aviation policy debates. All ANSPS are subject to safety regulation and some form of economic regulation because of their monopoly on ATC services.

In 2005, the Government Accountability Office (GAO) conducted a large-scale evaluation of the performance of commercialized ANSPs. The GAO collected extensive data from five major ANSPs—Australia, Canada, Germany, New Zealand, and UK. They found that after commercialization of ATC, safety had either been unaffected or even improved since implementation. They also found that all five of the systems studied had taken significant steps to invest in new technology and equipment and had taken meaningful steps to reduce operating costs. Similarly, a 2005 FAA study (Federal Aviation Administration, 2005) found that commercialized systems were more cost-effective in airspace with equivalent traffic density. In short, commercialized ATC has become the norm for most of the industrialized world and apparently has a solid record of improved efficiency with no decline in safety.

POLITICAL OBSTACLES TO REFORM

Most opposition to ATC commercialization centers around user fees. General aviation has been particularly fearful of user fee impact. However, it is very feasible, and probably politically necessary, to exempt piston aircraft flying under visual flight rules (VFR) from user fees. For those that sometimes fly instrument flight rules (IFR), reasonable accommodations can be made. In Canada, for example, they pay a modest annual fee based on aircraft weight. Moreover, the envisioned board of directors for a US government corporate ATC system would include representatives of general aviation that would have to approve any changes in user fees.

Business jets would have to pay some user fees, but experience shows that these can be reasonable. Both Canada and Europe have experienced strong growth in business aviation after ATC commercialization (Poole, 2006).

One might expect public employees to be potentially strong opponents of commercialization. However, The National Air Traffic Controllers Association, the main FAA union, supported the Clinton Administration's proposal to divest ATC to a government corporation, structured along the lines discussed here. Moreover, commercialization could readily include no-layoff guarantees for all current controllers and technicians.

Government reform often proceeds at a glacial pace—no matter how inefficient existing institutions are there are always interest groups who perceive a vested interest in maintaining the status quo. However, in the case of ATC perhaps the US has reached a stage where the problems are so severe that the political logjam blocking major reform may soon be broken.

Recently, the industry itself has mounted a strong lobbying campaign to change the situation. This effort has been led by the Airlines for America (formerly the Air Transport Association of America) and it has involved a political campaign to influence Congress to change the ATC system. Basically, the intent of the effort is to replace the existing ground-based radar and voice communication system with a more technologically advanced system. The new system will be based on a much more accurate surveillance technology, namely, the Automatic Dependent Surveillance Broadcast or ADS-B that is intended to replace the existing radar-based system. This system will also provide a more open standards-based architecture to replace the existing national airspace system software and it will also include airborne collision avoidance and shared intent information within the cockpit. The present radio communication system is supposed to be replaced by a data link system that will allow direct exchange of messages between controllers and pilots. Finally, new navigation systems will allow direct routing and replace the current airway system.

SOLUTIONS TO AIR TRAFFIC CONTROL PROBLEMS

The FAA expects flight operations to increase by over 35 percent in the next decade and this increases the cost of delay discussed above. Whether or not meaningful reform will actually take place is problematical, since it is difficult to believe that an organization with a set of incentives and goals that are fundamentally different from the aviation industry will be able or willing to implement any of these changes. The following quote is taken from the Airlines for America "Smart Skies" initiative:

> Without dramatic change in the way our airspace is managed, congestion and resulting delays will be overwhelming for passengers, shippers, consumers and businesses. Failure to meet future airspace demand could cost the US economy $40 billion annually by 2020.

Regulation

In response to numerous instances of passengers experiencing lengthy tarmac delays, the Department of Transportation (DOT) issued a final ruling entitled "Enhancing Airline Passenger Protections," also referred to as the "Three-hour Tarmac Rule," effective April 29, 2010. The new rule requires that at large and medium hub airports, the aircraft must depart or passengers given the opportunity to deplane no later than three hours after the cabin door has been closed. Exceptions to this rule are when returning to the gate or tarmac poses a safety or security risk, or would severely disrupt airport operations. Particularly applicable to regional operations is the requirement that food and potable water must be made available no later than two hours after push-back or touch-down with operable lavatory facilities while the aircraft remains on the tarmac. Fines for violating up to $27,500 per passenger may be levied against the airline. Foreign flag carriers were initially exempt from the ruling but in 2011 amendments to the ruling require all foreign carriers with at least one aircraft with 30 or more seats to adopt and adhere to tarmac delay contingency plans. International flights are not allowed to remain on the tarmac at a US airport for more than four hours without allowing passengers to deplane subject to safety, security, and ATC exceptions.[5] In addition to these restrictions, the ruling also prohibits airlines from scheduling chronically delayed flights.

Those opposed to the implementation of regulations including, but not limited to the three-hour tarmac rule, point to a surge in flight cancellations. In the eight months directly after the rule went into effect, cancellation rates at US airports rose 24 percent. Proponents of the ruling counteract this rise in cancellations with the fact that the data does not take into account variables like weather (one of the common causes for cancellations).

Air traffic control charges

Another attempt to control delays and congestion in the air traffic system revolves around changing the way in which charges are levied against airlines operating at these airports from a uniform-pricing structure (or one based on weights) to a congestion/peak-load pricing method. This method of pricing was originally developed for roads where peak road usage is excessive, because individual users do not take into account the delays imposed on other users. Economists have long argued for a similar pricing of airport resources according to each user and the marginal cost imposed on the other users of the system. This pricing system takes into account the differences in willingness to pay

---

5   Department of Transportation, Rules and Regulations: Federal Register Vol. 76, No. 79, April 25, 2011.

as well as the incremental costs of meeting demand during peak times when delays and congestions occur. European airports tend to follow a weight-based system of charges with variations for day and night periods combined with a slot coordination system. However, none of the major European airports makes use of a true congestion pricing method. The differences in charges between night time and day are in accordance with strict noise abatement policies and not based on peak/off-peak periods.

Supporters of congestion pricing for airports point out that mid-sized airlines will redistribute their flights to off-peak hours. By doing this, hub airlines can operate more efficiently because of the reduced number of flights during peak hours and the airport benefits from increased revenues. Despite these economic benefits, peak-load pricing has been met with heavy criticism and has yet to be implemented in a North American airport. North American airports' landing fees are currently a function of landing weights and unrelated to time of day or airport conditions. Instead of congestion pricing, some of the more congested airports are moving toward slot-controlled airports where capacity is fixed to reduce congestion and delays.

AIRPORT OWNERSHIP AND MANAGEMENT

The last part of this chapter analyzes and discusses the appropriate ownership of airports and the use of pricing mechanisms to allocate and improve the scarce resources at the airports. According to Airports Council International (ACI), in 2010, ACI members handled 5 billion passengers, 91 million metric tons of freight, and 74 million aircraft movements. There are approximately 1,670 commercial airports serving more than 900 airlines worldwide. According to the FAA, there are 546 commercial airports in the US. Of these, 422 have more than 10,000 annual enplanements and are grouped as commercial service airports in the US. Furthermore, there were 33 large hubs, with 464,486,847 enplanements, 35 medium hubs, with 115,177,169 enplanements, 68 small hubs, with 50,202,980 enplanements, and 410 non-hub airports. At the present time, almost all of the airports in the US are under some type of government control. One of the byproducts of this control is typically a pricing system (landing fees) that is fixed over the entire day. Consider Figure 5.5 below. If the airport authority sets a price below equilibrium for a particularly advantageous period of the day, then it is easy to see that the quantity demanded will exceed the available supply by the amount of $Q_E - Q_D$ and some other rationing device must be found. This typically takes the form of delay for some or all of the aircraft and/or, in some cases, a rationing of the available landing times (slots) for the airport in question. Both of these solutions have inherent efficiency problems when considered from an economic point of view. And, in the case of slot controls, the airlines that are awarded the slots benefit from an economic rent.[6] Therefore, the issue becomes the appropriate ownership of airports. Many economists would maintain that private ownership of the airports would provide a better set of incentives for the long-term viability of the industry. The next few sections discuss this question in more detail.

---

6   An economic rent, also known as Ricardian theory of rent, is the payment over and above that needed to keep a factor of production in its current use.

Trends in airport privatization

> In Europe it's driven by money and regulatory requirements. In developing nations, is also driven by need for expertise.

David Feldman[7]

Increasingly, many countries around the world are rethinking the appropriate role for government in operation and ownership structure of aviation infrastructure. In 2012, the Brazilian government raised a total of $14.3 billion through the sale of a controlling interest in São Paulo Guarulhos International (GRU), Campinas Viracopos (CPQ), and Brasília International airports (BSB), three of the country's largest air travel facilities.[8] Many countries have privatized airports, concluding that the private sector can run airports more efficiently, just as they run airlines more efficiently or produce better aircraft than would a state-owned manufacturer.[9]

In 1987, the British government led the way when it completely privatized seven major airports, selling British Airport Authority (BAA) to the public for $2.5 billion. In 2006, the company was purchased by the Spanish firm Ferrovial with a market value of $18 billion. BAA currently owns and runs five airports in the UK, including Heathrow, Stansted,[10] Southampton, Glasgow, Edinburgh, and Aberdeen in Scotland since they were forced

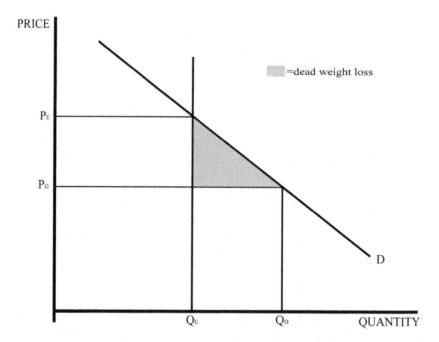

Figure 5.5   Demand for Airport Services

---

7   *Airline Business*, April 2012.
8   Dow Jones Newswires, February 3, 2012.
9   See Chapter 2 for a discussion of why efficiency tends to be enhanced by reducing government's role.
10   As of August 20, 2012, the Competition Commission of Britain ruled that BAA must sell London Stansted airport, although they will continue to operate for the time being until a buyer is found. See: After Antitrust Ruling, BAA to Sell a London Airport, NY *Times*, 20 August, 2012.

by the Competition Comission of Britain to sell Gatwick and Edinburgh in late 2011 on anti-competitive grounds. Following the apparent success of BAA, many countries have followed suit. BAA, also operated Indianapolis International Airport from 1995–2007 and had retail-management agreements at Baltimore-Washington, Boston Logan, and Pittsburgh international airports, which were eventually spun off into a new company, AirMall Inc.

As of early 2007, over 100 major airports worldwide have been at least partially privatized, including those at Belfast, Brussels, Budapest, Copenhagen, Düsseldorf, Frankfurt, Hamburg, Rome, South Africa, Argentina, Chile, Colombia, Mexico, Auckland, Brisbane, Melbourne, Sydney, and many others, with Hong Kong and Tokyo in the works. Roughly a dozen global airport companies are in the business of running airports.[11] The financial company, Macquarie, has created a privatized airports mutual fund for global investors (Poole, 2007). Figure 5.6 summarizes the number of airport privatization transactions from 2000 to 2009. From the figure, it can be determined that Budapest (BUD) was the most expensive while Hamburg (HAM) was the cheapest. This was determined by calculating the value (price paid) of the airport divided by EBITA (Earnings before interest, tax, depreciation, and amortization). The figure illustrates that airports of all economic levels are being transferred from government entities to private operators. The average cost was about 15 times annual earnings, but as can be seen from the figure, there is a large degree of variation in the different airports.

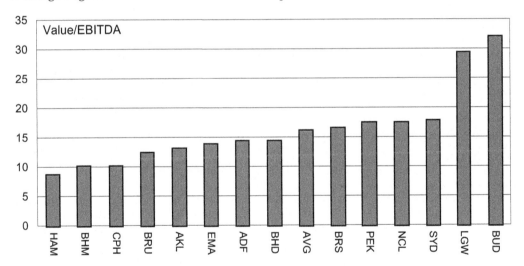

Figure 5.6   Recent Examples of Airport Privatizations (2000–2009)

Source: EBITDA: Air Transport Research Society, Global Airport Benchmarking Report

Value: Price paid for ownership at point of sale, converted into $US using same year exchange rate. See appendix for the airport code

Many other factors such as government regulation, availability of credit, and economic growth impact privatization initiatives. Furthermore, due to global credit problems, many

---

11   In 2005, BAA acquired a 75 percent stake in Budapest Airport (£1.2 billion), which it subsequently sold to Hochtief AirPort in 2007 after BAA's buyout by Ferrovial in 2006. Hochtief's parent company is currently looking to sell its AirPort division, thus emphasizing the premise that although privatization can lead to flexibility and better access to capital, it can also be more susceptible to the instabilities of the market.

privatizations have been suspended or aborted. For example, Spain's attempt to divest its El Prat airport in Barcelona and Barajas airport in Madrid were put on hold when German airport operator Fraport, Spain's Fomento de Construcciones Contratas SA, as well as Singapore's Changi Airports International all pulled out of the bidding citing the difficult market situation (Dominguez and Perez, 2012).

Reasons for privatization

While there is no conclusive empirical evidence to support the proponents of selling or long-term leasing of airports to the private sector, there are at least four perceived principle advantages:

- greater efficiency of operations, particularly in developing the non-aviation side of the airport;
- capital infusion: open up non-traditional sources of capital;
- lower labor costs resulting from either lower wages or less labor input;
- conversion of a public airport into a tax-paying corporate entity.

In addition to being generally more efficient, private companies can readily raise funds for needed airport projects without entanglement in the political problems and delays that often plague government airports looking for grants to expand or renovate. Moreover, these companies can engage in equity financing while government is only able to issue debt. Of course, there may be potential problems with privatization (Vasigh and Gorjidooz, 2006). Some worry that, even with continued government regulation, a private company may not be motivated to properly maintain infrastructure. However, two decades of experience seem to indicate that there are really not many problems in this regard. Most economists would probably argue that profit motives provide strong incentives for proper maintenance of airport infrastructure. Since consumers tend to be hyper-sensitive to safety concerns in air travel, any hint of corner-cutting in this regard is likely to depress demand and sink profits. Also, private airports are more likely to be held accountable by liability laws since it is generally easier to sue a private party for damages than to sue the government.[12] In essence, a private airport appears to have the same regulatory incentive for safety as a government airport since there is no change in safety regulations with privatization, plus the added incentives of stricter legal liability and of a profit motive. Indeed, these added safety incentives for private companies may explain why one seems sometimes more likely to encounter dangerous infrastructure failures of government levies, bridges and roads than on private roads, parking lots or other structures.

Opposition to privatization

Airlines and passenger dissatisfaction with privatization most often occurs when the service level deteriorates and airport users face sky rocketing prices. Subsequent to a

---

12   Of course, overly harsh liability laws can be an impediment to economic efficiency and consumer well-being, and many would argue that tort reform that would render liability laws less harsh. However, tort reform is not a necessary pre-condition to airport privatization—many companies are clearly interested in buying airports under the existing tort system, whatever its faults.

complaint by British Midland International of discriminatory pricing, the UK Civil Aviation Authority launched a formal investigation into passenger and landing charges at London Heathrow during April 2011 (Buyck, 2011). Another fear sometimes voiced is that private airports might sometimes go bankrupt. Although this is a distinct possibility, bankruptcy is more a financial disaster for stockholders than an operational problem for air travelers. Just as airlines have (routinely, unfortunately) continued to operate normally in bankruptcy, so too would viable airports. Furthermore, if management is markedly at fault then a bankruptcy judge might well eject such management. Thus, bankruptcy provides a new channel, one not available in the case of mismanaged government airports, for eliminating poor management.

Another common objection to airport privatization is that airports have monopoly power and, if private, will raise landing fees to extremely high levels. However, it is not certain that this is all bad for the industry or air travelers. Recall from the economic analysis in the introduction to this section that artificially low landing fees can result in high costs from dead weight losses that are worse than the cost of a higher landing fee. Very low prices for airline operations result in excess fuel burn, maintenance costs, and time wasted for both crew and travelers. Moreover, any business that is unable to choose its price is fundamentally hampered. In the case of airports, major benefits from airport expansion, for instance, may not be affordable with price held artificially low.

Also, the danger of excessively high monopoly pricing may not be as significant as it first appears. In the end, airports can only charge airlines higher fees if airlines are able to pass on those higher costs to customers. It is probably safe to say that the leisure traveler will not bear such costs. That is, leisure travelers will likely either travel by other modes of transportation or fly via more distant airports if prices go up substantially at their home airport. Business travelers are perhaps more likely to pay higher prices but even in their case, the monopolist airport must consider competition from other transportation modes, secondary airports, modern telecommunications, corporate jets, and the developing "air taxi" competition from the very light jets that are able to operate out of smaller airports. The situation is similar to that discussed under "contestability" theory in Chapter 9; that is, the threat of competition may be sufficient to preclude extreme monopoly pricing behavior on the part of privatized airports. Given these considerations, some experimentation with free pricing of landing fees may be warranted.

Although the theoretical grounds for supporting airport privatizations seem solid to most economists, actual experience may be the most persuasive evidence. The very fact that so many different governments are abandoning control of their airports is a strong statement in favor of privatizing. After some initial skepticism, there now seems strong support for the argument that divestiture can enhance the efficiency of airport operations (Truitt and Michael, 1996). Vasigh and Hamzaee (1998) emphasize the benefits of privatization of airports in Western Europe, Latin America, and Asia that should inspire officials in search of new economic opportunities in transforming airports from publicly run into private businesses (Vasigh, Yoo and Owens, 2004).

Types of privatization

The techniques used to privatize airports vary in terms of the scope of responsibility and the degree of ownership transferred to the private sector. A traditional privatization tool involves the contracting of selected services (restaurants, parking, security services, cargo, baggage handling, and/or fueling services) to the private sector while the government retains overall operating responsibility for the airport. More comprehensively, under the contract management approach, the government transfers all responsibility for all airport operations and implementation of strategy to the private sector, while retaining the ownership and investment responsibilities.

Several US airports currently are operated under management contracts. These include Westchester County Airport in New York, Orlando-Sanford in Florida, and Bob Hope Burbank Airport, which is owned jointly by the cities of Burbank, Glendale, and Pasadena, California. The Burbank Airport has been managed by for over a decade by TBI Airport Management Inc. which receives a fixed management fee, plus expenses for the services it provides for the airport. The airport authority is responsible for capital improvements. Burbank Airport, which ranks fifty-ninth in size among US airports (as measured by annual passenger enplanements), often is held up as a viable model of public-private partnerships in airport operations (Ashford and Moore, 1992). However, in October 2011, airport employees voted to affiliate themselves with the International Union of Operating Engineers due to insufficient pay below industry averages.

In addition, in October of 1995, the BAA took over the management of Indianapolis International Airport promising to raise non-airline revenues by $32 million within the ten-year period of the contract. The contract was renegotiated in 1998 and extended until 2008. Between 1995 and 1999, costs per passenger were reduced from $6.70 to $3.70 and in spite of a moderate passenger annual growth rate of 3.5 percent, non-airline revenue per passenger more than doubled by 2003 (Vasigh and Haririan, 2003). Private management of the airport ended on December 31, 2007 and control was transitioned back to Indianapolis International Airport management.

A long-term lease approach allows the government to legally (and politically) retain ownership but to transfer investment, operational, and managerial responsibilities to a private tenant, with the lease long enough to motivate the private tenant to more or less behave as an owner. This method may be used to allow the financing of the construction of the airport or associated project by the private sector, which must then relinquish control at the end of the lease term. Several recent examples of this type of public–private partnership exist at Sydney Airport (99-year lease), and Brazil which has recently awarded three concession contracts of 20, 25, and 30 years at its busiest airports. One well-known example of such a lease arrangement is Teterboro Airport in New Jersey. The lease to operate Teterboro was established in 1970, when Pan American World Airways (now known as Johnson Controls World Services) secured a 30-year lease with Teterboro's owner, the Port Authority of New York and New Jersey. The lease expired in 2000 and the Port Authority resumed operation as scheduled. The company believed that it could relieve some congestion if general aviation aircraft could be lured away from the city. Johnson Controls secured a 30-year lease to operate Teterboro Airport.

Finally, using a full divestiture/sale of shares, the government transfers full (or partial) ownership to the private sector either through an Initial Public Offering of shares (IPO) or a competitive bidding process. An IPO can be part of a full sale, as in the case of BAA in 1987, or a partial sale, as in the cases of Bangkok in 2002 (30 percent) and Hong Kong (49 percent). It should be noted that even in the case where the airport is sold the government still retains substantial regulatory control in many areas, including safety and, usually, the regulation of landing fees.

Privatization in the US airport industry[13]

In many countries around the world, except the US, governments own and operate aviation infrastructure including gates. The gates are assigned dynamically to airlines as needed (common-use gates). Airlines pay landing fees and space rentals, at pre-set rates, based on how much of the facilities they use. This same model has continued under privatization, meaning very little change for the airlines.

In contrast, the typical US approach is one in which the anchor-tenant airlines sign long-term lease agreements with charges based on "residual cost," that is, the cost of operating the airport would first be covered by revenue sources other than the airlines—such as parking, concessions, and so on. Whatever costs that are not covered by these revenues would then be assessed on the airlines via landing fees and space rentals.

In effect, these signatory airlines became joint owners of the airport. In good years the airports would take in more of non-aviation revenues and the airlines would enjoy lower fees. Of course, in bad years with fewer passengers and therefore lower airport revenues, the airlines would have to pay higher fees. Thus, bad years for the airlines become even worse.

One might expect the airlines to refuse such an arrangement that make their profits even more intensely cyclical, but two considerations make the arrangement worthwhile for them. First, the federal funding system for airport expansion is often so cumbersome that joint airport ownership/funding by airlines is the only viable way for the timely expansion of the airport. Second, the airlines gain the ability to veto airport spending that they see as wasteful and might lead to more fees and charges. Thus, US legacy carriers tend to have a vested interest in maintaining the status quo, opposing airport privatization (as discussed below) because it might indirectly open their markets to more competition.

As airport privatization gained momentum in the 1990s, there was a call for legislation to eliminate federal regulations hostile toward private airports. A key problem was that regulations were interpreted to imply that any local government that sold an airport would then have to repay all previous federal airport grants and, of course, that airport would be ineligible for any future grants. In an attempt to address these problems the 1996 Airport Privatization Pilot Program was passed and supervised by the FAA. Under this program, cities whose airports were accepted for the pilot program would not have to repay previous grants.

However, political pressure, largely from the airlines, resulted in a provision that rendered the program essentially useless. In order to make use of lease or sale proceeds, a city has to get the approval of 65 percent of the airlines serving the airport. Otherwise, all profits must be reinvested in the airport, making the whole exercise from the viewpoint of the airport owners not worth the trouble.

The only airport actually privatized under the 1996 law—Stewart Airport in Newburgh, NY only lasted six years of the 99-year lease before being bought out by the Port Authority of New York and New Jersey; ironically, to initiate a $500 million expansion project that the firm wasn't willing to commit. Furthermore, the project did not get the needed airline approval, which in this case meant the city could not have gained from the profit. As such, the sale or lease of US airports is likely to remain politically unfeasible for the foreseeable future unless airline opposition weakens. Contractual privatization, in part or in whole, remains the only viable alternative.

---

13   This section draws heavily from Poole (2007) and from Vasigh, Yoo and Owens (2004).

SUMMARY

This chapter introduces the reader to some more practical aviation applications of the theoretical constructs of supply and demand that were introduced in the previous chapters. The unique and unprecedented control that government exerts over the industry in the form of an absolute monopoly over all facets of air traffic is discussed and analyzed. The chapter uses the previously introduced models of supply and demand to analyze the costs of delay imposed by the regulatory agency as an external tax that has been levied on the industry through the failure of the regulatory agency to modernize and use effective technologies. The effects of these costs are not only large in the short run but also extend to the long-run viability of the industry. Therefore, these costs are large and likely to grow larger over the foreseeable future. The chapter also uses supply and demand analysis to introduce the reader to the concept of airport privatization and the use of the market price system to allocate scarce resources at the airport. Various approaches to privatization, both internationally and domestically, are discussed and critiqued in the half of the chapter.

DISCUSSION QUESTIONS

1. In what sense can it be said that the government controls the means of production for the airline industry?
2. In what ways are the effects of the regulatory delay tax discussed in the chapter more burdensome than an ordinary tax?
3. Why can it generally be assumed that the supply curve for the airline industry is relatively inelastic?
4. What are some of the benefits of privatization that are mentioned in the chapter?
5. What are some of the ways in which the effects of an airport monopoly might be countered in the marketplace?

APPENDIX

Airport	Airport Code
Auckland, New Zealand	AKL
Beijing, China, Capital	PEK
Belfast City Airport, Ireland	BHD
Berlin, Germany, Schoenefeld	SXF
Berne, Switzerland, Belp	BRN
Birmingham, AL, USA	BHM
Bologna, Italy, Guglielmo Marconi	BLQ
Boston, MA, USA, Logan International Airport	BOS
Brasilia, Distrito Federal, Brazil	BSB
Bristol, UK	BRS

Brussels, Belgium, National	BRU
Budapest, Hungary	BUD
Charlotte, NC, USA, Charlotte/Douglas Intl Airport	CLT
Chicago, IL, USA, O'Hare International Airport	ORD
Cincinnati, OH, USA, Greater Cincinnati Intl Airport	CVG
Copenhagen, Denmark	CPH
Dallas/Ft Worth, TX, USA	DFW
Denver, CO, USA	DEN
East Midlands, UK	EMA
Frankfurt, Germany	FRA
Hamburg, Germany	HAM
Haneda, Japan, Tokyo International Airport	HND
Helsinki, Finland	HEL
Houston, TX, USA	IAH
Leeds/Bradford, UK	LBA
Lisbon, Portugal, Lisboa	LIS
London, UK, Gatwick	LGW
London, UK, Luton	LTN
Madrid, Spain, Barajas	MAD
Narita International Airport, Japan	NRT
Newcastle, UK	NCL
Paris. France, Charles De Gaulle Airport	CDG
Rio De Janeiro, Brazil	GIG
Rio De Janeiro, Brazil, Santos Dumont	SDU
Rome, Italy, Ciampino	CIA
Rome, Italy, Leonardo Da Vinci/Fiumicino	FCO
Singapore Airport, Singapore	SIN
Sydney, Australia	SYD
Toronto, Ontario, Canada	YYZ
Vancouver, British Columbia	YVR
Verona, Italy	VRN
Warsaw, Poland, Okecie	WAW
Zurich, Switzerland, Zurich	ZRH

REFERENCES

Ashford, N. and Moore, C. (1992). *Airport Finance.* Kluwer Academic Publishers.

Dominguez, P. and Perez, S. (2012, January 12). Spain Grounds Airport Privatization Plan, *The Wall Street Journal.* Retrieved May 6, 2012, from http://online.wsj.com/article/SB100014240529702044 09004577156633018644026.html.

Buyck, C. (2011, July 7), London Heathrow charges investigated by CAA after BMI complaint, *Air Transport World.* Retrieved on November 21, 2011, from http://atwonline.com/airports-routes/ news/london-heathrow-charges-investigated-caa-after-bmi-complaint-0706.

Federal Aviation Administration. (2005). *International Terminal Air Traffic Control Benchmark Pilot Study.*

Kent, R. (1980). *Safe, Separated, and Soaring: A History of Federal Civil Aviation Policy, 1961–1972.* Washington DC: DOT/FAA.

Poole, R. (2006). Bizjet Sales Cut by Commercialization. *ATC Reform News* (p. 31). Reason Foundation. Retrieved on 11 February, 2007 from www.reason.org/atcreform31.shtml.

Poole, R. (2007). *Will Midway Lease Re-Start U.S. Airport Privatization?* Reason Foundation, Public Works Financing.

Truitt, L. and Michael, J. (1996). Airport Privatization: Full Divestiture and its Alternatives. *Policy Studies Journal,* Vol. 24, No. 2, 100–124.

Vasigh, B. and Gorjidooz, J. (2006). Productivity Analysis of Public and Private Airports: A Causal Investigation. *Journal of Air Transportation,* Vol. 11, No. 3, 142–162.

Vasigh, B. and Hamazaee, R. (1998). A Comparative Analysis of Economic Performance of US Commercial Airports. *Journal of Air Transport Management,* Vol. 4, No. 4, 209–216.

Vasigh, B. and Haririan, M. (2003). An Empirical Investigation of Financial and Operational Efficiency of Private Versus Public Airports. *Journal of Air Transporation,* Vol. 8, No. 1, 91–110.

Vasigh, B., Yoo, K. and Owens, J. (2004). A Price Forecasting Model for Predicting Value of Commercial Airports: A Case of Three Korean Airports. *International Journal of Transport Management,* Vol. 1, No. 4, 225–236.

6

International Economics and Aviation

... when I cock my ear toward Mexico, I still hear that "giant sucking sound" of American jobs headed south of the Rio Grande.

Ross Perot, the Reform Party's 1996 presidential candidate

Although other opinions are addressed, the thrust of this chapter is to explain why economists largely agree that international trade for the most part benefits the economy. Aviation is generally not an exception to that rule, though it has some special complications. Anti-trade arguments are discussed and, with some slight exceptions, demonstrated to be mainly fallacious. This chapter lays the foundation for the next which details international aviation agreements and industry alliances. Allowing international competition in aviation, and elsewhere, produces a net gain to the economy rather than a destructive "concession," as politicians often suggest. This chapter will cover the following topics in order:

- Trade Globalization
  - World Trade Organization (WTO)
  - International balance of payments
  - Current account
  - Capital and financial accounts
- International Economics and Trade
  - Arguments for Free Trade
  - Arguments against Free trade
    o Job losses
    o National security concerns
    o Infant industry
  - Trade deficit and surplus
- Why Nations Trade
  - Production possibility curve
  - Absolute advantage
  - Comparative advantage
  - Trade protections and trade barriers
  - Tariffs
  - Quotas

- Other forms of trade protection
- Aircraft Manufacturing and Governmental Subsidies
  - Boeing versus Airbus
  - Recent World Trade Organization rulings against Airbus
- International Trade Policy in Air Travel–Optimality versus Political Realities
- Foreign Currency and Exchange Markets
  - Exchange rate Quotes
  - Exchange rate Regimes
  - Fixed (pegged) exchange rate
  - Floating (adjustable) rates
- Summary
- Discussion Questions
- Appendix: International Free Trade Agreements

TRADE GLOBALIZATION

All countries, regardless of their size, depend to some degree on other economies and are affected by trade and transactions outside their borders. In short, international trade remains the engine of world economic growth. The globalization of the international economy has occurred in almost in every country and has led to the development of many regional trade agreements. As a consequence of international trade, Japan and Korea became able to transfer themselves to become the world's largest car producer, the largest steel producer, and the largest fiber producer.

> We had the globalization of trade, we had the globalization of capital, and now we have the globalization of talent.
>
> London Free Press

World Trade Organization (WTO)

The World Trade Organization (WTO) is a global international organization dealing with the rules of trade between nations and was formed as the successor to the General Agreement on Tariffs and Trade (GATT) established after the Second World War. At the core of the WTO multilateral trading system are agreements that are negotiated and signed by the majority of the world's trading nations and ratified in their parliaments. The WTO's primary functions include:

- administering trade agreements;
- acting as a forum for trade negotiations;
- settling trade disputes;
- reviewing national trade policies;
- assisting developing countries in trade policy issues, through technical assistance and training programs;
- cooperating with other international organizations.

The 150-member WTO accounts for approximately 95 percent of global trade with rules (agreements) that are the result of negotiation between members. The current rules are the result of a round of negotiations called the Uruguay Round, which spanned eight years from 1986 to 1994 and revised the original GATT rules. The 30,000-page revision consists of 30 agreements and separate commitments (called schedules) made by individual members.[1] Several key aspects of the WTO, particularly settling trade disputes will be discussed further.

International balance of payments

The Balance of Payments (BOP) method is used by countries to summarize, for a specific time period, the economic transactions of an economy with the rest of the world. The BOP is a system of recording all of a country's economic transactions, exports, and imports, with the rest of the world during a given period of time and is divided into two main categories: current accounts and capital accounts. In the US, the trade deficit, as a percentage of GDP, increased dramatically in the 1980s and again in the late 1990s. During 2000, imports have continued to rise as a percentage of gross domestic product (GDP) and exports have fallen. If a country is experiencing persistent BOP deficits, it may signal that the country's industries lack productivity and efficiency.

Current account

The current account is the sum of the balance of trade (difference between a nation's total exports and its total imports), net factor income (dividends and so on) and net transfer payments (foreign aid and so on). The current account balance can therefore be either a surplus or a deficit and serves as one of the broadest measures of international trade. Put simply, if the current account balance is positive, the country is a net lender to the world and vice versa for a deficit. The US's current account has been in a deficit since 1992; this trend continued into 2010 where the US recorded a current account deficit of $470.9 billion (see Figure 6.1).[2] When a country runs a current account deficit, its purchases of goods and services from abroad exceed its sales of goods and services to foreign buyers. Contrary to the US, Japan has realized continuous current account surpluses. When a country runs a current account surplus, sales of goods and services to foreign buyers exceeds its purchases of goods and services from other countries.

Following China's accession to the WTO in 2001, its exports surged and generated tremendous growth in its trade surplus in the 2000s. In 2010, China recorded a current account surplus of $305.4 billion USD, obtaining its number one ranking in terms of current account surpluses.

---

1   World Trade Organization, 2012.
2   The World Bank Economic Policy and External Debt data, 2012.

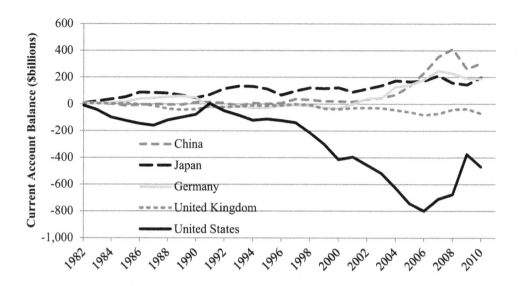

Figure 6.1   Current Account Balances, 1982–2010

Source: The World Bank, Economic Policy and External debt data

Capital and financial accounts

The capital account is the net result of public and private investment flowing into and out of a country including changes in foreign direct investment, stocks, bonds, loans, bank accounts, and currencies. The major components of the capital account are capital transfers and acquisition/disposal of non-produced, non-financial assets. Coupled with the current account, this forms the BOP; the current account reflects net income while the capital account reflects net change in national ownership of assets. The International Monetary Fund (IMF) divides the capital account into financial and capital components where the monetary flows related to investment in business, real estate, bonds, and stocks are recorded in the financial account and changes in capital are recorded in the capital account.

INTERNATIONAL ECONOMICS AND TRADE

Economists in general agree on most of the key policy issues relating to international economics. However, as is often the case in economics, this consensus has failed to break through many popular misconceptions and, therefore, public policy in international trade deviates substantially from the ideal. International trade allows countries to take advantage of other countries' resources through the theory of comparative advantage (explained later in the chapter). It has been successfully argued and empirically demonstrated that the overall production of the world increases through trade and partnership.

In 2010, the volume of world merchandise exports increased by 14.0 percent. China, Germany, US, and Japan have emerged as important exporters in world merchandise trade (Table 6.1). The US is the world's largest market for exporting countries. In 2005, it imported more than $1,969 billion worth of merchandise (Table 6.2). China alone exported more than

$1,578 billion worth of products to the rest of the world—10.4 percent of total exports and imported $1,395 billion worth—9.1 percent of total imports. Japan imported $770 billion during the same period. Among the ten leading exporters, the five most dynamic economies are China (10.4 percent), US (8.4 percent), Germany (8.3 percent), Japan (5.1 percent), and the Netherlands (3.8 percent). The five leading importers comprise the same countries as the group of the top five leading exporters, except that France replaces the Netherlands in fifth place with 3.9 percent.[3] Figure 6.2 categorizes the world merchandise according to regions, of which the Asian and European regions make up the largest collective exporters/importers.

Table 6.1   Top Ten Exporters in World Merchandise Trade, 2010

Rank	Exporters	Value ($billions)	Percentage Share	Annual Percentage Change
1	China	1,578	10.4	31.3
2	United States	1,278	8.4	21.0
3	Germany	1,269	8.3	13.3
4	Japan	770	5.1	32.6
5	Netherlands	573	3.8	15.2
6	France	521	3.4	7.4
7	South Korea	466	3.1	28.3
8	Italy	448	2.9	10.0
9	Belgium	412	2.7	11.5
10	United Kingdom	406	2.7	15.0

Source: World Trade Organization, *International Trade Statistics*, 2011

Table 6.2   Top Ten Importers in World Merchandise Trade, 2010

Rank	Importers	Value ($billions)	Percentage Share	Annual Percentage Change
1	United States	1,969	12.8	22.7
2	China	1,395	9.1	38.7
3	Germany	1,067	6.9	15.2
4	Japan	694	4.5	25.7
5	France	606	3.9	8.1
6	United Kingdom	560	3.6	16
7	Netherlands	517	3.4	16.6
8	Italy	484	3.1	16.6
9	Hong Kong	442	2.9	25.5
10	South Korea	425	2.8	31.6

Source: World Trade Organization, *International Trade Statistics*, 2011

---

3   World Trade Organization, *International Trade Statistics*, 2011.

The world trade market can also be broken down according to regions as shown in Tables 6.3 and 6.4. We see that the level of merchandise exports originating in Europe has remained fairly stable in the time period 1948 to 2010 with Europe maintaining a dominant position throughout. Moreover, the runner-up in percentages of global merchandise exports by region has shifted from North America to the Asian region. In 1948, exports by North America comprised 28.1 percent of the world while Asia was half of that at 14.0 percent. In 2010, we see near exact numbers in reverse: North America has 13.2 percent of global exports while Asia has more than the double the amount at 31.6 percent.

Table 6.3   World Merchandise Exports by Region (Percentage), 1948–2010

	1948	1953	1963	1973	1983	1993	2003	2010
North America	28.1	24.8	19.9	17.3	16.8	18	15.8	13.2
South and Central America	11.3	9.7	6.4	4.3	4.4	3	3	3.9
Europe	35.1	39.4	47.8	50.9	43.5	45.4	45.9	37.9
Commonwealth of Independent States (CIS)	-	-	-	-	-	1.5	2.6	4
Africa	7.3	6.5	5.7	4.8	4.5	2.5	2.4	3.4
Middle East	2	2.7	3.2	4.1	6.8	3.5	4.1	6
Asia	14	13.4	12.5	14.9	19.1	26.1	26.2	31.6

Source: World Trade Organization, *International Trade Statistics*, 2011

Table 6.4   World Merchandise Imports by Region (Percentage), 1948–2010

	1948	1953	1963	1973	1983	1993	2003	2010
North America	18.5	20.5	16.1	17.2	18.5	21.4	22.4	17.8
South and Central America	10.4	8.3	6.0	4.4	3.8	3.3	2.5	3.8
Europe	45.3	43.7	52.0	53.3	44.2	44.6	45.0	38.9
Commonwealth of Independent States (CIS)	-	-	-	-	-	1.2	1.7	2.7
Africa	8.1	7.0	5.2	3.9	4.6	2.6	2.2	3.1
Middle East	1.8	2.1	2.3	2.7	6.2	3.3	2.8	3.7
Asia	13.9	15.1	14.1	14.9	18.5	23.7	23.5	29.9

Source: World Trade Organization, *International Trade Statistics*, 2011

Further analysis can be conducted into who the largest trading partners are for a particular country. Table 6.5 shows the top ten trading partners with the US for 2011 (US Census Bureau: Foreign Trade Division, 2012).

Arguments for free trade

Trade is the natural enemy of all violent passions. Trade loves moderation, delights in compromise, and is most careful to avoid anger. It is patient, supple, and insinuating, only resorting to extreme measures in cases of absolute necessity. Trade makes men independent of one another and gives them a high idea of their personal importance: it leads them to want to

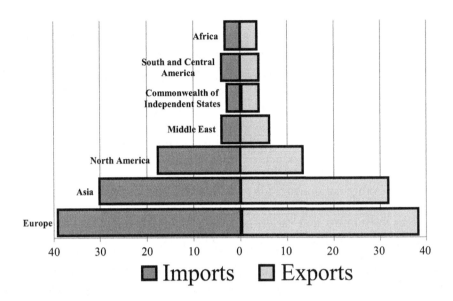

Figure 6.2   World Merchandise Trade According to Regions, 2010*

Source: World Trade Organization, *International Trade Statistics*, 2011

* The results do not reflect any one country's specific trade but all countries in their respective regions collectively

Table 6.5   Top Ten US Trading Partners in 2011 ($ billions)

Rank	Country	Exports (Year-to-Date)	Imports (Year-to-Date)	Total Trade (Year-to-Date)	Percent of Total Trade
1	Canada	258.3	290	548.2	16.20%
2	China	94.2	366.5	460.7	13.60%
3	Mexico	181.2	241.8	423.1	12.50%
4	Japan	60.8	116.9	177.7	5.30%
5	Germany	45.1	89.5	134.6	4.00%
6	United Kingdom	51.1	46.4	97.4	2.90%
7	Korea, South	39.5	52.1	91.7	2.70%
8	Brazil	39.6	27.9	67.5	2.00%
9	France	25.7	36.5	62.1	1.80%
10	Taiwan	23.6	38.2	61.8	1.80%
11	Netherlands	39.1	21.3	60.4	1.80%
12	Saudi Arabia	12.3	42.4	54.7	1.60%
13	India	19.7	33.6	53.2	1.60%
14	Venezuela	11.2	40.1	51.3	1.50%
15	Singapore	28.7	17.7	46.4	1.40%
...	Total, All Countries	1,037.3	1855.4	2892.7	100%

Source: US Census Bureau: Foreign Trade Division, Foreign Trade Statistics 2012

manage their own affairs and teaches them to succeed therein. Hence it makes them inclined to liberty but disinclined to revolution.

Alexis De Tocqueville, French Political Philosopher, 1805–1859

Proponents of free trade cite the following advantages:

- reduces the price of every item sold in the market;
- increases the supply of products in other markets and results in lower prices for those products;
- encourages other nations to trade more freely with their trading partners which helps the global economy;
- increases the number and variety of products for consumers to choose;
- is a driving force behind a high standard of living.

Economists generally argue that the world at large will benefit from free trade and that trade liberalization can promote development. Free traders claim that trade protection always harms all the trading partners. The main political objections to free trade are centered on the fears that cheaper imports will destroy jobs and reduce income. Labor unions and management may be against free trade if they believe that the competition from free trade will make them lose jobs and/or bankrupt the firm or industry. In the simple "two-good" case illustrated later on in the chapter it is easy to see that, for the economy as a whole, this fear has no basis in reality. Although some jobs may indeed be made obsolete (or lost) by free trade, other jobs are created. The new jobs are more productive so average wages and overall income rises. Again, this is easy to see in the two-good case (where the country lost jobs making one product but gained more productive jobs making another product).

The same comparative advantage principle holds when there is trade with many goods but the impact is not so obvious. Suppose, for instance, a country dramatically increases purchases of imported autos. Everyone will be able to see the resulting layoffs in the domestic auto sector, the downside of this international trade. However, the new jobs created by this trade are widely dispersed and not at all obvious. When consumers get cheaper autos some will then be able to spend more on computers, others will enjoy more air travel, some will buy new clothes, eat out at restaurants more, buy a nicer home, and so on. Thus, cheap auto imports allow us to produce more of other products with additional job creation to enable this new production. However, the job gains will be widely dispersed throughout many, seemingly unrelated, industries. The individual who obtains an airline job made possible by the availability of cheaper cars in the economy is unlikely to see the connection. In other words, even though the benefits of trade far outweigh the costs, free trade is often controversial because the costs are concentrated, visible to even the least discerning citizens, and therefore easy to politically exploit.

Arguments against free trade

Today, most countries around the world impose some form of trade restrictions such as tariffs, taxes, and subsidies. The arguments most often heard against free trade are:

- keeping jobs in the country;

- limiting imports to keep the wealth inside of the country;
- national security concerns;
- other nations don't treat their workers fairly or are not concerned about the environment;
- a nation may become too specialized and dependent on other nations.

Job losses   Even though free trade increases average wealth, some individuals probably will be made worse off. This, after all, is true of any advance in technology. For example, discovering a cure for cancer would destroy some jobs. In an advancing economy most people will be able to adapt to new job opportunities with little or no serious problems. However, in extreme cases some individuals may require substantial assistance to survive and adjust to technological advances in a growing economy. But with trade, as with improving technology, overall wealth is increasing in society so such assistance is more readily affordable. Low trade barriers inherently create hundreds of billions of dollars more in benefits than they impose in costs (Pugel, 2007). Hufbauer and Elliott (1994) state that the average protected job costs an economy $170,000. So, rather than using trade barriers to "protect jobs," it would be cheaper for consumers to allow free trade and then, as taxpayers, compensate all workers who have lost jobs to more efficient foreign competition.

National security concerns   In a very few special cases it may be true that national security justifies a particular trade barrier. It would not be prudent, for example, to allow aircraft from hostile nations to have "open skies" access to our airspace. But, more typically, free international trade enhances national security by raising income and promoting friendly relations. This is particularly true for aviation since free trade in air travel inherently "makes the world smaller" and promotes more economic integration and social interaction between nations. When economies are strongly integrated, many of each country's citizens have a large stake in other countries, and therefore a strong vested interest in avoiding the massive destruction and attendant loss of wealth associated with war. This is why wars between major trading partners are relatively unusual in world history. Indeed, it was the desire to promote peaceful interaction, even more so than economic development, which initially motivated European leaders to form the European Union (EU) (van den Berg, 2004).

Some try to argue that allowing foreign airlines, even those from friendly nations who are staunch allies, to compete freely in domestic markets somehow inherently jeopardizes national security. They insist that airlines must be domestically-owned just in case the government needs to use civilian aircraft in some emergency to, say, move military troops within the country. Most economists are very skeptical of such claims. All governments reserve the right, for example, to confiscate private property (hopefully with just compensation being paid at the appropriate time) in emergencies, regardless of who owns the needed property. Instances where government would need to confiscate or commandeer civilian aircraft are rare, possibly even non-existent in many cases, but if the situation ever arises, the government has the power to take what it needs.

In the US, the government has a contract in place with some airlines to provide troop transport if ever needed (US Air Force: Air Mobility Command, 2006). Given the relative efficiency of US airlines and the availability of military transport aircraft, even if the US domestic air travel market were completely open to foreign competition, it is unlikely in the near term that foreign carriers would so dominate the market that there would not be enough US aircraft to move troops. If, in the distant future, foreign carriers begin to achieve such dominance there seems to be no reason that the US could not arrange the same sort of contract with a foreign carrier; it could even be required as a condition of the carriers being allowed to compete in the US market. Other countries could do likewise. National security does not provide a reasonable argument against free international trade and competition in air travel.

Infant industry   Infant industries (emerging domestic industries) are often offered some limited and temporary protection by their governments through tariffs, quotas, and duty taxes from international trade competitors. This allows the infant industry sufficient time to take advantage of the learning curve and mature enough to sustain itself amid global competition without protection. This protection is also often labeled under "launch aid" in the aviation industry.

A significant number of the justifications offered for this temporary protection of infant industries stem from the learning curve that all industries experience. In the early production stage, higher costs exist versus those of foreign companies who have already taken advantage of their learning curve to reduce unit costs. Launch aid and infant industry protection, particularly in the aviation industry, have long been argued as an unfair trade practice, generally by the direct foreign competitor. Complaints brought forth to the WTO by the US government on behalf of Boeing against Airbus (which will be covered in greater detail further on in the chapter) continue to argue that Airbus has outgrown an infant industry's need for the aid it receives from the governments of the EU.

Trade deficit and surplus

It is useful to begin by pointing out that the term "trade deficit" is completely arbitrary and might just as easily be called a "trade surplus." If imports are greater than exports, more goods and services flow into a country than flow out, so that a deficit exists. Contrarily, if the difference in the value of a nation's exports over imports is positive the country enjoys a trade surplus. Suppose, to illustrate, that someone came to your house and brought a number of products to you; they took some of your goods in return but the value of what they gave you was greater than the value of what they took. Would it not be more natural to say you enjoy a surplus of trade with this individual, rather than following convention which would term this situation a trade deficit?

But is it harmful when money "leaves the economy?" To begin, remember that the world's leading currencies are no longer backed by gold or any other real assets. Dollars, yen, and euros are pieces of paper, valuable in exchange, but very inexpensive to print in virtually limitless quantities. Of course, printing too much currency results in devaluation of that currency, that is, inflation, but printing up currency to simply replace that which leaves the country will not be inflationary at all (as long as foreigners just hold the dollars). Suppose that the US were to experience the ultimate trade deficit—foreigners acquired dollars and simply collected them, refusing to buy anything from the US. If the government did nothing to offset the effect of dollars pouring out of the US then the country would experience deflation, pervasive falling prices. So, to keep the value of the currency stable the government needs to create new currency and put it into circulation to replace the currency leaving the country.

To put money into the system, the authorities (the Federal Reserve System in the US) buy existing US Treasury debt. Since interest is paid on this debt the government is now paying interest to itself; in effect this debt is retired and no longer a burden to taxpayers. Thus, the more money that leaves the country the better![4] The US is able to trade currency, cheap pieces of paper, for real goods plus retire substantial portions of debt. For this reason the US, and all governments, generally encourage other countries to use their currency. Another way of looking at this is the obvious fact that the principal place where dollars can be spent is in the US. This obviously will increase the demand for US goods and services or, as is more likely to be the case, investment in productive resources in the US. There are currently more dollars circulating outside the US than inside, something the US welcomes (Elwell, 2004).

But what if those outside dollars are suddenly returned and spent in the US? While this would not be catastrophic, this would be somewhat costly. Consumption would have to fall as foreigners traded their dollars for goods and services; in other words, rather than having pieces of paper leave and goods and services flow in, the US would now see the reverse. To avoid inflation with the returning currency, the US will have to take some currency out of circulation. Some of that US treasury debt will be resold, and taxpayers will now be charged to make the outside interest payments to the buyers of the US treasury debt. The currency gained from reselling that debt would be held out of circulation.

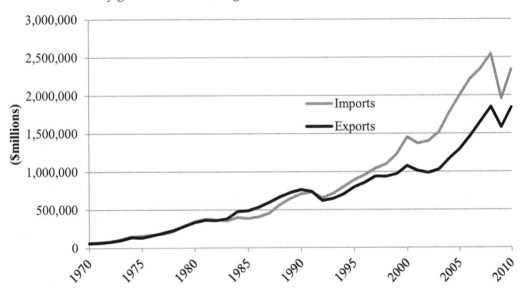

Figure 6.3   Imports and Exports in the United States, 1970–2010

Source: US Department of Commerce, Bureau of Economic Analysis, 2010

Of course, foreigners willingly hold dollars because they believe that it is in their interest to do so, and thus are very unlikely to suddenly change their minds en masse and flood the US with returning currency. But it is comforting to know that if such a thing should somehow happen, the negative impact is not at all overwhelming.

As indicated in Figure 6.3 above, the US has generally experienced a perpetual, growing trade deficit since about 1980. A few trade surpluses have occurred but only when the

---

4   Recall the original assumption that foreigners prefer to hold the dollars.

economy weakened in recessions. Although some currency did flow out of the US, the trade deficit is basically balanced by a surplus of capital inflows. The capital account surplus (capital inflows minus outflows) is essentially the mirror image of the trade deficit. In other words, foreigners who wish to invest in the US have largely outbid foreigners who wish to consume US goods. When the US "sends dollars out of the country" they largely "come back" in the form of capital flows. For instance, foreigners directly build factories in the US or buy stocks and bonds from US companies (US Department of Commerce, Bureau of Economic Analysis, 2007). The US is seen as an ideal place to invest—international investors are attracted to the world's largest economy, political stability, and a substantial degree of economic freedom. As long as this holds true, the US is likely to continue to see capital surpluses/trade deficits for the foreseeable future.[5]

WHY NATIONS TRADE

In this section, we begin to explore the reasons why nations choose to engage in international trade and, particularly, the advantages to engaging in international trade. The baseline reasoning behind international trade is the fact that nations are not equally able to produce all goods; one nation may be particularly abundant in the natural resources needed to produce a particular good while another may have the wherewithal to capitalize on those resources. First, we need to explain the concept of a production possibility curve (PPC).

Production possibility curve

Whenever the structure of production is addressed within an economy of any scale, it is helpful to understand the concept of a PPC. A PPC shows all different combinations of output which an economy can produce by using all the available resources. The curve is plotted along a two-dimensional axis in which the y and the x axes each signify a quantity of a good, making it a "two-good" analysis. The curve itself is seen as a frontier outside of which production of the two goods is impossible. Any combination of outputs within the curve would prove to be inefficient and maximum efficiency lies only along the curve. Therefore, any point within the PPC represents inefficiency and any point outside the PPC represents something unattainable, given available technology and other resources. The PPC of Figure 6.4 shows the tradeoff in production between, for example, "Food and Clothing" and any two combinations of these two products such as A, B, C or any others could be chosen.

> A PPC can be used to represent a number of different economic concepts, such as the term of trade opportunity cost, and economies of scale.

For example, at point A where a quantity of 600 units of food can be produced along with 300 units of clothing, or at point B where 200 units of food can be produced along with

---

5   Even though the so-called trade deficit is not harmful one can occasionally observe negative reactions to it in financial markets. This happens because investors fear that rising trade deficits/capital surpluses may eventually trigger trade barriers that will harm the economy. Also, under certain circumstances, a rising trade deficit/capital surplus can be an early indicator of currency depreciation which can also spook investors.

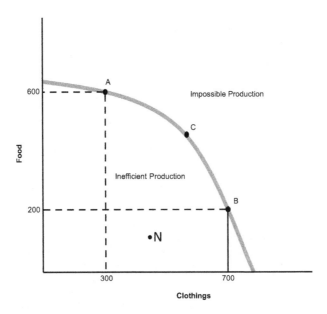

Figure 6.4   Production Possibility Curve

700 units of clothing. Hence, the opportunity cost to create 400 more units of food than at point B can be said to be 400 units of clothing. Point M represents a level of production that can't be achieved with the current level of resources and lies outside the production possibilities curve, while point N represents a level of inefficient production.

The slope at any point on the curve describes the marginal rate of substitution or how much of one good must be sacrificed to produce one more unit of the other. For example, if the slope at point C is 0.75, then that is equal to the opportunity cost at point C: to produce one more unit of clothing, 0.75 units of food will be taken out of the production schedule, or inversely, to produce one more unit of food, 1.33 units of clothing will not be produced.

Absolute advantage

A country is said to have an absolute advantage over another in the production of a good if it can produce the good with less resources (Das, 2008). Adam Smith, a Scottish social philosopher, first described the principle of absolute advantage in the context of international trade, using labor as the only input. Absolute advantage is determined by a simple comparison of labor productivities, so it is possible for a country to have no absolute advantage in anything, and therefore according to the theory of absolute advantage, no trade will occur with the other party. If a country can produce more output per unit of productive resources than its trading partner, then that country is said to have an absolute advantage in the terms of trade. Through specialization, different countries can produce and export goods where they have a natural or acquired absolute advantage and import those goods they don't specialize in. Table 6.6 shows an absolute advantage situation for the US with respect to China in aircraft and China with respect to the US in car production. Assume that, as shown in Table 6.6, China can produce 100 aircraft or

Table 6.6   Production Possibilities

	Aircraft per Unit of Input	Cars per Unit of Input
China	100	100,000
US	160	80,000

100,000 cars per one unit of productive resources, while the US can produce 160 aircraft or 80,000 cars per unit of productive resources. Clearly, the US has an absolute advantage in aircraft production while China has an absolute advantage in car production.

Using this information we can calculate the opportunity cost of producing each product in each country as demonstrated in Table 6.7. To produce one aircraft means China must forego 1,000 cars giving an opportunity cost of 1,000 cars. Furthermore, to produce one aircraft, the US must forego 500 cars giving an opportunity cost of 500 cars.

With an absolute advantage a country can charge a lower price than a competing trading country since more of the good with the absolute advantage can be produced with less resources. In the absence of free trade, in China one unit of aircraft will exchange for 1,000 cars and in the US one unit of aircraft will exchange for 500 cars. With the introduction of free trade, both the US and China can gain benefits. If the US can get more than 500 cars per unit of aircraft, then they will be better off than they would have been without trade. On the other hand, China has been giving up 1,000 cars to get one aircraft so if they can get an aircraft for less than 1,000 cars, then they will be better off. Therefore, the terms of trade should fall somewhere between 500 and 1,000 cars per unit of aircraft. Let's suppose that the terms of trade are 750 cars per unit of aircraft, we arrive at the following possibilities.

Table 6.7   Opportunity Cost

China	Opportunity cost
cost of producing 1 aircraft	1,000 cars
cost of producing 1 car	0.001 aircraft
US	
cost of producing 1 aircraft	500 cars
cost of producing 1 car	0.002 aircraft

In terms of the PPC introduced above, trade has effectively moved the curve to the right thereby making available more of both goods, as seen in Figure 6.5 and 6.6. Clearly, this simple example shows that there are significant gains to trade when each country has an absolute advantage in the production of one of the goods. However, what happens when one of the countries has an absolute advantage in the production of both goods? This situation is discussed in the next section.

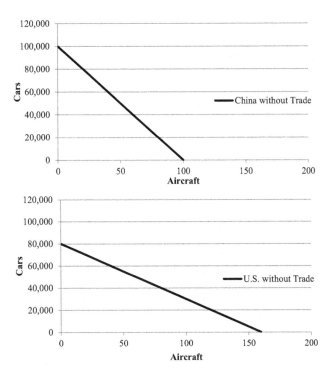

Figure 6.5   China/US Production Capabilities without Trade

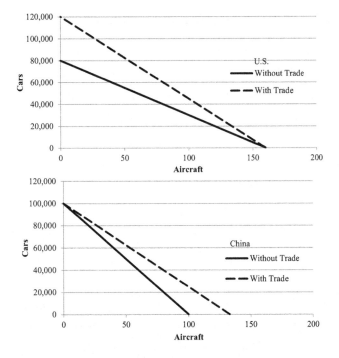

Figure 6.6   China/US Production Capabilities with/without Trade

Comparative advantage

In 1817, David Ricardo outlined the theory of comparative advantage; this theory shows how the gains from trade can still come about even if one country has an absolute advantage in the production of both goods (Miller, 2009). A country has a comparative advantage in that product with the lowest opportunity cost of production, so China should specialize in cars and the US should specialize in aircraft. To show this let us alter the numbers in the example we presented above to provide a simple proof of the benefits of international trade when one country as an absolute advantage in both goods. Suppose now that the US is a superior producer of both cars and airplanes.

Table 6.8 shows that the US has an absolute advantage in both products. Nonetheless, Ricardo argued that both countries can still benefit from trade. In the US, the opportunity cost of 1,000 cars is one aircraft. This includes the opportunity cost of production, meaning that if labor, energy, material, and other resources are reallocated away from car production to aircraft, then US firms can produce another aircraft but will lose 1,000 cars in its output. Likewise, if we reverse the reallocation and devote more resources to making cars then US firms can produce another car but will have to sacrifice 1/1000th of an aircraft. Meanwhile, in China the opportunity cost of an aircraft is 1,600 cars, with the opportunity cost of a car equal to 1/1600th of an airplane. Thus, even though the US has an absolute advantage in the production of both cars and aircraft, it has a comparative advantage only in the production of airplanes, since each plane costs 1,000 cars compared to a cost of 1,600 cars in China. On the other hand, China's firms have a comparative advantage in the production of cars since their cost is only 1/1600th of an airplane compared to 1/1000th of an airplane opportunity cost in the US.

Table 6.8   China/US Comparative Production Possibilities

	Aircraft per Unit of Input	Cars per Unit of Input
China	50	80,000
US	100	100,000

If unrestricted trade is now permitted then firms in each country will naturally shift production into the product with the comparative advantage. Given the numbers in Tables 6.9 and 6.10, it is clear that the terms of trade will fall somewhere between 1,000 and 1,600 cars for each aircraft. Thus, US firms will produce aircraft and "convert" each aircraft into (say) 1,300 cars through trade. Note that, without trade, the US could gain only 1,000 cars for each aircraft given up. By trading aircraft for cars the US is able to acquire more cars and US citizens can afford to consume both more cars and more aircraft.

China's wealth also increases since they can now trade 1,300 cars for an aircraft rather than having to give up 1,600 cars for each aircraft they directly produce. This simple mathematical proof confirms common sense—when a society produces goods at the lowest possible cost of resources, it is possible to produce more. Allowing free international trade unambiguously increases overall wealth. Likewise, it follows that international trade barriers reduce wealth—such as tariffs (special taxes on imports), import quotas, discriminatory regulation, or outright import bans.

Table 6.9   **Production Costs without Trade**

Production Costs *Without* International Trade	
US	China
1 aircraft costs 1,000 cars	1 aircraft costs 1,600 cars
1 car costs 1/1000th of an aircraft	1 car costs 1/1600th of an aircraft

Table 6.10   **Production Costs with Trade**

Production Costs *With* International Trade	
US	China
1 car costs 1/1300th of an aircraft	1 aircraft costs 1,300 cars
(as opposed to 1/1000th without trade)	(as opposed to 1,600 without trade)

Trade protections and trade barriers

Underlying most arguments against the free market is a lack of belief in freedom itself.

Milton Friedman, Chicago School of Economics, 1912–2006
Awarded Nobel Prize for Monetary Theory of Economics

Trade barriers are attempts by the government to regulate or restrict international trade. They all work on the same common principle of imposing an additional cost on the imported good that will result in an increased price for that good. A country can protect domestic industry by imposing a trade *tariff*, a *quota* or a trade *subsidy*.

Tariffs   A tariff, a common trade barrier, is an additional tax on imported goods. Some tariffs are intended to protect local industries from cheaper foreign goods, while others are an attempt by that government to generate revenue off the imported good. Tariffs are the easiest trade barrier to impose and can successfully reduce free trade (Husted and Melvin, 2007). Assume Figure 6.7 shows supply (S) and demand (D) for automobiles in the US. In the absence of trade restriction, cars will be imported at the prevailing market price of (P_{FT}). In this example:

- Q_4 = Total consumption in US
- Q_1 = Total domestic production
- $Q_4 - Q_1$ = Total import
- $Q_4 = Q_1 + (Q_4 - Q_1)$

P_{NT} and Q_{NT} are the unit price and quantity available in conditions where there is no international trade and hence no international competition. P_{FT} and Q_4 are the unit

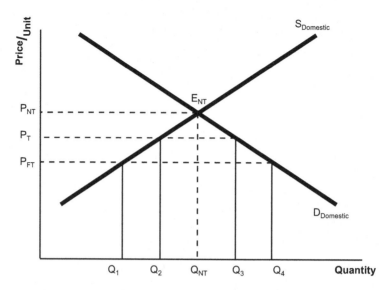

Figure 6.7   The Effects of International Trade Barriers to a Domestic Market

price and quantity available in conditions where there is free international trade and no international trade barriers.

Suppose the US imposes a tariff on imported cars. As a result of this, the price of cars will rise by the amount of the tariff, to P_T. In the absence of any retaliation by other countries, the increase in price of the car reduces consumption and increases domestic production. This would change the values to:

- Q_3 = Total consumption in US
- Q_2 = Total domestic production
- $(Q_3 - Q_2)$ = Total import
- $Q_3 = Q_2 + (Q_3 - Q_2)$

Quotas   The objective of an import quota, like a tariff, is to protect domestic producers from outside competition. Import quotas limit the quantity of various commodities that can be imported into a country during a specified period of time. In Figure 6.8, import quotas on cars will induce domestic manufacturers to expand production from Q_1 to Q_2, Q_{QU} being the total supply counting both domestic production as well as the quota. Dairy products in the US are a good example of this as they are subject to annual import quotas administered by the Department of Agriculture.

Other forms of trade protection   Another more controversial trade practice is so-called "dumping." Dumping supposedly occurs when a manufacturer in one country exports their product and sells it at an unreasonably low price, usually claimed to be below the cost of production, in another country. It is then usually alleged that workers in the second country may become unemployed because of this "unfair competition." However, free market advocates see dumping, if it actually occurs, as beneficial to consumers since it obviously lowers the price of the product in question. According to WTO regulations, a government may act against dumping when material injury to the domestic industry has

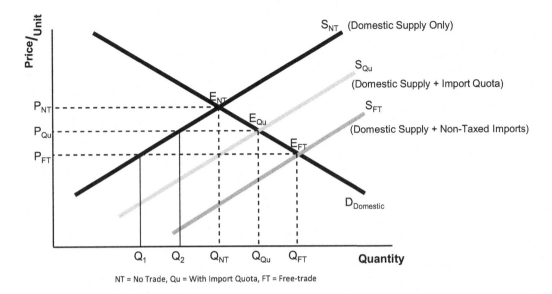

NT = No Trade, Qu = With Import Quota, FT = Free-trade

Figure 6.8   The Effects of Trade Quotas on a Domestic Market

occurred. This directive is intentionally vague so as to allow each government to decide on its own to what extent dumping will be permissible (van den Berg, 2004).

Another common practice is a subsidy, when a government gives financial aid to a company to help produce or purchase a product. Subsidies have been used to aid a new or failing industry if that industry cannot generate enough revenue to maintain itself and it is in the interest of the general public (Carbaugh, 2007). Amtrak, for example, has received government subsidies in the form of loans to keep its service running. In 2007, the Bush Administration agreed to pay $900 million of Amtrak's estimated $3.1 billion budget. One of the more controversial subsidies is that which allegedly exists between Airbus and its associated countries and this will be covered in detail in a subsequent section.

AIRCRAFT MANUFACTURING AND GOVERNMENTAL SUBSIDIES

Much of the discussion of government subsidies as it pertains to aircraft manufacturing tends to focus on Airbus and Boeing's Large Commercial Aircraft (LCA) manufacturing; however the smaller regional jet manufacturers are also the recipients of government aid and subsidies. The main regional jet manufacturers; Embraer out of Brazil and Bombardier from Canada, have been engaged in their own battle over unfair subsidies since the 1990s. Several rulings between 1999 and 2002 by the WTO found that both countries ran illegal subsidy programs for their aircraft manufacturing companies. In a 2000 ruling, the WTO found that Embraer has received illegal subsidies and ordered Brazil to modify its Proex export subsidies program. In 2001, the WTO ruled against Canada, just as it had ruled against Embraer, over low-interest loans from the Canadian government designed to aid Bombardier in gaining market share.

Boeing versus Airbus

It is commonly alleged that Airbus and Boeing are unfairly subsidized by their respective governments, allowing them to charge less than full price for their aircraft at a disadvantage to other manufacturers. These disputes originate from the first Airbus aircraft manufactured in the 1970s, the A300. In the case of Boeing, partisans argue that the US government should impose some sort of tariff on Airbus to offset the subsidy and afford Boeing a level playing field. Airbus disputes this analysis, but economists point out that even if the subsidy exists as alleged, it would be harmful to the US to impose a tariff. Throughout the history of aircraft manufacturing, a number of truces and agreements have been signed between Boeing, Airbus and their respective governments, including a 1992 bilateral agreement that established limits on the direct and indirect subsidies used for financing new aircraft. In 2005 (after a 2004 complaint against the EU in the WTO citing breaches of the 1992 agreement), both the governments of the US and the EU agreed (again) to stop subsidizing Boeing and Airbus for a short period of time while they try to resolve the decades-old dispute over billions in subsidies to the aircraft makers outside of the WTO.

If people in the US can buy aircraft more cheaply then they will have more resources to produce and consume additional aircraft and whatever else they might desire—total wealth clearly increases. The impact would generally be the same if the European taxpayers mailed checks directly to US consumers rather than giving money to Airbus. If there is a loser in this case then it would be the European taxpayers who see some of their wealth transferred from them to the US.

Boeing does suffer some lost sales from the subsidy, but Boeing's loss is less than the combined gain to airlines, air travelers, and the economy in general. Notice that the general effects of trade are not importantly affected by the existence or lack of a subsidy. Boeing loses less than the general economy gains if Airbus provides a better aircraft for the price. Regardless of whether that better price comes from a subsidy, hard work, luck, better technology, or whatever, the impact is the same.

Recent World Trade Organization rulings against Airbus/Boeing

The WTO requires proof of damage (a "measurable injury to competitors") before any government assistance can be termed a subsidy. The initial WTO report released in 2010 stemming from the 2004 complaint brought forth against Airbus included rulings on over 300 separate instances of alleged subsidization covering a period of almost 40 years. The subsidies in question relate to the entire family of Airbus products (A200 through to the A380) and fall under five general categories:

1. *Launch aid or "member state financing"*: the provision of financing by the Governments of France, Germany, Spain, and the UK (hereinafter, the "member States") to Airbus for the purpose of developing the A300, A310, A320, A330/A340, A350, and A380.
2. *Loans from the European Investment Bank (EIB)*: 12 loans provided by the EIB to Airbus companies between 1988 and 2002 for aircraft design, development, and other purposes.
3. *Infrastructure and infrastructure-related grants:* the provision of goods and services, as well as grants, to develop and upgrade Airbus manufacturing sites.

4. *Corporate restructuring measures:* the provision to Airbus by Germany[6] and France of equity infusions, debt forgiveness, and grants through government-owned and government-controlled banks.
5. *Research and technological development funding:* the provision of grants and loans undertaken by Airbus.

These WTO rulings were appealed by the EU. On May 18, 2011 the Appellate Body upheld the Panel's finding that certain subsidies provided by the EU caused serious prejudice to the interests of the US. The principal subsidies covered by the ruling include financing arrangements (known as "launch aid" or "member state financing") provided by France, Germany, Spain, and the UK for the development of the A300 through to the A380 aircraft. The Appellate Body found that the effect of the subsidies was to displace exports of Boeing single-aisle and twin-aisle aircraft from the EU, Chinese, and Korean markets and Boeing single-aisle aircraft from the Australian market. The Appellate Body found the amount of illegal subsidies and aid to be $18 billion for Airbus. It should be noted that a separate dispute brought by the EU against the US for subsidies allegedly provided to Boeing is currently before the Appellate Body.[7]

INTERNATIONAL TRADE POLICY IN AIR TRAVEL— OPTIMALITY VERSUS POLITICAL REALITIES

Economists generally agree that free trade is the best policy in aviation and most everything else. Ideally, foreign airlines from friendly nations should be allowed to freely compete; indeed, there is no reason for government policy to favor domestic airlines over foreign airlines. Implementation of this policy would maximize competition and efficiency in air travel—prices would be lower and there would be more variety of service and consumers would have more choices. Overall wealth would increase because air travel would be more efficient in its use of resources and the improved transportation system would help virtually all industries to be more efficient. A more efficient air travel industry has impacts analogous to a more efficient road system—it allows firms to expand into more output markets, to gather resources from more input markets, and, as appropriate, take more advantage of economies of scale in production.

We mentioned earlier that it has been estimated that establishing free international trade in 2004 would have increased annual world income by about a half trillion dollars. Formal estimates for the impact of free competition in air travel alone are not available but income would certainly increase by many billions.

Given the huge benefits of free trade in air travel, why have politicians in most nations failed to implement it? Think back to the public choice principles discussed in Chapter 2.

---

6   Panel Report, para. 2.5(e). The United States challenged specific transactions arising from the German Government's restructuring of Deutsche Airbus in the late 1980s, including the 1989 acquisition by the German Government, through the development bank Kreditanstalt für Wiederaufbau ("KfW"), of a 20 percent equity interest in Deutsche Airbus (ibid., paras. 2.5(e) and 7.1250), the 1992 sale by KfW of that interest to Messerschmitt-Bölkow-Blohm GmbH ("MBB"), the parent company of Deutsche Airbus (ibid., paras. 2.5(e) and 7.1253), and the forgiveness by the German Government, in 1998, of debt owed by Deutsche Airbus in the amount of DM 7.7 billion (ibid., paras. 2.5(d) and 7.1308).
7   World Trade Organization Report of the Appellate Body Measures Affecting Trade in Large Civil Aircraft, 18 May, 2011.

Most citizens are "rationally ignorant," and are unaware of the benefits of free international trade; there is no strong consumer movement clamoring for free trade in air travel or anything else. However, all firms, including airlines, like to avoid increased competition as much as possible. Consequently, governments have a tendency to act in the interests of airlines rather than in the interests of the nation as a whole. In other words, absent a well-informed public, politicians tend to give into the special interests of the domestic airline industry.

This is reflected in the typical language of trade politics. Any time foreign firms are allowed to compete in a market, that is to reduce prices and increase the importing nation's wealth, politicians refer to this as a "trade concession," something politicians appear to reluctantly agree to in exchange for rights for domestic firms to go compete in the other country. The entire attitude of politicians is that increasing import competition is an awful result that must be tolerated in order to negotiate export rights for domestic firms. This, understandably, is the perspective of the domestic firms these politicians seek to please—access to more markets is welcomed, more competition and lower prices are not welcomed, and even though national wealth increases, the wealth of companies facing more competition usually does not. But it is ironic that politicians who increase national well-being by reducing trade barriers feel compelled to cloak these good steps by calling them concessions.

The optimal policy favored by economists would be for a country to unilaterally open its own market to foreign competition. We would want to see increased competition and lower prices as soon as possible, with no negotiations necessary.

It is useful to summarize the likely results of such a policy in the US market. If the US allowed *cabotage*, foreign carriers handling domestic traffic within the country, the immediate results, though beneficial, are unlikely to be spectacular. US airlines have been deregulated for a long time and are quite efficient by world standards—the market is not an easy one to make a profit in. Thus, foreign airlines would not be anxiously pouring into this market. Some, Virgin America Inc. for instance, would enter in a major way. Others might enter a few markets they already have some link to. For example, there are some *blind routes* that are already being flown by foreign carriers where they are currently prohibited from handling domestic traffic. That is, an airline might already fly, for example, from London to New York to Los Angeles but is currently prohibited from picking up passengers in New York and dropping them off in Los Angeles. With the prohibition lifted they could freely market that segment.

From the viewpoint of US domestic carriers' self-interest, the disadvantage of increased competition would be at least partially offset by the injection of foreign capital. In abolishing the laws prohibiting cabotage the US would inherently also be abolishing the laws prohibiting foreign controlling investment in US airlines. Some cash-strapped carriers would welcome some sort of partnership, perhaps even a formal merger, with a wealthier foreign airline. Likewise, the best strategy for a foreign airline looking to break into the US market might often be to team up with a US partner. Current airline alliances achieve only a very limited amount of this sort of cooperation.

Since there would be lower prices and greater efficiency there would be more airline passengers and therefore higher demand for labor in the airline industry; employment in the US airline industry would rise. Some of the new jobs might go to foreign workers brought in by foreign-based airlines but net airline employment for US workers is still likely to rise since airlines, like most service industries, prefer to hire locally in order to promote better customer relations.[8]

The impact on average wages in the airlines is more complex. In competitive labor markets, rising labor demand would normally bid up the wages. However, much of the US airline industry is dominated by unusually powerful labor unions. Essentially these unions band workers together to bargain as a labor monopoly and thereby raise wages above competitive levels. Since increased competition tends to erode union monopoly power, the effect from cabotage could theoretically reduce average wages. That is, rising labor demand tends to raise wages while increasing competition tends to reduce union power and union wages so the overall wage effects of cabotage are not immediately clear.

It is beneficial to review the US experience since airline deregulation occurred in 1978. In this case, the effect of increased labor demand swamped the effect of eroded union monopoly power, and wages generally rose after deregulation. Even after major union concessions following the 9/11 terrorist attacks, US airline employees were compensated at a level almost twice the average for all US industries (Eldad, 2005).

Since the move from regulation to deregulation probably impacted airline competition far more dramatically than cabotage would, it seems reasonable to conclude that airline wages would not be driven down following the opening of the US market. Of course, the effects on the US economy overall are unambiguously beneficial, though, as mentioned earlier, these effects are unlikely to be dramatic (at least in the short run) given the relatively efficient state of the US airline industry and the likelihood of only moderate initial new entry.

The same general impact, naturally, would occur in any country that opened its airline market, though the effects would often be more intense. In a number of cases it is quite possible that efficient foreign carriers would drive prices so low that flag carriers could not survive. To most economists this loss would not be at all tragic, the gains for the broader economy in a more efficient air transport system would easily exceed the sentimental regret at the loss of an inefficient flag carrier—there is no more reason to insist that air travel be supplied internally than there is to insist on locally grown pickles.

But does "national pride" justify preserving an inefficient flag carrier? One response is to note that if the market is opened, the flag carrier is free to market itself to travelers on the basis of national pride—if consumers feel it is important to support the flag carrier they can do so. Naturally, there are likely to be limits on how much of premium consumers would be willing to pay—a flag carrier that is vastly less efficient than the competition is probably doomed. But if so this clearly implies that the people don't value "national pride" that much; in some sense, this argument in defense of the flag carrier is inherently invalid if consumers won't freely support it.

It may also be possible to compromise with and overcome the politics of protectionism in this sort of case by requiring foreign carriers to exclusively employ native-born employees, and maybe use aircraft painted in the home nation's colors with appropriate insignia. This would preserve much of the feel of having a flag carrier while still enjoying at least some of the benefits of open competition.

It is ironic that government interference has rendered air travel, an industry that should naturally be more global than most, far less global in its operations than virtually any other large industry. The pace of government reform is mostly very slow but the direction

---

8   Although most economists would prefer to avoid added regulations, if employment politics are an obstacle to establishing cabotage then government regulation could stipulate a certain level of "native employment."

of the trend is at least somewhat encouraging. Deregulation of air travel within the EU has been impressive. The US and Europe seem to be slowly finding their way to establishing a Trans-Atlantic Common Aviation Area (European Commission, Directorate General, Energy and Transport, 2004). A large step in this direction occurred when the EU–US Open Skies Agreement was signed in 2007 and became effective in 2008. Perhaps, open skies in lieu of cabotage will eventually become routine and allow a true flourishing of the air travel industry; one day it may truly be a very small world. Foreign ownership of US airlines is another restriction that prevents a full open market for airlines in the US airline industry. The US law limits the amount of foreign ownership in its domestic airlines to a maximum of 49 percent, with a maximum of 25 percent control. Nonetheless, many other countries have similar protection of their domestic markets.

FOREIGN CURRENCY AND EXCHANGE RATES

International trade (imports and exports) requires foreign currency in order to complete the transactions. When goods and services are bought in a country, they are bought using that country's currency. To obtain foreign currency, one must trade in one's own local currency via the currency exchange rate. The exchange rate or the price of one nation's currency in terms of another nation's is a central concept in international finance. Exchange rates are influenced by a wide range of different factors, and the importance of each differs from country to country. For example, one factor affecting the exchange rate between currencies is the rate of inflation. As a general rule, the currency from countries with lower inflation rates rise in value, while the currency from the countries with higher inflation rates fall in value. Therefore, the products from countries with high inflation rates become more attractive than the products from countries with lower inflation rates because of the relative increase in purchasing power of the lower inflating currency.

Another factor affecting the exchange rate between currencies is the interest rates. Everything being equal, a higher interest on US securities (compared to say Canadian securities) would make investment in US securities more attractive. Therefore, an increase in the US interest rate raises the flow of Canadian dollars into US securities, and decreases the outflow of American dollars to Canadian securities. This increased flow of funds into the US economy would increase the value of the US dollar and decrease the value of the Canadian dollar. Hence, the ratio of US dollar to Canadian dollar, as it is represented in the foreign exchange market, would decrease. Finally, the balances of payments of a country with the rest of the world influence the country's exchange rate. Demand for foreign currency arises from the import of foreign merchandise, the payment for foreign services, or from the redemption of foreign capital obligations. The supply of foreign currency, on the other hand, comes from the export of goods and services, or from an inflow of foreign capital.

Exchange rate quotes

An exchange rate quotation is the value of one currency in terms of another. For example, a quotation of 1.06 CAD/USD signifies that 1 Canadian dollar will be needed to acquire

1.06 US[9] Dollars. In this quotation, the price currency is CAD (Canadian Dollars) and the unit or base currency is USD (US Dollars). When the base currency is the home currency (the US in this example), it is known as a direct quotation. Using direct quotation, the exchange rate decreases when the home currency appreciates (strengthens) and increases when the home currency depreciates.

Exports generally increase the exchange rate of the domestic currency (appreciation) since they cause more foreign currency to come into the domestic country. For example, if Japan Airlines bought large numbers of wide-body aircraft from the US-based Boeing Company, it would have to convert its Japanese yen to US dollars to complete the transaction. This would result in the US banks receiving the Japanese yen and exchanging them for the requisite amount of US paper dollars at the current exchange rate. This increased demand for US dollars acts like any increase in demand; that is, the price of dollars in yen is increased. This increase in price is pictured in Figure 6.9 below.

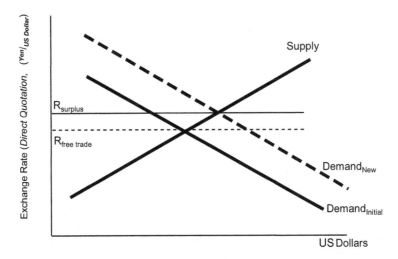

Figure 6.9   Demand for Dollars Causes the Value of US Dollar to Rise and an Increase in the Exchange Rate

Imports reduce foreign currency reserves causing a decrease in the foreign currency supply (supply shifts to the left), driving up the value of foreign currency relative to the local currency (depreciation), and decreasing the exchange rate. For example, suppose Wal-Mart buys consumer electronics from the Sony Corporation of Japan. To complete the transaction, Wal-Mart must convert its US dollars to Japanese yen. The Bank of Japan would take the US dollars and issue the requisite amount of Japanese yen to Wal-Mart at the current exchange rate. The supply of yen relative to the dollar would be depleted so that the supply curve for yen would shift to the left. Alternatively, this would cause a relative increase in the supply curve for the dollar, depreciating the dollar, and increasing the dollar price for yen as shown below (Figure 6.10).

---

9   On June 22, 2012, the direct quote was .9695 CAD/USD; within the 52-week range of 0.9384-1.0630.

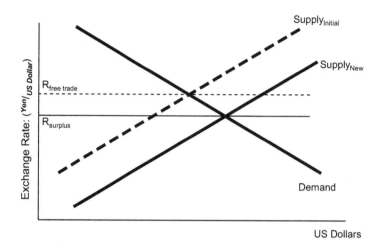

Figure 6.10   **Surplus in Dollars Causes the Value of US Dollar to Drop, Exchange Rate Decreases**

Exchange rate regimes

An exchange rate regime is how a country manages its currency in relation to other currencies. There is a spectrum across which the price of a currency can be determined against another: fixed exchange rates are at one extreme end of the spectrum, and floating (adjustable) rates are at the other.

Fixed (pegged) exchange rate   The concept behind fixed exchange rates is simple: a government or central bank ties its official exchange rate to another country's currency (or the price of gold). This allows the currency value to fluctuate within a very narrow range and is also called a pegged exchange rate. With this exchange rate, a government or central bank determines its currency's worth in terms of either a fixed weight of gold or a fixed amount of another currency. Prior to 1971, most countries operated under a form of this system known as the Bretton-Woods system where the exchange rates of member countries were fixed against the US dollar which was in turn fixed against a fixed amount of gold.

The Gold Standard is a form of fixed exchange rates where that currency can be converted into gold at a fixed rate. The US switched to the Gold Standard *de jure* in 1900 when Congress passed the Gold Standard Act but had already adopted the Gold Standard de facto in 1834. Between 1880 and 1914, the majority of countries stuck to the Gold Standard as a means of valuing their currency. The Gold Standard period broke down in the First World War, but dissatisfaction with high inflation rates in the 1970s and 1980s renewed moderate interest in the Gold Standard, as inflation was not a problem associated with it. Because the government of a country can only print as much money as it has in gold, inflation is discouraged as well as high debt levels.

The Gold Standard, because of the limited stock of gold, (and the fact that any new discoveries would be relatively small compared to the accumulated stock) ensured that the money supply, and hence the price level, would not vary much and caused price levels around the world to move together. There also existed several problems with the Gold Standard: governments seeking to protect their reserves created large economic fluctuations. For example, the US, during the period from 1890 to 1905 when it adhered to the Gold Standard, suffered five major recessions directly tied to this. However, the Gold Standard effectively vanished in 1971 when the last formal link (the Bretton-Woods System) was ended.

Floating (adjustable) rates   As the Bretton-Woods System collapsed in 1971, a new regime of exchange rates was implemented: floating exchange rates. With a floating exchange rate, the country's currency is allowed to vary according to the foreign exchange market (that is, the interactions of banks, firms and other institutions that sell and buy currency). Despite the move to floating rate systems, only a few of the world's currencies are classified as true floating exchange rates. This includes the US, Canada, Australia, Britain, and the European Monetary Union.

SUMMARY

This chapter discusses the international aspect of the airline industry. The international character of aviation as an industry is shown to be particularly relevant given the characteristics of the industry; that is, its ability to easily cross international borders more quickly than any other mode of transportation. Economists generally argue that the world at large will benefit from free trade and that trade liberalization can promote development. Trade protectionists claim that trade protection in most cases benefits all the domestic economies, whereas the majority of economists argue that trade protection only benefits individual industries and is probably not sustainable in the long run. Other arguments for and against free trade are explored along with a presentation of the theories of absolute and comparative advantage. These theories are illustrated with quantitative examples. The determination of foreign exchange rates is analyzed using standard supply and demand models along with historical methods for controlling exchange rates. Finally, the topic of free trade agreements is discussed.

DISCUSSION QUESTIONS

1. Explain why it is particularly important to understand the international aspects of the aviation industry.
2. Consider the following situation: Country A takes 50 hours to produce ten aircraft or five jet engines. Country B takes 50 hours to produce five aircraft or ten jet engines. Which country has an absolute advantage in which product? What are the domestic terms of trade for each country? Draw the production possibilities frontier for each country if they do not engage in trade. Draw the production possibilities frontier if the two countries engage in trade.
3. Now assume that Country A takes 100 hours to produce 20 aircraft or ten jet engines and Country B takes 100 hours to produce 15 aircraft or five jet engines. Which

country has an absolute advantage in which product? Does either country have a comparative advantage in one of the products? If so, which product is it? Can trade take place between the two countries and what would be the limits of the terms of trade if it were to take place?

4. Why do nations trade? Briefly explain the arguments for and against free trade.
5. Name and describe the rationale behind the three types of barriers to trade that a country may impose.
6. How have the effects of free trade impacted the aviation industry?
7. Generally speaking, what happens to the value of a country's currency when imports exceed exports and how, in terms of supply and demand, does this take place?
8. What is the difference between a free-floating currency and a pegged exchange rate?

APPENDIX: INTERNATIONAL FREE TRADE AGREEMENTS

Labor and Regional trade agreements (RTA) among different countries are the framework by which most of the world's economy is organized. Over the last few years, the number of RTAs has significantly increased but their actual effectiveness is controversial. While the main purpose of many RTAs is to reduce trade barriers and encourage free trade, an increasing number of agreements also deal with other trade-related issues, such as environment and labor issues. This section reviews and discusses several different regional trade agreements such as North American Free Trade, Dominican Republic-Central America Free Trade, the US, Jordan Free Trade Agreement, and US Australia Free Trade Agreements.

- Dominican Republic-Central America Free Trade Agreement (CAFTA-DR).
  On August 5, 2004, the US signed the Dominican Republic-Central America-United States Free Trade Agreement (CAFTA-DR) with five Central American countries (Costa Rica, El Salvador, Guatemala, Honduras, and Nicaragua) and the Dominican Republic. The CAFTA-DR is the first free trade agreement between the US and a group of smaller developing economies. CAFTA aims to create a trade free zone amongst its member nations like that of NAFTA. Eighty percent of tariffs on US exports were eliminated immediately, with the remaining 20 percent to be phased out over time.

- North American Free Trade Agreement (NAFTA).
  NAFTA was originally signed on January 1, 1994 as a trade agreement that would link Canada, the US, and Mexico. The immediate effects of NAFTA were an end to tariffs on goods (immediately for some, and over time for others). As a result of NAFTA, imports and exports from the US to Canada and Mexico increased from one-quarter to one-third. One of the major fears with the implementation of NAFTA was the potential loss of jobs in the US to Mexico. And, while manufacturing jobs in the US did decrease, this loss was compensated for by the creation in the US economy of more than 2 million jobs a year from 1994 to 2000 in other industries.

Destination	NAFTA			Other Origin/Destination		
	Value*	Annual Percentage Change		Value*	Annual Percentage Change	
	2010	2009	2010	2010	2009	2010
Exports						
Canada	294	-33	22	94	-22	24
Mexico	250	-20	29	49	-28	35
United States	956	-24	24	1,009	-18	21
NAFTA	413	-19	24	866	-17	20
Imports						
Canada	229	-22	22	174	-20	21
Mexico	159	-26	28	152	-23	29
United States	900	-26	25	1782	-25	22
NAFTA	512	-27	26	1457	-26	22

* Billions of US dollars

- The US–Jordan Free Trade Agreement (USJFTA).
  The agreement between the US and Jordan was fully implemented on January 1, 2010. This was the first agreement the US made with an Arab nation, though it was the third free trade agreement implemented by the US. The goal of this agreement is to eliminate tariff and non-tariff barriers on industrial goods and agricultural products over the following ten years of implementation. In addition, the Qualifying Industrial Zones (QIZs), established by Congress in 1996, allow products to enter the US duty-free if manufactured in Israel, Jordan, Egypt, or the West Bank and Gaza.

- US–Australia Free Trade Agreement (AUSFTA).
  FTA is an agreement (modeled after NAFTA) between Australia and the US that was implemented on January 1, 2005. The agreement, like others, sought to reduce barriers to trade; however, following its inception, trade from Australia to the US declined, while trade from the US to Australia has continued to increase.

- European Union (EU).
  The EU was established in 1993 by the Treaty on European Union (The Maastricht Treaty) and is the successor to the six-member European Economic Community (EEC). The EEC was an organization established by the Treaty of Rome in 1957 between Belgium, France, Italy, Luxembourg, the Netherlands, and West Germany, known informally as the Common Market (the Six). The EU is a confederation run by 27 member nations and mostly located in continental Europe. Representatives of these nations make decisions partly by unanimity, partly by majority vote and partly by delegation to lesser bodies. It has its own flag, anthem, central bank, currency, elected parliament, Supreme Court, and common foreign and security policy.[10]

---

10   As of today, the EU has 27 members including; Austria, Belgium, Bulgaria, Cyprus, Czech Republic,

184 INTRODUCTION TO AIR TRANSPORT ECONOMICS

Citizens belonging to EU member states are also EU citizens. They are allowed to invest, live, travel, and work in all member states except for temporary restrictions on newly inducted member states. With a few exceptions, systematic border controls were mostly abolished by the Schengen Agreement in 1985. The EU economy relies on a complex web of multilateral trade agreements, international rules, and standards that cover products, markets, investment, health, and environmental issues. There are still concerns about the nature of the union being intergovernmental (unanimous voting only) or supra nationalist (majority votes imposed on all members); however, the EU has proved to be a mix of both. In the last five decades the EU has shown remarkable success in achieving economic prosperity and stability on a continental scale. It now accounts for about 30 percent of global GDP and 20 percent of global trade flows and the euro has become an important international currency. The example of the EU is now considered a working model for regional integration.

- Union of South American Nations (UNASUL).
Loosely modeled on the EU, the Union of South American Nations (UNASUL) will combine the free trade organizations of MERCOSUR (Southern Common Market) and the Andean Community, plus the three countries of Chile, Guyana, and Suriname by the end of 2007. The Union's headquarters will be located in Quito, the capital of Ecuador. Formerly known as the South American Community of Nations, it was renamed at the First South American Energy Summit on April 16, 2007. The foundation of the Union was formally announced at the Third South American Summit, on 8 December, 2004. Representatives from 12 South American nations signed the Cuzco Declaration, a two-page statement of intent. An important operating condition is the use of institutions belonging to the pre-existing trade blocs (MERCOSUR and Andean Communities) to establish the union. So far, most of the countries within the union have waived visa requirements for travel and there is an established consensus for a single South American currency.

- AFTA (ASEAN Free Trade Area).
The Association of Southeast Asian Nations (ASEAN) is an organization of ten countries located in Southeast Asia. The ASEAN Free Trade Area (AFTA) agreement was signed on January 28, 1992 in Singapore by the member nations. When the AFTA agreement was originally signed, ASEAN had six members, namely, Brunei, Indonesia, Malaysia, the Philippines, Singapore, and Thailand. New member nations were required to sign the agreement upon entry into ASEAN and were given timeframes in which to meet AFTA's tariff reduction obligations. Beginning in 1997, ASEAN began creating organizations within its framework with the intention of accelerating south East Asian integration to include the People's Republic of China, Japan, South Korea, India, Australia, and New Zealand.

- African Union (AU).
The African Union (AU) is an organization of 54 African states created in 2001 from the amalgamation of various pre-existing regional blocs. The AU preserved the

Denmark, Estonia, Finland, France, Germany, Greece, Hungary, Ireland, Italy, Latvia, Lithuania, Luxembourg, Malta, Netherlands, Poland, Portugal, Romania, Slovakia, Slovenia, Spain, Sweden, and the United Kingdom.

free trade areas established by these pre-existing blocs and will be combining and expanding them under the banner of the African Economic Community. The AU aims to have a single currency and a sustainable economy by bringing an end to intra-African conflict and creating an effective common market.

- Greater Arab Free Trade Area (GAFTA).
  The Greater Arab Free Trade Area (GAFTA) came into existence on January, 2005. Similar to ASEAN, GAFTA was an agreement that was initially signed by 17 Arab League members; the agreement aimed at decreasing the customs on local production and the creation of an Arab Free Zone for exports and imports between members. The GAFTA rules involve member nations coordinating their tariff programs, maintaining common standards for specifications and restrictions on goods, promoting the private sector across all member countries, maintaining a base of communication, and decreasing customs duties. The members participate in 96 percent of the total internal Arab trade, and 95 percent with the rest of the world. Overall, the agreement would tie the pre-existing African-based Agadir Agreement and Middle Eastern GCC organizations together to form one large free trade area.

- South Asian Free Trade Area (SAFTA).
  Born out of the efforts of the South Asian Association for Regional Cooperation (SAARC), the South Asian Free Trade Area was an agreement reached on January 6, 2004 for the creation of a free trade area involving India, Pakistan, Nepal, Sri Lanka, Bangladesh, Bhutan, Maldives, and Afghanistan. Its influence is the largest of any regional organization in terms of population with almost 1.5 billion people. The SAARC members have frequently expressed their unwillingness to sign free trade agreements. Though India has several trade pacts with Maldives, Nepal, Bhutan, and Sri Lanka, similar trade agreements with Pakistan and Bangladesh have been stalled due to political and economic concerns on both sides. However, even with this slow progress, the foreign ministers of the member countries have signed a framework agreement to bring their duties down to 20 percent by the end of 2007 and zero customs duty on the trade of almost all products in the region by the end of 2012.

- Trans-Pacific Strategic Economic Partnership (TP SEP).
  The Trans-Pacific Strategic Economic Partnership (TP SEP) is a free trade agreement between Brunei, Chile, New Zealand, and Singapore which was signed on June 3, 2005. The TP SEP was previously known as the Pacific Three Closer Economic Partnership (P3-CEP). Despite cultural and geographical differences, the four member countries share the similarities of being relatively small countries (as compared to some of their trading partners) and are members of the Asia-Pacific Economic Cooperation (APEC). It aims to reduce all trade tariffs by 90 percent by January 2006 and to completely eliminate them by 2015. Because of an accession clause within the agreement, it has the potential to include other nations as well. Countries belonging to the 21 member APEC have shown some interest in this agreement.

- Pacific Regional Trade Agreement (PARTA).
  The Pacific Regional Trade Agreement (PARTA) was founded in 1971 and is aimed at increasing trade between the island nations of the Pacific. Australia, New Zealand, New Caledonia, and French Polynesia are associate members of PARTA. Most of the member island countries are smaller in population and some are quite poor. Australia and New Zealand have much larger populations and are significantly wealthier. Australia's population is around twice that of the other 15 members combined and its economy is five times larger. Because of their position, the poorer countries are awarded concessional tariff deals to ease their exports.

- Caribbean Community (CARICOM).
  The Caribbean Community was originally called the Caribbean Community and Common Market and was established in 1973. Currently its membership has grown to a total of 20 countries (15 members and five associate members), the majority of which have joined the CARICOM Single Market and Economy (CSME) and the CARICOM Common Passport. Moreover, CARICOM is representing all its members as one single entity for bilateral agreements with the EU, members of NAFTA, and members of UNASUL.[11] Twelve of the CARICOM countries have signed an oil alliance with Venezuela (Petrocaribe) which permits them to purchase oil on conditions of preferential payment.

- Central American Common Market (CACM).
  The Central American Common Market (CACM) is an economic trade organization that was established in 1960 between the nations of Guatemala, El Salvador, Honduras, and Nicaragua. Costa Rica joined the CACM in 1963. The organization collapsed in 1969 due to a war between Honduras and El Salvador, but was reinstated in 1991. Because of its inability to settle trade disputes, the CACM has not been able to achieve all the goals of unification that were espoused in its founding. But despite its shortcomings, the CACM has succeeded in removing duties on most products traded between its members, unifying external tariffs, and increasing trade between its member nations.

REFERENCES

Carbaugh, R. (2007). *International Economics* (11th ed.). Mason, OH: South-Western.

Das, M. (2008). Absolute and Comparative Advantage. In W. Darity, *International Encyclopedia of the Social Sciences* (pp. 1–2). Detroit: Macmillan Reference USA.

Eldad, B. (2005). *Evolution of the US Airline Industry*. New York: Springer, p. 251,

Elwell, C. (2004). *The US Trade Deficit: Causes, Consequences, and Cures* (Order Code RL31032). Washington, DC: The Library of Congress–Congressional Research Service.

European Commission–Directorate General, Energy and Transport. (2004). *International Aviation Agreements: Opening the Market for Efficient Air Travel*. Luxembourg: Office for Official Publications of the European Communities.

---

11   Caribbean Community (CARICOM) Secretariat, 2012.

Hufbauer, C. and Elliott, A. (1994). *Measuring the Costs of Protection in the United States.* Washington, DC: Peter G. Peterson Institute for International Economics.

Husted, S. and Melvin, M. (2007). *International Economics* (7th ed.). Boston, MA: Addison-Wesley.

Miller, R. (2009). *Economics Today* (15th ed.). Boston, MA: Pearson Education, p. 559.

Pugel, T. (2007) *International Economics* (13th ed.). New York: McGraw Hill.

US Air Force–Air Mobility Command. (2006). *US Air Force Fact Sheet: Civil Reserve Air Fleet.* Retrieved June 13, 2007 from US Air Force, Air Mobility Command Library: http://www.amc.af.mil/library/factsheets/factsheet.asp?id=234.

US Census Bureau: Foreign Trade Division. (2011). *Top Trading Partners—Total Trade, Exports, Imports: Year-to-date December 2011.* Retrieved June 23, 2012 from US Census Bureau, Foreign Trade Statistics: http://www.census.gov/foreign-trade/statistics/highlights/top/top0612.html.

US Department of Commerce–Bureau of Economic Analysis. (2007). *US International Transactions.* Retrieved June 13, 2007 from Bureau of Economic Analysis: International Economic Accounts.

Van den Berg, H. (2004). *International Economics.* New York: McGraw Hill.

7

Open Skies and Global Alliances

It is probably not love that makes the world go around, but rather those mutually supportive ALLIANCES through which partners recognize their dependence on each other for the achievement of shared and private goals.

Fred A. Allen, American Humorist (1894–1956)

Forces in the global marketplace increasingly require airlines to collaborate with other airlines from different countries for market efficiency, to increase profitability, to reduce costs, and to counter competition. It can also be argued that alliances amongst airlines may lead to more itinerary choices for their passengers. In addition, one of the great benefits of aviation has been its ability to make the world smaller and promote globalization. Also, from its very roots the international regulatory environment has been critical to the success of the aviation industry. Therefore, this chapter will explore two major themes encompassed in international aviation: open skies and global alliances. "Open skies" incorporates the legal framework surrounding the rights granted to airlines with its roots tracing back to the original air transportation agreements between countries. (While the true ideal of open skies has yet to be achieved, it is the goal for international aviation.) "Global alliances" refer to the arrangements that airlines have made with one another to expand their scope on a global basis. The specific topics that are covered in the chapter are listed below:

- Chronology of International Air Transport Agreements
- Chronology of International Air Transport Agreements
- Bilateral and Multi-lateral Air Service Agreements
  - Freedoms of air transportation
  - Bermuda Agreement
  - Open skies agreements
  - Characteristics of open skies
  - Benefits of open skies
- Open Skies in Europe
- Open Skies in Asia
- Global Airline Alliances
  - History of global airline alliances
  - Global alliances and competitiveness
  - Benefits of global alliances
  - Disadvantages of global alliances
  - Future for global alliances

- Summary
- Discussion Questions

CHRONOLOGY OF INTERNATIONAL AIR TRANSPORT AGREEMENTS

The first international agreement concerning air transportation occurred shortly after the end of the First World War in Paris. With the tremendous leap in aviation that occurred during the First World War, delegates from 26 countries drew up the Convention relating to the Regulation of Air Navigation (US Centennial of Flight Commission (USCOF), 2006). The Convention voted to give each nation, "complete and exclusive sovereignty over the airspace above its territory" (USCOF, 2006). This was the first time countries were provided with an internationally recognized legal authority over their airspace, enabling them to allow or disallow aviation access into their country. At the end, neither Russia nor the US signed the Paris Convention of 1919 (USCOF, 2006).

The US signed its first international aviation agreement at the Havana Convention on Civil Aviation of 1928. This agreement guaranteed the innocent right of passage as well as the formulation of rules concerning such issues as aircraft navigation, landing facilities, and pilot standards. The Havana Convention also provided the right for each country to set the route to be flown over its territory. In total, the US and 20 other western hemisphere countries signed and ratified the Havana Convention of 1928 (USCOF, 2006).

The Convention for the Unification of Certain Rules Relating to International Carriage by Air was convened on October 12, 1929 in Warsaw, Poland. One of the major results of this convention was a formal definition of "international carriage." Article 1 of the Warsaw Convention states:

> International carriage means any carriage in which, according to the contract made by the parties, the place of departure and the place of destination whether or not there be a break in the carriage or a transshipment, are situated either within the territories of two High Contracting Parties, or within the territory of a single High Contracting Party, if there is an agreed stopping place within a territory subject to the sovereignty, suzerainty, mandate or authority of another Power, even though that Power is not a party to this Convention. A carriage without such an agreed stopping place between territories subject to the sovereignty, suzerainty, mandate or authority of the same High Contracting Party is not deemed to be international for the purposes of this Convention (Warsaw Convention (WC), 2006).

The Convention also established a general set of guidelines for the operation of the commercial air transportation industry for international flights. For example, Article 3 describes the requirements for a passenger ticket, while Article 4 outlines what needs to be included on a luggage tag (WC, 2006). One of the more practical outcomes of the Warsaw Convention concerned air carrier liability. Article 17 states that the air carrier is liable for:

> ...damage sustained in the event of the death or wounding of a passenger or any other bodily injury suffered by a passenger, if the accident which caused the damage so sustained took place on board the aircraft or in the course of any of the operations of embarking or disembarking (WC, 2006).

This article states that the air carrier is liable for death or bodily injury suffered by a passenger on an air carrier's flight. However, Articles 20 and 21 provide escape clauses for the airlines if it is determined that they took all measures necessary to avoid the loss or there was some contributory negligence on behalf the person (WC, 2006). Such issues are not as important today since aviation is extremely safe, but it was important at the time of this convention.

Eventually, the Warsaw Convention was completely overhauled by the Montreal Convention of 1999 which is the current convention that governs international carriage liability. Article 21 of the Montreal Convention states that an air carrier has unlimited liability; that is, there is no maximum cap on the payment, and that in the event of death, the minimum the airline must compensate is 100,000 Special Drawings Rights (SDRs) (Montreal Convention, 2006).[1] This translates into roughly $150,000.

The next major international agreement concerning air transportation was the Chicago Convention of 1944 held near the end of the Second World War and hosted by US President Franklin D. Roosevelt. Roosevelt's goal for this convention was revolutionary in that he wanted an agreement that would allow any airliner from any country to fly to any other country with little or no restriction (Phillips, 2006). What Roosevelt was pushing for was a true open skies agreement, whereby there would be few if any restrictions on international flying. Unfortunately, few of the 54 delegations attending the Chicago Convention actually backed him on his goal for open skies (Phillips, 2006). Instead, Article 6 of the Chicago Convention created a system of bilateral air service agreements between countries for all scheduled international flying. The article states:

> No scheduled international air service may be operated over or into the territory of a contracting State, except with the special permission or other authorization of that State, and in accordance with the terms of such permission or authorization (Chicago Convention (CC), 2006).

A major outcome of the Chicago Convention was the creation of the International Civil Aviation Organization (ICAO) with the objective of, "developing the principles and techniques of international air navigation and to foster the planning and development of international air transport" (CC, 2006). The Chicago Convention superseded the previous Paris and Chicago conventions (CC, 2006) and it still remains the major basis for all international aviation law.

BILATERAL AND MULTILATERAL AIR SERVICE AGREEMENTS

The bilateral and multilateral air services agreements allow designated airlines of participating countries to operate commercial flights and legally cover the transport of passengers and cargo between their countries. After the Chicago Convention, bilateral air service agreements between countries became the predominant method of regulating international air transportation. These agreements controlled market access, market entry,

---

1   The SDR is an artificial currency unit based upon several national currencies. SDR is used by the International Monetary Fund (IMF) for internal accounting purposes and by some countries as a peg for their own currency, and is used as an international reserve asset.

and in many cases market pricing (Rowell, 2002). In granting market access, countries allow various degrees of freedom of air transportation. There are up to eight degrees of freedoms that may be granted in bilateral air service agreements.

Freedoms of air transportation

The *first freedom* provides the right for an airline to fly over another country without landing (ICAO, 2004). An example of a first freedom right would be an international non-stop flight from Los Angeles to London that over flies Canada on the way to England. In order for this to occur, Canada must grant the US first freedom rights. Today, with a few exceptions, almost all countries grant unilateral first freedom rights. Moreover, countries may charge airlines for the permission to over fly their country, essentially placing economic barriers on the first freedom. For example, the over-flight fee for the US is $38.44 per 100 nautical miles over the continental US (FAA, 2012). Russia is notorious for charging high over-flight fees, especially on new polar flights from North America to Asia. This provides an economic hindrance for airlines flying such routes.

> The EU and the US Open Skies Agreement which was signed in 2007 became effective on March 30, 2008. The agreement allows any airline of the EU and any airline of the US to fly between any point in the EU and any point in the US. The Agreement superseded previous bilateral agreements between the US and individual European countries.

The *second freedom* of the air is the right to make a landing for technical reasons (that is, refueling) in another country without picking up or setting down revenue passengers (ICAO, 2004). An example of a flight requiring the second freedom would be Cathay Pacific's flight from Hong Kong to Toronto, Canada with a refueling stop in Anchorage, Alaska. In order for this flight to occur, the US would have to grant Hong Kong second freedom rights. The second freedom right is usually granted since the airline provides revenue to the granting country in terms of landing fees and fuel purchase but does not compete with the domestic airlines. With today's modern aircraft, the requirement for refueling stops has diminished greatly, but just a few years ago the second freedom was important to many operators. Today, cargo carriers utilize second freedom rights the most.

The *third and fourth freedoms* of the air are essentially two sides of the same coin. The third freedom grants the right to carry revenue traffic from your own country to another country, while the fourth freedom provides the right to carry revenue traffic from the other country back to your own country (ICAO, 2004). These rights are usually granted together in order to allow an airline to operate a return air service. Third and fourth freedoms may only be granted for certain city pairs in air service agreements, and this puts limitations on air travel. Any international flight that carries passengers between two countries requires third and fourth freedoms for that particular flight.

Fifth freedom rights enable an airline to carry revenue traffic from their own country to another country, and then pick up and drop off traffic from the intermediate country to a third country. For these rights to be useable, the third country must also agree to the right. A prime example of fifth freedom rights in action is Cathay Pacific's flight from Hong Kong to Vancouver and then onto New York. On this flight Cathay Pacific is allowed to carry traffic from Hong Kong to Vancouver and also from Vancouver to New York. In order for Cathay Pacific to operate this flight, Canada and the US must grant Hong Kong fifth freedom for the route. Fifth freedom rights are rarely granted since the foreign airline is now competing with domestic airlines for the same traffic. However, there are other examples of fifth freedom rights such as Northwest's inter-Asia operation from Tokyo Narita and EVA Airways operating the Taipei–Bangkok–London flight with full traffic rights from Bangkok to London. Fifth freedom rights are highly desirable to airlines, as segments can be tagged on to an existing flight, and this increases its profitability.

The *sixth freedom* of the air allows an airline to carry traffic between two other countries by using its home base as a transit point. A prime example of this is an airline that flies a passenger from Europe to North America and then transfers that passenger onto a flight to Mexico or Central America. Usually sixth freedoms are not granted explicitly, but are given implicitly when third and fourth freedoms are granted. A slightly modified form of the sixth freedom, or "modified sixths" would allow an airline to transfer a passenger through its hub from two points in the same foreign country (Field, 2005). An example of a "modified sixths" would be a passenger flying on a Canadian airline from Boston to Seattle that uses Toronto, Canada as its transfer point. Currently "modified sixths" are not allowed, but there is movement in North America to possibly allow this to happen (Field, 2005).

The *seventh freedom* allows an airline to carry revenue traffic between points in two countries on services which lie entirely outside its own home country. The liberalization of European airspace allowed seventh freedom rights as airlines are now allowed to fly throughout Europe. For instance, Ryanair, an Irish airline, can fly from Germany to Portugal. The tremendous access provided by seventh freedom rights enables increased competition; this has lowered air fares and increased the quantity of air travel demand. While Europe has allowed seventh freedom rights, they are rarely granted by other countries.

Finally, the *eighth freedom* is probably the most controversial freedom. Also referred to as cabotage, the eighth freedom allows a foreign airline to fly between two domestic points in a country. For instance, cabotage rights would need to be granted if Qantas wanted to continue its Los Angeles flight onto New York with local traffic between Los Angeles and New York. Cabotage is controversial because it allows foreign competition on domestic routes. Few countries have granted cabotage rights, but they are actively sought during negotiations. True open skies between two countries would require cabotage rights from both countries. Europe's liberalization of air transport has allowed cabotage rights since the entire European Community is considered domestic from an air transportation perspective. Figure 7.1 graphically summarizes the eight freedoms of air transportation.

Bermuda Agreement

The first bilateral air service agreement was signed between the US and the UK on February 11, 1946 (shortly after the Chicago Convention) in Bermuda (Department of Transportation (DOT), 1978). While this agreement holds the distinction of being the first

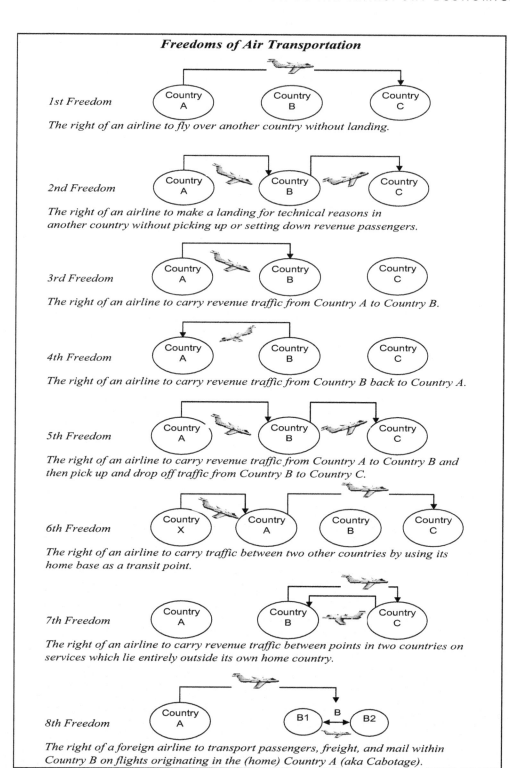

Figure 7.1   Freedoms of Air Passage

bilateral air service agreement to be signed, it is also one of the longest standing and most important bilateral agreements in aviation. The agreement had been updated numerous times and was overhauled in 1977, creating the Bermuda 2 agreement. This agreement was the standing body of airline regulation between the UK and the US until the Open Skies agreement between the EU and the US was finalized in 2000. Even so, the Bermuda 2 agreement is a good example of bilateral air transport agreements between countries; therefore, it is instructive to take a closer look at some of the more important provisions of the agreement.

One of the major themes of most bilateral agreements is regulatory approval on air fares. Both countries must approve the pricing of tickets for all carriers operating between the two countries. This "double approval" of tariffs is usually related to a cost plus profit formula, ensuring a profitable operation while keeping airfares artificially high. Other possibilities for tariff regulation are "dual disapproval," zone pricing, or free pricing (or no pricing regulation).

The tariff approval system in the Bermuda 2 agreement was based on Article 11 which states in part that, "the designated airlines of one Contracting Party shall have a fair and equal opportunity to compete with the designated airlines of the other Contracting Party."[2] (DOT, 1978). This stipulation governed all actions by the airlines, including tariffs, and therefore it acted to create some level of price fixing. The article contains stipulations against capacity dumping, which could severely impact the profits of the other operators. Today, with greater liberalization of competition laws, the tariff approval mechanism largely rubber-stamps airlines' requests for changes in fares.

The Bermuda 2 agreement also dealt with other issues concerning security, airworthiness, dispute resolutions, and customs issues; however, from an economic standpoint, and from an airline standpoint, the greatest impact of the Bermuda 2 agreement was the granting of air freedoms and route authorities. While at the time the rights granted by the Bermuda 2 agreement were considered quite liberal, it is now considered very bureaucratic. The Bermuda 2 agreement placed numerous restrictions on air transportation; these are detailed below:

First, Article 2 of the Bermuda 2 agreement granted the rights bestowed on airlines of both the UK and the US Airlines of the provisions therein:

a) the right to fly across its territory without landing; and
b) the right to make stops in its territory for non-traffic purposes" (DOT, 1978).

Part (a) provided the first freedom to airlines from both countries, while part (b) granted second freedom rights to all airlines. Article 2 goes on to grant fourth and fifth freedom rights as long as they were a part of the agreed upon routes (DOT, 1978).

One of the major restrictions placed in the Bermuda 2 agreement was that only two airlines from each country were allowed to operate scheduled passenger services from London Heathrow to the US (Competition Commission, 1999). Under this agreement, only British Airways, Virgin Atlantic, American Airlines, and United Airlines were permitted to fly from Heathrow to the US. This effectively created a government-enforced cartel that considerably inhibited competition, especially considering the fact that Heathrow is London's most desired airport for passengers. American and United received these rights from Trans-World and Pan Am respectively, and in doing so, received considerable

---

2   *Flight International*, Bermuda 2 initialed, July 2, 1977.

windfall profits. As Heathrow was the only airport that had such restrictions, all other carriers operating between the US and London were forced to do so from secondary UK airports, such as London Gatwick. With Open Skies now in effect, any airline can now legally operate from Heathrow to the US; granted of course that there are slots available.

Under the Bermuda 2 agreement, there were only certain cities that could be served by US and UK airlines from either London Heathrow or London Gatwick. US airlines could only serve a select number of US cities from these two airports, and alternatively, UK airlines had a slightly different list of US cities that they were permitted to serve from London. While although some of these cities were "switchable," the archaic system not only impacted airlines, but communities as well. The Bermuda 2, like many other bilateral air service agreements, placed tremendous restrictions on international travel.

The general thrust of bilateral air service agreements has been to protect national interests and provide support for national airlines. While such protectionism helps carriers who receive the benefits, it is frustrating for airlines looking from the outside-in. By its protectionist nature, bilateral agreements curtail a market solution to international air travel and replace it with government regulation. Generally speaking, the artificial restrictions that are imposed by bilateral air transport agreements raise costs, create inefficiencies in the market, and allow rent-seeking behavior on the part of the favored airlines. Opening up of the international skies would be similar to the deregulation movement that occurred domestically in the US in 1978. Open skies would not only benefit consumers and the economy, but also increase the airline's profit and reduce cost.

> Open Skies denotes to multilateral agreements between states in order to liberalize the airline industry and minimize governmental regulations.

Open skies agreements

In 1992, the US and the Netherlands signed the first open skies agreement. This was followed by similar open skies agreement between the US and Canada in 1995 (Field, 2005). As of 2012, the US has signed 105 open skies agreements with countries ranging from Germany to Chad, from Chile to Uzbekistan (DOS, 2012). The general trend in international aviation is to do away with complicated and restriction-laden bilateral agreements and move toward "open skies." In March of 2007 the US and the EU entered into an Open Skies agreement. Phase 2 of this agreement was signed in June 2010 and further eliminates the restrictions on air service between the US and EU; airlines from both sides can fly any route without limitations on the number of carriers that can fly or the number of flights that they can operate.

Characteristics of open skies

US open skies agreements generally contain eight key provisions (DOS, 2012). The first, and probably the most important, provision contained in all open skies agreements is the absence of restrictions on international route rights (DOS, 2012). This means that carriers from either country are free to fly between any two cities they wish, with whatever size aircraft they wish, as many times a day/week as they want. This lack of restrictions lowers barriers to entry for airlines, but does not entirely eliminate them as carriers as they may still require landing slots at foreign airports in order to initiate a new flight. However, open competition will allow airlines to bid for these rights and ultimately they will be assigned to their highest-valued economic bidder.

The second major provision included in open skies agreements is that airline pricing should be determined by market forces (DOS, 2012). While under true open skies agreement, the governments would play no role in airline pricing, the US model does include a "double-disapproval" stipulation, whereby a fare can be disallowed if both countries agree (DOS, 2012). In practice, carriers are allowed to set whatever fares they want, but the presence of a "double-disapproval" stipulation could possibly prevent some low-cost carriers (LCCs) from entering certain international markets and offering deeply discounted fares, if disallowing such fares happened to be politically attractive to the two governments involved.

The third major provision contained in US open skies agreements is a clause ensuring fair and equal opportunity to compete (DOS, 2012). This clause covers a wide variety of issues, such as non-discriminatory airport slot allocations or user fees. This provision also covers issues such as availability of ground handling and establishing sales offices. In essence, countries should allow airlines of both countries equal opportunity to be able to compete fairly.

The fourth major provision contained in open skies agreements allows airlines to enter cooperative marketing agreements (DOS, 2012). As will be pointed out later in the chapter, airline alliances are critical to the success of airlines. Prior to open skies agreements, restrictions could be placed on the air carriers' ability to enter alliance agreements with airlines of both countries. For instance, while both British Airways and American Airlines are both founding members of the Oneworld alliance, they were not permitted to code share on each other's flights under the previous Bermuda 2 agreement (Button, 2002). Ideally under open skies, airlines are permitted to enter whatever code share agreements that they wish. Thanks to the open skies agreement between the US and the Netherlands, Northwest Airlines and KLM were able to enter a strong alliance that included revenue sharing between the airlines. In order for this cohesive agreement to occur, the open skies agreement had to be in place, and the extensive code share agreement between the airlines was given anti-trust immunity from the US Department of Justice. While US open skies agreements allow full code share agreements, they do not address issues pertaining to foreign ownership of airlines.

Other provisions that are frequently contained in open skies agreements are mechanisms for dispute settlement, consultation pertaining to unfair practices, liberal legal charter agreements (whereby carriers can choose to operate under the charter regulations of either country), and agreements pertaining to the safe and secure operation of flights between the two countries (DOS, 2012). In open skies agreements the US also seeks the provision that there be the seventh freedom rights for all-cargo flights (DOS, 2012). This permits cargo flights to operate between the other country and a third country, via flights that are not linked to its homeland. This stipulation enables airlines like FedEx and UPS to operate cargo hubs in foreign countries. Currently only about half of the open skies agreements signed by the US contain this optional eighth provision (DOS, 2012). Fifth freedom rights are rarely provided for in open skies agreements. However, one agreement that does allow this is the new US–Canada open skies agreements whereby Canadian carriers received unilateral fifth freedom rights from the US in exchange for seventh freedom all-cargo rights for US carriers.

Benefits of open skies

The benefits achieved from open skies agreements are similar to the benefits obtained from domestic deregulation. As a result of open skies agreements, airlines are able to fly more routes, which ultimately results in increased competition, resulting in lower average air fares. Open skies also enables new city pairs (domestic to foreign) to be flown that were previously not possible. In general, consumers benefit from open skies as they receive more frequent service and lower prices. This is a result of more competition between the airlines. On the other hand, it may also result in greater fluctuations in the profitability of the airlines. Airlines also benefit from open skies agreements, but those benefits vary depending on the airline's position in the market. For example, the airline that already has extensive rights to the foreign country is currently receiving some windfall profits from the protection it is receiving in the market. With open skies, that airline would no longer receive the protection and would face more competition. Carriers that are currently excluded from a market (or have limited service) gain more from open skies agreements than carriers that already have extensive route rights. This is the major reason why airlines such as Continental and Delta lobbied hard for an open skies agreement between the US and the UK, while carriers such as American, which has extensive London access rights, remained relatively quiet in its lobbying.

> "As our international travel market continues to grow, we see even more growth and expansion as Open Skies truly opens up even more air travel for our local and connecting passengers. And for those who do not travel, there's good news for you too. These new flights will mean hundreds of millions of new dollars for the North Texas economy that will create new opportunities for many businesses."
>
> J. Fegan, CEO of DFW Airport, 2008

As mentioned above, the US signed one of the first open skies agreements with Canada. In a study conducted by the US Department of Transportation three years after the signing of the open skies agreement, it was found that trans-border traffic averaged an 11.1 percent yearly growth rate compared to 1.4 percent per year for the three years prior to the agreement (DOT, 1998). Moreover, the number of non-stop markets with over 50,000 annual passengers increased from 54 in 1994 to 77 in 1997 (DOT, 1998). It is estimated that 38 new city pairs were opened up between Canada and the US as a result of the agreement. (DOT, 1998). While this tremendous growth rate in the market will not continue, the figures clearly show the large latent demand, from both business and tourism, which was being suppressed before the open skies agreement. The original open skies agreement between the two countries has been recently amended in 2005 to provide both countries with increased freedoms, but some, including Air Canada President Robert Milton, want to see the North American market resemble the European market where a Canadian carrier would essentially be granted US cabotage rights, and vice versa (Field, 2005). However, the likelihood of such rights being granted is probably slim.

OPEN SKIES IN EUROPE

Europe has had a successful experience of open skies with the creation (in 1997) of a single European aviation market. (Kinnock, 1996). Under this single European aviation market, European carriers are free to fly routes throughout Europe. For instance, British Airways could fly from Paris to Frankfurt or Amsterdam to Rome. This granting of seventh freedom rights also included the granting of eighth freedom rights, or cabotage. This further enabled a British airline to be able to offer Frankfurt to Munich flights or Barcelona to Madrid flights. In fact, British Airways created a German subsidiary to fly domestic German routes. National ownership has become irrelevant for intra-European flights, and this is the primary reason that LCCs such as Easyjet and Ryanair have been able to expand rapidly. While European liberalization has increased competition and helped lower airfares throughout Europe, making aviation a viable competitor to train travel, it has also caused many airline bankruptcies; this, of course, is similar to what occurred in the US after deregulation (Kinnock, 1996).

While the European Union (EU) has successful liberalized intra-Europe travel, global travel from Europe is still largely dominated by each country's respective flag carriers (de Palacio, 2001). German airlines cannot fly from London to the US while British Airlines cannot fly from Paris to Japan. This is a result of the current bilateral agreements between individual European countries and other countries around the world; these agreements limit international flights to airlines with full national ownership (de Palacio, 2001). This problem is made more complicated by the fact that individual European countries have signed bilateral agreements with foreign countries, when the European community is attempting to become a single market. Because it has had stronger leverage over each individual nation than over the collective whole, the US currently has open skies agreements with most European nations. However, these nationality clauses have been deemed illegal in a 2002 European Court of Justice (ECJ) ruling; this has placed pressure on the EU to create multilateral aviation agreements with foreign countries (Baker, 2005). The largest such multilateral agreement involved the EU and the US.

The second-stage of the open skies agreement between the EU and US was adopted on June 24, 2010 as part of the mandate set forth in the first stage to reach a balanced agreement in 2010.[3] The ultimate objective of the agreement is a single air transport market between the EU and the US with no restrictions on air services, including access to the domestic markets of both parties. Both sides have agreed to remove the remaining barriers which include legislative change in the US regarding foreign ownership. Currently foreign ownership in US airlines is limited to 25 percent of voting rights and the EU has stated its intentions of reciprocally allowing majority ownership of EU airlines by US nationals.

OPEN SKIES IN ASIA

While most of the world's aviation industries have adopted liberalized aviation markets, the Asia-Pacific region has not implemented open skies; this has been largely due to perceived national interests (Oum and Yamaguchi, 2006). These are the same national interests that kept the US and Europe from liberalizing their air space. One reason for this is the fact that the aviation industry in the Asia-Pacific region is still developing compared to the mature North American and European markets. However, as these markets have shown, the benefits of open skies far outweigh the benefits of protectionism, but this may take some period of time before it is fully appreciated in this region. Two of the more successful airlines in the region are Emirates Airlines and Singapore Airlines.

---

3   Business Wire, A Berkshire Hathaway Company, March 22, 2011.

These carriers play a large role in promoting their relatively small countries and have been successful, through the adoption of open skies agreements, in the creation of global aviation hubs.

Australia is another country that has implemented open skies by creating a single aviation market with New Zealand. Australia has also eliminated foreign ownership restrictions (Oum and Yamaguchi, 2006). This removal of foreign ownership restrictions has enabled Richard Branson to start-up Virgin Blue in Australia (now Virgin Australia). On the other hand, while Australia has liberalized aviation considerably, it also denied Singapore Airlines's request to fly from Sydney to Los Angeles. This action was undoubtedly to protect Qantas Airways from competition on this route.

India, a long-time highly regulated aviation market, has slowly become more liberalized as the economy has grown; however, much of the liberalization effort has only occurred in the domestic market. This liberalization has spurred tremendous growth in domestic air travel and the industry has created several successful LCCs. Only recently has the Indian government permitted private Indian carriers to fly internationally, and many of the country's bilateral agreements have severe capacity restrictions placed on them.

In northern Asia, Japan and Korea still are highly regulated with most international routes containing capacity restrictions. Liberalization of the Japanese market is even more difficult due to airport restrictions at congested airports such as Tokyo Narita. Only recently, since 2010, has Japan had an open skies agreement in place with the US. Besides this, very few countries in the Asia-Pacific region actually have open skies agreements with the US.

Probably the most attractive country for foreign carriers to fly to in the Asia-Pacific region is China. However, for political reasons, China has also historically been one of the most restrictive countries in the region. One of these political sanctions that affect the aviation industry was the prohibition of direct flights between China and Taiwan. While this hindered both Chinese and Taiwanese carriers, the regulations benefited Macau which was used as a transiting point between China and Taiwan. While China has opened up its country to foreign business ownership, the liberalization of the aviation sector has progressed at a much slower pace.

Domestically, the Chinese industry was previously exclusively government-owned, but operated by multiple small carriers. Recently, the domestic industry has consolidated forming three large Chinese carriers, Air China, China Southern Airlines, and China Eastern Airlines, which are based at Beijing, Guangzhou, and Shanghai respectively (Francis, 2004). This consolidation has enabled the Chinese carriers to be stronger internationally, and allowed slow, progressive reform in international aviation. Additionally, China has allowed private ownership of airlines, including a foreign ownership cap of 49 percent, which possibly could be raised in the future (Francis, 2004).

While China has liberalized some of their international aviation agreements, these agreements are still quite restrictive. For instance, China and Singapore reached an open skies agreement, but the open skies agreement forbids Singapore LCCs from flying into Shanghai and Beijing (Francis, 2005). The agreement did, however, permit full open skies to all other Chinese cities. In this respect it is just a further example of certain open skies agreements not truly being open skies.

China is also experimenting with other liberalized air policies such as creating an entirely open aviation policy for the Hainan region of China (Francis, 2004). Under this open skies policy, foreign carriers are permitted unlimited access to the Hainan region, have full fifth freedom rights, and limited cabotage rights to other Chinese cities other than Beijing, Guangzhou, and Shanghai (Francis, 2004). This policy is an effort to open up aviation markets outside China's three dominant cities, and also is probably protectionist in nature since these three cities also happen to be the hubs of China's three largest airlines (Francis, 2004).

Probably the most ambitious economic liberalization project to occur in the Asia-Pacific region is the multilateral open skies agreement between the ASEAN (Association of Southeast Asian Nations) countries. The ten members of the ASEAN are Brunei, Singapore, Thailand, Cambodia, Indonesia, Laos, Malaysia, Myanmar, the Philippines, and Vietnam. As part of the broader ASEAN Air Transport Liberalization Plan, the agreement to allow unlimited flights between members was initiated in 2008 and is expected to be fully implemented by 2015. These reforms are a giant step forward for air transport liberalization in the region, since many of the ASEAN countries have historically had very protectionist viewpoints toward aviation. Such air transport liberalization would not only benefit individual countries, but also the economic region as a whole in both the short term and long term (Forsyth King, and Rodolfo, 2006). Although the ASEAN open skies agreement shows promise, the Asia-Pacific region still remains a heavily regulated industry in comparison to Europe and North America.

GLOBAL AIRLINE ALLIANCES

In order to help overcome restrictive barriers to entry in international markets, many airlines have formed alliances with foreign carriers. The purpose of these alliances has generally been an attempt to introduce service in those countries and regions where there have been legal or financial restrictions. While the initial global alliances started as simple code share agreements between airlines, they have evolved into global alliances with multiple airlines that span the globe. The introduction of global alliances has also magnified the prominence of global or regional hubs. Today nearly 60 percent of the world's total air traffic flies on some sort of a global alliance (Baker, 2006).

History of global airline alliances

In 1986, the first international airline alliance was signed between Air Florida and British Island whereby Air Florida provided a passenger feed for British Island's London–Amsterdam route (Oum, Park and Zhang, 2000). This basic form of an alliance between two carriers, commonly called a code share, enables an airline to place passengers onto the flight of another carrier. Code share agreements can be signed that cover a few particular routes or flights, or they can also cover almost all of the airline's flights. While this original code share agreement was quite simple, code share agreements have evolved to the point that they may even include some blocked seats for the code sharing airline. For instance, in 1993 Air Canada and Korean Air entered into a code share agreement whereby each airline purchased 48 seats per departure on the other airline's flight. In this example, the inventory assigned to the airline is fixed, while in other cases it can be variable, depending on demand. After this initial code share agreement was signed, multiple carriers signed code share agreements. Some examples of these are: Japan Airlines and Thai Airways in 1985; American and Qantas on Qantas's transpacific flights in 1986; and Air France and Sabena with a blocked space agreement on the Paris to Brussels route in 1992 (Oum, Park and Zhang, 2000).

In 1992, Dutch carrier KLM and Northwest formed a major transatlantic airline alliance whereby a broad code share agreement was put in place. In 1993 the alliance received ATI from the US Department of Transportation (DOT), thereby enabling the two airlines to closely coordinate their flights across the Atlantic. This led to a joint operation venture where revenues were divided between the two carriers, regardless of the operating airline. Joint ventures were not a new phenomenon to the airline industry, as both Braniff and Singapore Airlines operated a quasi-joint venture with the Concorde aircraft. Because it covered so many flights, not only across the Atlantic but also in Europe and North America, the KLM/Northwest deal was precedent-setting. The KLM/Northwest alliance was also precedent-setting since there were equity investments for both airlines. In the agreement, KLM purchased 25 percent of Northwest's voting rights, and 49 percent of Northwest's total equity share (Oum, Park and Zhang, 2000). While equity investments bring the carriers closer together and enable the investing airline to help shape overall strategy, it may also limit the flexibility of the alliance. For instance, Gudmundsson and Lechner (2006) argue that one of the downfalls of the Qualiflyer alliance was that Swissair's equity investments made it difficult for the alliance partners to enter new agreements.[4] While Swissair's equity investment highlights the pitfalls involved in such an agreements, they have also proven successful, such as British Airways's 25 percent investment in Qantas in 1993, Air Canada's 27.5 percent investment in Continental also in 1993, and Singapore Airlines's current 49 percent investment in Richard Branson's Virgin Atlantic (Oum, Park and Zhang, 2000).

The next development in the airline alliances model was the creation of global alliances. While Delta, Swissair, and Singapore had the initial roots of a global alliance in the late 1980s, the first truly global alliance was formed in 1997 between United, Lufthansa, SAS, Air Canada, and Thai Airways (Baker, 2001). This alliance was called the Star Alliance, and it was shortly followed by similar global alliances such as: the Qualiflyer in 1998 that originally included Swissair, Sabena, Turkish Airlines, Air Liberte, and TAP Air Portugal; the oneworld alliance in September 1998 between American, British Airways, Qantas, and Cathay Pacific; and the SkyTeam alliance in September 1999 originally between just Delta Air Lines, Air France, and Aeromexico (Baker, 2001). Since then, the global alliances have added additional airlines in an attempt to reach all corners of the globe. Airlines were more than willing to join these alliances for the undoubted benefits, and they also did not want to be left behind in the alliance game. However, several large carriers continue to remain unaligned with any of the "Big Three" alliances including Emirates (UAE), Southwest Airlines (US), Ryanair (UK), EasyJet (UK), Qatar Airways (Qatar), JetBlue Airways (US), and Virgin Atlantic Airways (UK).

Many airlines choosing not to align with an alliance feel that despite having code share agreements, such alliances would not provide them with any additional benefits. Another often cited reasons from these airlines are concerns of declining service standards that might result from joining a global alliance. In the US, Alaska Airlines has chosen not to join a global alliance, but has decided to have multiple broad code share agreements with

---

4   The Qualiflyer alliance was a European alliance led by Swissair and Sabena.

carriers from both oneworld and SkyTeam. On the other hand, Aer Lingus decided to back out of the oneworld alliance as it redefined itself as a LCC.

Today, three major global alliances exist: Star, SkyTeam, and oneworld. Combined, these alliances generated 57.9 percent of global revenue passenger kilometers (RPKs) and 64.2 percent of revenue for 2010. Star Alliance is comprised of 27 airlines and is the largest global alliance with a 29.6 percent global market revenue share. The Star Alliance also carries approximately 513 million passengers a year. SkyTeam obtained a 19.0 percent global market share in 2010 with 477 million passengers annually. Oneworld is the smallest alliance in terms of revenue with only 15.6 percent global market share. The three major alliances are compared in Table 7.1.

All three alliances have expanded to provide geographical coverage to all areas of the world. Star Alliance has solidified its coverage in Africa in 2012 when Ethiopian Airlines joined its ranks. SkyTeam added its first Middle-Eastern airline in 2012 with the addition of Saudi Air, but its real focus has been on the Asian Markets with the joining of China Southern as well as China Eastern and its subsidiary, Shanghai Airlines (which was previously aligned with Star Alliance). Figures 7.2 through 7.4 provide various comparison statistics for the alliances; these include number of countries served, number of destinations served, and total aircraft fleet.

Table 7.1   Global Airline Alliance Statistics

	Star Alliance	SkyTeam	oneworld
Member airlines	27	17	11
Revenue ($millions)	174,368	112,043	91,933
RPKs (millions)	1,210,278	921,557	718,383
Passengers (millions)	513	477	307
Countries served	189	173	148
Employees	414,498	399,496	268,147
Fleet	4,386	3,542	2,194

Source: Airline Business Airline Alliance Survey, September 2011

Two major trends that have occurred in the global alliance game is the push for Eastern Europe members and the creation/acceptance of smaller regional airlines. While Czech Airways was the first Eastern European carrier to join a formal global alliance, numerous other Eastern Europe carriers have joined alliances as the region has prospered in the past few years. Also, both Star and SkyTeam have created regional/associate members for niche airlines that serve a particular need. These regional/associate members must have a sponsoring carrier for entry into the alliance.

Global alliances and competitiveness

It should be noted that it is the strategy of every airline to weaken competition on certain routes and to gain market share; therefore, a global alliance helps achieve that particular goal. Since legacy carriers on both sides of the Atlantic have been faced with substantial and growing competition from LCCs on short and medium-haul routes, they have expanded their global networks where they have an important comparative advantage

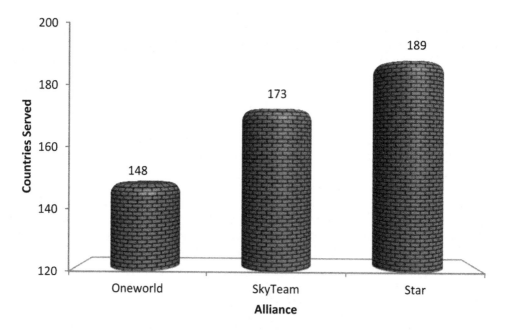

Figure 7.2   Global Airline Alliances: Countries Served 2011

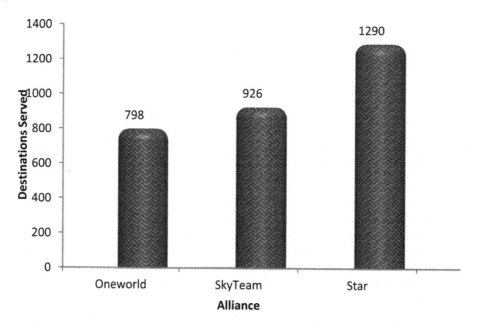

Figure 7.3   Global Airline Alliances: Destinations Served 2011

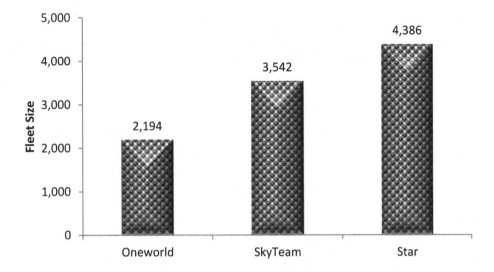

Figure 7.4   Global Airline Alliances: Aircraft Fleet 2011

versus LCCs. This makes their overall costs more competitive with the growing LCCs. The global alliances have facilitated the achieving of these competitive advantages.

Table 7.3 and Figure 7.5 display the impact of global alliances on the top ten airports in the world in terms of available seat miles (ASMs). The market share for each airline is calculated based on second quarter 2012 ASMs published by the Official Airline Guide (OAG). In terms of concentration, the Star alliance commands the greatest market share, averaging 30 percent of all ASMs originating from the world's ten busiest airports. Oneworld is a close second, averaging 28 percent market share, while SkyTeam has been able to achieve a 21 percent share of these markets. On the other hand, Dallas-Fort Worth airport is the most concentrated airport, where the oneworld alliance, largely driven by American Airlines, commands an 86 percent market share of the airport.

Almost all the airports are highly concentrated when grouped by global alliances — much more so than when grouping the airlines individually. Of the top ten airports, Los Angeles is the least concentrated with all three alliances and non-member carriers having a fairly even market distribution. Even market distribution results in greater competition and lower airfares, while greater concentrations generally lead to reduced service and higher air fares — a negative effect for consumers, but a positive effect for the airlines.

It can be argued that global alliances may decrease competitiveness on specific routes. If two airlines with hubs in large cities at either end of a city-pair combine their networks, their cooperation as part of the alliance may grant them market power to raise prices, alter capacity and/or reduce the quality of service on that route. When an airline alliance is granted antitrust immunity (ATI) the number of independent competitors decreases significantly. The US DOT and the European Competition Commission (ECC) carry out competitive assessments before granting ATI to these alliances, albeit utilizing different methods, to determine whether global alliances generate unfair competitive advantages within their respective countries. Under EU competition rules, the Commission conducts an analysis of possible negative competitive effects whereas the DOT weighs the potential efficiencies or benefits before it grants an ATI.

Table 7.2   Global Alliance Membership, 2011

oneworld	SkyTeam	Star Alliance
Air Berlin	Aeroflot	Adria Airways
American Airlines	Aeromexico	Aegean Airlines
British Airways	Air Europa	Air Canada
Cathay Pacific	Air France	Air China
Finnair	Alitalia	Air New Zealand
Iberia	China Airlines	All Nippon Airways
Japan Airlines	China Eastern Airlines	Asiana Airlines
LAN	China Southern Airlines	Austrian Airlines
Qantas	Czech Airlines	AviancaTaca
Royal Jordanian	Delta	Blue1
S7 Airlines	Kenya Airlines	Brussels Airlines
	KLM	Copa Airlines
	Korean Air	Croatia Airlines
	Middle East Airlines	Egyptair
	Saudia	Ethiopian Airlines
	Tarom	LOT Polish Airlines
	Vietnam Airlines	Lufthansa
		SAS Scandinavian Airlines
		Singapore Airlines
		South African Airways
		Swiss
		TAM
		TAP Portugal
		Thai Airways
		Turkish Airlines
		United Airlines
		US Airways

Source: Airline Business Alliances Survey, September 2011

Table 7.3   Concentration of Top Ten Global Airports by Alliance in Terms of ASMs

Market Share	ATL	PEK	LHR	ORD	HND	LAX	CDG	DFW	FRA	HKG
Star	2%	43%	19%	48%	54%	24%	11%	5%	78%	12%
oneworld	2%	1%	49%	37%	34%	21%	4%	86%	5%	39%
SkyTeam	78%	18%	6%	5%	2%	17%	66%	4%	4%	9%
Non-members	18%	38%	26%	9%	11%	38%	19%	5%	13%	40%
Herfindahl Index	6,374	3,582	3,474	3,812	4,144	2,761	4,806	7,398	6,315	3,345

Source: Compiled by the author from 2012 OAG data

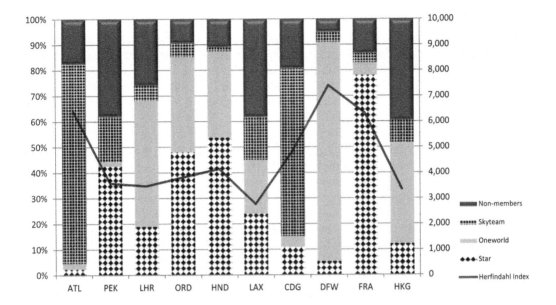

Figure 7.5   **Concentrations of Top Ten Global Airports by Alliance in Terms of ASMs**

Source: Compiled by the author from 2012 OAG data

Benefits of global alliances

The major benefit of global alliances to both airlines and passengers is an expanded and optimized route network. Through global alliances, passengers can, in theory, easily travel from one destination to another destination, anywhere in the world, with one ticket from one airline. Without global alliances or code share agreements, international travel was more complex as passengers might have to purchase multiple tickets on multiple airlines to fly to their desired destination. Through global alliances the airlines are now able to sell tickets to destinations that they previously could not serve for one reason or another. For example, Delta Airlines and Air France are both SkyTeam members, and technically there is no difference which one a passenger is booked on, the passenger will earn miles, check-in all the way through to the final destination, and access their lounges anywhere in the world.

Global alliances provide a traveler with greater flight options and help make the traveling experience more enjoyable. The various routings that are available from the three alliances reduce the international travel time for passengers. For the airlines, one might not find it profitable to fly directly to a certain city, but it can still carry passengers to that city on connecting flights. Therefore, alliances enable airlines to achieve benefits from economies of scope, and this is the reason that the alliances are seeking a strong presence in every major market in the world.

To be successful, the connections between carriers in an alliance must be seamless and easy. Information technology is critical to success in this area. Alliance members all operate different information technology platforms, but they still need to interact effectively. Each of the three alliances are addressing the information technology and ticketing structures differently. Star is creating a common platform that all carriers may choose to use, while SkyTeam is not pursuing a common information technology approach (McDonald, 2006). Oneworld is the first airline alliance to have full e-ticketing between all its members. SkyTeam undoubtedly will have difficulties achieving full e-ticketing connections since Russian law requires that Aeroflot provide paper tickets (McDonald, 2006). Full e-ticketing is probably critical to the future success of these alliances since it is more convenient for the passengers and cheaper for the airlines.

While the exact traffic and revenue benefits of joining an alliance are difficult to quantify, oneworld estimated that the alliance generated almost $400 million in additional revenues for the eight alliance members. Oum, Park and Zhang (2000) compared traffic increases on alliance routes to non-alliance routes. In the short-lived USAir/British Airways alliance, British Airways was able to increase traffic by 8.3 percent on alliance routes over non-alliance routes (Oum, Park and Zhang, 2000). Iatrou and Skourias (2005) compared the traffic difference on alliance routes between the pre- and the post-alliance periods for all three major alliances. On average, they found that traffic increased by 9.4 percent as a result of airline alliances; however, the greatest increases in traffic were experienced by SkyTeam and Star, while traffic for oneworld actually decreased (Iatrou and Skourias, 2005). They also found, not surprisingly, that traffic increased the greatest when an alliance received ATI.

Another major benefit of airline alliances is cost reduction in maintenance and operational activities as a result of bulk purchasing and sharing of resources. As mentioned above, joint information technology could greatly reduce costs among alliance members. One area that the Star Alliance has pioneered is joint purchasing; this is in an effort to receive greater volume discounts. In an effort to reduce fuel costs for alliance members, the Star Alliance launched Star Fuel Co. in December 2003 (Mecham, 2004). Using volume discounts, Star Fuel Co. was able to reduce fuel costs for Star Alliance members at Los Angeles, San Francisco, and London Heathrow by $50 million in 2004 (Mecham, 2004). Star Alliance members Air Canada, Austrian, Lufthansa, and SAS all explored a joint regional aircraft purchase in the same year, but the initiative did not occur as Air Canada and Austrian made independent decisions (Field and Pilling, 2004). While the airlines were able to develop common specifications for the aircraft, the failure of the joint purchasing agreement highlights the difficulties that may be encountered when separate alliance members have unique and different requirements and objectives. While the Star Alliance intends to pursue joint aircraft purchasing again in the near future, the best strategy is probably to focus on the joint purchasing of commodities such as fuel and aircraft parts (Field and Pilling, 2004).

The other major area where alliances are looking at reducing costs is at the airports. While airlines commonly use personnel from alliance members in order to help reduce costs, alliances are seeking to extend this further by hosting operations all under one roof. Under this scenario, not only would the carriers be able to share airport resources, but they would also provide passengers with easier connections. An airport where all the alliances would appreciate dedicated facilities is London Heathrow. With the opening of Terminal 5 in 2008, airlines began shuffling all over the airport in order to consolidate operations (Thompson, 2004).

The creation of dedicated alliance terminals such as that described at Heathrow would not only help reduce costs, but enable alliances to have shared business lounges, self-service check-ins, and possibly ground service. The Star Alliance is actively pursuing similar "one roof" initiatives in such airports as Paris Charles de Gaulle, Nagoya, Tokyo

Narita, Miami, Bangkok, and Warsaw (Thompson, 2004). Additionally, in an effort to improve customer service, the Star Alliance has initiated a program at three airports, where Star Alliance customer service teams meet incoming flights and ensure that passengers and their baggage make their connecting flight (Thompson, 2004).[5] While the program may seem more like a customer service idea, the Star Alliance claims the program saves approximately $3.3 million per year by reducing passenger misconnects and lost baggage claims (Thompson, 2004).

Having a strong global presence also enables the alliances to obtain global corporate travel agreements for corporations that require global travel (Field and Pilling, 2004). SkyTeam estimates that there are about 75–100 organizations that require global corporate travel agreements (Field and Pilling, 2004). This additional source of revenue can be substantial for alliances, and the attractiveness of their products over competing alliances depends greatly on the alliances' global market coverage.

Another benefit that frequent travelers enjoy is the fact that global alliances enable travelers to accrue and redeem miles on a variety of airlines. No longer do frequent flier programs limit travel to just one airline, but they may now include 20 or even 30 airlines. Additionally, depending of the passengers' frequent flier status, they can receive reciprocal benefits such as upgrades, priority boarding/check-in, and/or lounge access. This is also one of the reasons why alliance members have to adhere to a certain level of quality, and it is also one of the major reasons why no LCC has yet to join an alliance.[6]

Disadvantages of global alliances

The greatest cost involved in joining global alliances is bringing information technology (IT) systems in line with other alliance members. These costs can be considerable, especially if the system needs a complete overhaul. The time needed for harmonization of IT systems is one of the reasons why alliance members usually announce their intention to join an alliance six to 12 months before they actually join. In the case of South African Airlines, it took nearly two years for the airline to finally join the Star Alliance (Star Alliance, 2006).

There are additional costs associated with global alliances, such as increased overhead costs. The Star Alliance is the most structured alliance with a full-time staff of around 75 people (Field and Pilling, 2004). The reason for this is the fact that the Star Alliance is so large that it needs some oversight, but it is also because the Star Alliance is the most aggressive of the three alliances in creating a master brand (Field and Pilling, 2004). Oneworld is slightly less structured with a full-time staff of 23 people, but the majority of these people are in sales for the various oneworld ticket packages (Field and Pilling, 2004). Finally, although SkyTeam was originally structured with no true overhead group, the alliance is now overseen by a team of nine core managers in addition to support staff.[7] Regardless of the alliance's structure, such coordination efforts raise overhead costs to some degree.

One area of potential concern for various alliance partners is that as new airlines are added to the alliance, the importance of an existing airline may be reduced. That is, carriers can be in competition with their own alliance partners. An interesting example of this

---

5   The airports include: Chicago O'Hare, Frankfurt, and Los Angeles.
6   Note that both Westjet in Canada and Gol in Brazil have expressed an interest in joining a global alliance (Baker, 2006).
7   skyteam.com.

was Delta's and Continental's battle on trans-Atlantic routes, even though both airlines were SkyTeam members and should technically have been working together. Because of these issues, Gudmundsson and Lechner (2006) argue that airlines will drop out and switch alliances in order to achieve the greatest benefits. On June 19, 2008, Continental announced that it would be leaving SkyTeam and began participating in Star Alliance on October 27, 2009 as part of a code sharing agreement with Star Alliance charter member United Airlines.[8] This has already begun with Aer Lingus and Mexicana dropping out of oneworld and Star Alliance respectively. While switching costs may prohibit a high degree of alliance movement by airlines, there will be situations where airlines may feel that it is in their best interests to switch.

Future for global alliances

Since the late 1990s, oneworld, SkyTeam, and Star Alliance have emerged as the dominating airline alliances with about two-thirds of the world's air traffic operations. This trend is expected to continue with these three dominant alliances seeking to fill in any gaps in coverage offered by their current members. There are currently no plans for the creation of a fourth global alliance as the current alliances balance between expanding their global networks and revenue generation and the risk of inefficiency due to the larger size. New and non-aligned carriers will likely join one of the three current alliances; however, simplified alliances and cooperation between LCCs may continue to occur within the domestic networks.

Global alliances are expected to continue to be driven by economies of scale, scope, and density, marketing advantages, circumvention of restrictive bilateral agreements, and market power. However as the potential number of alliance members begins to decrease, more strategic decisions will be made by the big three global alliances to include members who can fill in any network gaps.

SUMMARY

Open skies allow unrestricted access by the airlines of each country to, from, and beyond the other's territory, without limitations on where airlines fly and the prices they charge. This chapter presents an instructional discussion of aviation agreements and the concept of open skies. Bilateral and multilateral open skies agreements provide a framework that encourages competition, and expands international air services to benefit passengers and global trade. A detailed presentation of the various aviation freedoms that are contained in international agreements is presented as well as an overview of recent and vital transnational aviation agreements including the EU–US open skies agreement. The future of these aviation agreements is discussed and highlighted. The chapter also covers the three major global airline alliances and the history of such large alliances. It also covered the associated benefits and costs of belonging to a global alliance.

---

8   Star Alliance, Alliance Customer Benefits Remain in Place until 31 May, 2012. Press Release 2012.

DISCUSSION QUESTIONS

1. List and briefly explain the eight provisions of open skies agreements.
2. Global airline alliances were created with the intention of growing the networks of airlines and providing greater access to more parts of the world. Recently, three major global alliances started to accept smaller, regional airlines to join ranks. What are the benefits to the alliance of the membership of the regional airlines?
3. While global alliances have benefits for travelers and airlines, there are some negative concerns. Identify and describe some of the costs associated with joining a global alliance.
4. Define Cabotage Law and explain the impact of cabotage agreements on international travel.
5. How did the open skies agreement between the EU and the US reduce the barriers to entry?
6. How were LCCs in Europe affected by the EU open skies agreement?
7. What are the main provisions of open skies agreements and why have some countries avoided such agreements?
8. Which freedom is not likely to be extended to foreign airlines and why?
9. Name the three major world alliances. What are some of the benefits and costs associated with these alliances?
10. Why is aviation particularly suited to open skies agreements and what are some gains for individual airlines in such agreements?

REFERENCES

Baker, C. (2001). The Global Groupings. *Airline Business,* July. Retrieved on October 12, 2006 from Air Transport Intelligence.

Baker, C. (2005). Back to the Table. *Airlines Business,* September. Retrieved on September 28, 2006 from Air Transport Intelligence.

Baker, C. (2006). Stellar Orbit, *Airline Business,* September.

Baker, C. and Field, D. (2004). Europe Rules out US Open Skies Offer, *Airline Business,* April.

Button, K.J. (2002). Toward Truly Open Skies. *Regulation,* Vol. 25, No. 3.

Chicago Convention. (2006). *Convention on International Civil Aviation.* Signed at Chicago, on 7 December 1944.

Competition Commission. (1999). British Airways Plc and CityFlyer Express Limited: A report on the proposed merger, Appendix 4.2 *Bermuda 2.*

de Palacio, L. (2001). Open Skies: How to Get the Airlines Airborne Again. *Wall Street Journal (Europe),* September 11.

US Department of State (DOS). (2012). *Open Skies Partners* Retrive on August 15, 2012 from: http://www.state.gov/e/eb/rls/othr/ata/114805.htm.

Department of Transportation (DOT). (1978). *Air Services Agreement between the Government of the United States of America and the Government of the United Kingdom of Great Britain and Northern Ireland*. Washington, DC: DOT.

Department of Transportation (DOT). (1998). *The Impact of the New US–Canada Aviation Agreement At Its Third Anniversary*. Retrieved on August 15, 2012 from http://ostpxweb.dot.gov/aviation/intav/canada2.pdf.

Federal Aviation Administration (FAA). (2012). *Overflight Fees*. Retrieved on August 15, 2012 from http://www.faa.gov/air_traffic/international_aviation/overflight_fees/.

Field, D. (2005). True Open Skies? *Airline Business*, March. Retrieved on September 28, 2006 from Air Transport Intelligence.

Field, D. and Pilling, M. (2004). Team Spirit. *Airline Business*, September.

Forsyth, P., King, J. and Rodolfo, C. (2006). Open Skies in ASEAN. *Journal of Air Transport Management*, Vol. 12.

Francis, L. (2004). Liberal Values. *Flight International*, October.

Francis, L. (2005). Singapore–China "Open Skies" has Restriction on LCCs. *Air Transport Intelligence News*, December 2.

Gudmundsson, S.V. and Lechner, C. (2006). Multilateral Airline Alliances: Balancing Strategic Constraints and Opportunities. *Journal of Air Transport Management*, Vol. 12, No. 3, 153–158.

Iatrou, K. and Skourias, N. (2005). An Attempt to Measure the Traffic Impact of Airline Alliances. *Journal of Air Transportation*, Vol. 10, No. 3, 73–99.

International Civil Aviation Organization, (ICAO). (2004) *Manual on the Regulation of International Air Transport* (Doc 9626, Part 4).

Kinnock, N. (1996). The Liberalization of the European Aviation Industry. *European Business Journal*, Vol. 8, No. 4, 8–13.

McDonald, M. (2006). When to Tie the Knot. *Air Transport World*, August. Retrieved on October 12, 2006 from Proquest.

Mecham, M. (2004). Fueling Star. *Aviation Week & Space Technology*, Vol. 161, No. 17. Retrieved on October 18, 2006 from Proquest.

Montreal Convention. (2006). *Convention for the Unification of Certain Rule for International Carriage by Air*, Montreal 28 May, 1999.

Oneworld. (2006). http://www.oneworld.com.

Oum, T. H., Park, J. H. and Zhang, A. (2000). *Globalization and Strategic Alliances: The Case of the Airline Industry.* Oxford, UK: Elsevier.

Oum, T. H. and Yamaguchi, K. (2006). Asia's Tangled Skies. *Far Eastern Economic Review,* Vol. 169, No. 1, 30–33.

Phillips, D. (2006). "Open Skies" Reality Still Proves Elusive. *International Herald Tribune,* June 4, 2006.

Rowell, D. (2002). Freedom of Air. *The Traveller Inside,* November 15, 2002.

Thompson, J. (2004). Come Together. *Airline Business,* November. Retrieved on October 18, 2006 from Air Transport Intelligence.

US Centennial of Flight Commission (USCOF). (2006). *International Civil Aviation.* Retrieved on October 3, 2006 from http://www.centennialofflight.gov/essay/Government_Role/Intl_Civil/POL19.htm.

Star Alliance (2006). *Star Alliance Welcomes South African Airways,* Press Release, April 10, 2006. Retrieved on August 15, 2012 from http://www.staralliance.com/en/press/southafricajoins-prp/.

Warsaw Convention (WC). (2006). *Convention for the Unification of Certain Rules Relating to International Carriage by Air, Signed at Warsaw on 12 October 1929.* Retrieved on October 3, 2006 from http://www.jus.uio.no/lm/air.carriage.warsaw.convention.1929/doc.html.

8

Competitive Market Structure and Monopolistic Markets

And while the law of competition may be sometimes hard for the individual, it is best for the race, because it ensures the survival of the fittest in every department.

Andrew Carnegie

The airline industry has been in persistent instability since the passage of deregulation in 1978. Many airlines such as Aloha Airlines, Braniff, Eastern, Hungarian airline Malev, Pan Am, TWA, People Express, Sabena the national airline of Belgium, and others have all left the market. The next two chapters of the text deal with market structure. Market structure refers to the competitive environment that surrounds the firm. It generally can be described in terms such as barriers to entry and exit, numbers of buyers and sellers competing in the market, individual seller's control over price, extent of product substitutability, and the degree of mutual interdependence between firms. Market structure determines the type of competition that is found in the industry.

This chapter has the following format:

- Perfect Competititon
  - Conditions of perfect competition
    - Homogenous product
    - Many buyers and sellers
    - Full dissemination of information
    - No barriers to entry
  - Perfect competition in the short run
  - Perfect competition in the long run
- Monopoly
  - Economic conditions of monopoly
  - No close substitutes/unique products
  - Lack of information
  - High barriers to entry
    - Legal/government barriers
    - Capital requirements
    - Technology
    - Labor unions
    - Project risks

Table 8.1 is a depiction of the market continuum displaying the four major market structures. Perfect competition occurs when there are many buyers and sellers who have very little or no control over their price. At the other extreme of the continuum are monopolies. Here the market contains only one seller who has practically complete control over the output or price. These two market structures are the focus of this chapter. Monopolistic competition and oligopolies, otherwise known as hybrid market structures, are the focus of Chapter 9.

Table 8.1   Market Continuum

	Perfect Competition	Monopolistic Competition	Oligopoly	Monopoly
Number of Sellers	Large	Many	Few	One
Type of Product	Homogenous	Differentiated	Homogenous or differentiated	Unique
Control over Price	None	Very little	Strong	Very strong
Entry Condition	Very Easy	Easy	Difficult	Impossible
Example	Agriculture	Retail	Airlines	Public utilities

PERFECT COMPETITION

A perfectly competitive industry is one in which there are a large number of small buyers and sellers who can enter and exit the industry with no restrictions. This creates a situation where the individual firm has little or no power on the price of their good. The price is dictated by the market and this makes the individual firms price takers. Based on this, each firm has a simple decision: to sell at the market bearing rate or not to sell at all. In the real world, perfectly competitive industries are a rare occurrence as most products tend to have some degree of differentiation and thus, do not satisfy the requirement of being a homogenous product. Agriculture is the closest example to a perfectly competitive industry.

Conditions of perfect competition

In order for perfect competition to exist, four conditions must be met. Since very few markets satisfy all four conditions exactly, examples of perfectly competitive markets are limited. Nonetheless, since many industries approximate these conditions, the idea of a perfectly competitive market is a useful construct when comparing market structures. The four conditions are:

- homogenous (identical) product;
- many small buyers and sellers;
- perfect dissemination of information;
- very low barriers to entry.

Homogenous product   The first condition for perfect competition is that the product must be homogenous (indistinguishable for the users). This condition flows naturally from the idea that buyers of the product should feel that they are receiving essentially the same product (regardless of which seller they purchase the product from). A homogeneous product is one in which a product sold by one seller is not distinguishable from the same product sold by another seller. The more heterogeneous the product, the more sellers can take advantage of their market position. Therefore, we should expect and in the real world indeed see sellers attempting to differentiate their product regardless of the market. However, there are many types of markets, notably those for agricultural products, where it is difficult for sellers to differentiate their product. These markets are generally thought to be good examples of competitive markets.

Many buyers and sellers   The second condition of perfect competition is that many small buyers and sellers exist in the industry. Multiple buyers and sellers enable the market to dictate the price of the product through competition between buyers and sellers. As mentioned above, a good example of perfect competition is agriculture. In agriculture there are multiple individual producers and numerous buyers so that no one buyer or seller can control the market and artificially raise (or lower) the price of the commodity. A counter example of an industry where sellers have strong influence over price is the commercial aircraft manufacturing industry. Since Boeing and Airbus are the only two major sellers of large commercial aircraft, they have a strong influence on price; the commercial aircraft manufacturing industry is clearly not a perfectly competitive market.

Full dissemination of information   The third condition of perfectly competitive markets is the generally available and full dissemination of information. In order for markets to react efficiently, information needs to be available to all buyers and sellers. If information is not available, then distortions in price may occur within the market. An example of a perfectly competitive market with full dissemination of information is the stock market. Due to the demands of investors and requirements set by the Securities Exchange Commission (SEC), companies disseminate information relating to the company's financial well-being. Using this information, investors buy or sell and this adjusts the stock price, making the market efficient. Similar scenarios occur for other perfectly competitive markets, such as agriculture, where commodity markets provide readily available information on prices. But many times we have seen the most important information about the company or firm that has not been released to the public is used illegally by an insider for personal profit.

No barriers to entry   The final condition for perfect competition is that the barriers to enter the market must be low. In order for the market to act efficiently and adjust prices accordingly, firms must be able to easily enter and exit the industry. With low barriers to entry, the market can bear many small sellers. While barriers to entry can vary by market, common barriers are financial requirements, economies of scale, market power, customer loyalty, technology, and government regulations. For example, the barriers to enter the airline industry are thought to be quite large. Airlines generally require huge capital investments, economies of scale, economies of scope, and strong customer loyalty in order to be profitable. These massive barriers to entry are the primary reason why almost all start-up airlines since deregulation have failed. These issues are discussed in more depth in the chapter on low-cost airlines.

Perfect competition in the short run

In the short run in a perfectly competitive market, firms are considered as price takers. Individual firms are not able to affect the price for their product, and must accept the market price. Since the market determines the price for the product, the market price is set where market demand equals market supply. This is displayed graphically in Figure 8.1, where the equilibrium point is the market clearing price.

While the equilibrium point between market supply and demand determines the market clearing price, individual firms perceive this price as fixed for all quantities. The reason for this is the fact that the individual firms do not produce a sufficient quantity of the homogeneous product to affect the market price. Because of this, price is displayed as a horizontal line in Figure 8.2, which equates to the equilibrium point determined in Figure 8.1. Additionally, since the price is constant for all quantities, price also equals average revenue and marginal revenue. Using a hypothetical example, Table 8.2 displays how price equals average revenue and marginal revenue for price takers.[1] The horizontal price line in Figure 8.2 can be considered the demand for a firm in a perfectly competitive industry.

Since price is held constant in the short run, a firm's output decision is to produce where marginal revenue equals marginal cost (MC). Given that the price line is also the marginal revenue curve, a firm's optimal output (Q* in Figure 8.2) is where the marginal revenue line intersects the MC curve, which is point B in Figure 8.2. At this optimal level of output, the firm is producing Q* goods at price P. When the firm's average revenue is compared to the firm's average cost, the difference between points B and C in Figure 8.2

---

1   The formulas and explanations of total, average, and marginal revenue are all contained in the cost chapter (Chapter 4).

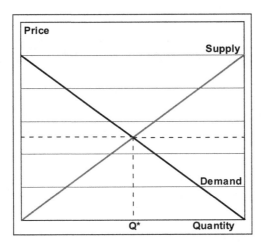

Figure 8.1   Market Equilibrium; Industry

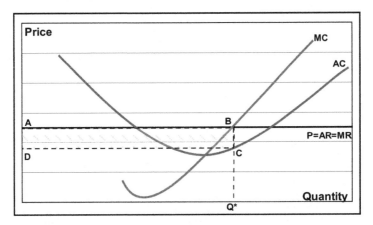

Figure 8.2   Market Equilibrium; Firm

represent the contribution margin of the firm. Therefore, the shaded box ABCD represents the total profit for the firm in a perfectly competitive market. It should also be noted that since demand is perfectly elastic, price is equal to marginal revenue and also average revenue.

However, the average cost curve may vary from firm to firm. For example, the firm in Figure 8.3 has a much higher cost structure, and instead of making a profit, the company is losing money, which is represented by the shaded region ABCD. Since the firm is a price taker, the firm has two options. The first option for the firm is to lower their cost structure to the point where they can actually make a profit or the second option is simply to shutdown. Since a firm can only continue to produce if its revenues exceed its variable costs (that is, it must meet its wage and supply bills), then the firm must shut down if it cannot cover its average variable costs. In the short run average variable costs do not include sunk costs since these must be paid regardless of output level.

Table 8.2   Total Revenue, Average Revenue, and Marginal Revenue

Quantity	Price	Total Revenue	Average Revenue	Marginal Revenue
0	$ 100	$0	$0	
5	$ 100	$500	$100	$100
10	$ 100	$1,000	$100	$100
15	$ 100	$1,500	$100	$100
20	$ 100	$2,000	$100	$100
25	$ 100	$2,500	$100	$100
30	$ 100	$3,000	$100	$100
35	$ 100	$3,500	$100	$100
40	$ 100	$4,000	$100	$100
45	$ 100	$4,500	$100	$100
50	$ 100	$5,000	$100	$100

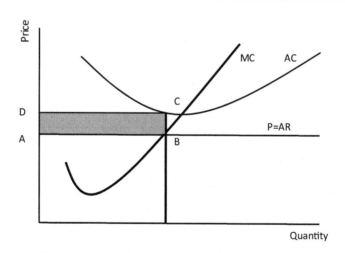

Figure 8.3   Short Run Loss in Perfect Competition

Perfect competition in the long run

The situations presented above highlight the price/output decision for firms competing in a perfectly competitive market in the short run. However, markets are not static, but continually evolving. Since the barriers to enter and exit in perfectly competitive markets are low, new firms will enter the market when the industry is profitable and firms will exit the market when the industry is sustaining losses. So, it is not the high number of firms per se (although more firms encourage more competition) that keep profits low but rather the entry of new firms.

To see the effects that market entry and exit have on firms in a perfectly competitive market, consider the situation where the firm is making an economic profit, or where the intersection between marginal revenue and MC is greater than average total cost.[2] Since firms are making a short-run profit, new firms easily enter the market (due to low barriers to entry) in hopes of also obtaining a profit. New entry ultimately increases supply in the market, causing a shift in the supply curve, which is represented in Figure 8.4. The shift in the supply curve also creates a new market clearing price, which all firms in the industry must accept.

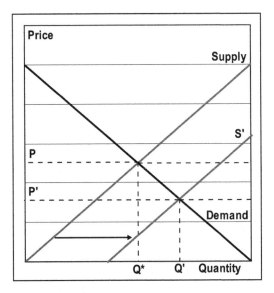

Figure 8.4   Supply Increase: Industry

Over time as more and more firms enter the market, the supply curve will continue shifting to the right. Every rightward movement of the supply curve increases market supply and resultantly decreases both the price and the quantity demanded for an individual firm. In order to maximize profits in the face of decreasing demand, the firm reduces output. These movements and market reactions occur until marginal revenue equals MC and average cost for individual firms in the market. At this point, the economic profit for the individual firm equals zero. Figure 8.5 displays the change in market output and individual firm output from a short-run scenario to a long-run scenario.

Based on these market adjustments, firms in perfectly competitive markets have zero economic profits in the long run. Since this is a long-run phenomenon, the market will continuously adjust in response to firms freely entering and exiting the market.

Market adjustments do not just occur in one direction. If some firms are losing money in the industry, then these firms will exit the market causing a leftward shift in the market supply curve. This reduces the total quantity supplied by the market, but increases the

---

2   Recall that by definition the average total cost curve contains a normal rate of return (profit) on investment. Anything above this is called an above average rate of return or economic profit.

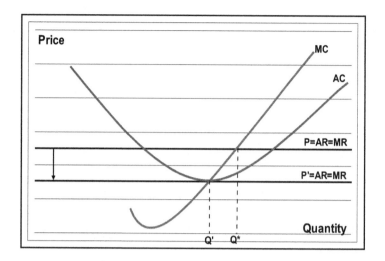

Figure 8.5   Supply Increase: Firm

demand (and the price) for an individual firm. Over time, the individual firm will return, to the point where marginal revenue equals MC and average cost and economic profit (or loss) equals zero. Figures 8.6 and 8.7 display the market reaction when firms that are losing money exit a competitive industry.

In order to change this situation, firms want to be able to differentiate their products so that they will gain some market power. Because of this, there are very few markets where all the conditions of perfect competition hold, but there are numerous markets where the perfect competition model described above is a good approximation of reality.

Suppose there is a firm serving a market and the market demand curve is equal to:

$P = \$105$

The total cost function is given by the equation:

$TC = 8,000 + 5Q + \frac{1}{2} Q^2$

The MC can be found by calculating the first order derivative of the TC function over the relevant range of production:

$MC = 5 + Q$

Given the market price, the purely competitive firm must decide what level of output to produce to maximize the profits. The profit is maximized when the market price is equal to MC.

$MC = P$

$5 + Q = 105$

Or $Q = 100$ units

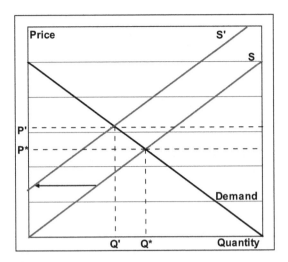

Figure 8.6   **Supply Decrease: Industry**

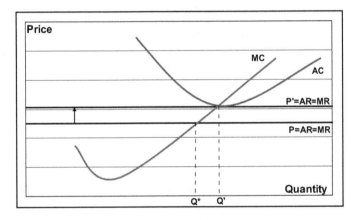

Figure 8.7   **Supply Decrease: Firm**

Therefore, the total revenue is $10,500 ($105 x 100), and the total cost of:

$$TC_{(Q=100)} = 4,000 + 5(100) + \tfrac{1}{2} (100)^2$$

$$TC_{(Q=100)} = \$9,500$$

This firm is enjoying a short run profit of:

Profit = $10,500 - $9,500 = $1,000

Since the typical firm in this industry is making abnormal profits ($1,000) in the short run there will be an expansion of the output of existing firms and we expect to see the entry of new firms into the industry. The entry of new firms shifts the market supply curve to the right and drives down the market price and the abnormal profit disappears. Perfect competition is used as a benchmark to compare with other market structures since it displays high levels of economic efficiency.

> "... Delta does not have a monopoly position at MSP. The airport already had low-fare competition with flights on Southwest, AirTran, Frontier and Sun Country before Frontier began the MCI flight. And it also has competition from American, Continental, United and Alaska."
>
> Kevin Mitchell, the Business Travel Coalition (BTC)

MONOPOLY

At the opposite end of the market continuum are monopolies. By definition, a monopoly is a market where there is only one seller. However, monopolies are all judged with respect to a certain market or geographical area. For example, the only fixed-base operator (FBO) in a small airport may be considered to have a monopoly power with respect to the airport, yet when the geographical scope is widened to include other airports, the county, or state, there are invariably other FBOs competing in the market. Thus, the market structure for providing aeronautical services such as fueling, tie-down and parking, aircraft rental, aircraft maintenance, and flight instruction would be considered a monopoly with respect to just the small airport; however, the market would be an oligopoly on a state level. At most commercial airports, car parks are monopolies and that is why people around the world complain about parking fees. Therefore, depending on the level of analysis, a monopoly can exist in many different ways. Since monopolies can exist in industries, especially when it is narrowly defined, it is important to understand the economic principles of monopoly.

Economic conditions of monopoly

A pure monopoly is an industry in which:

One Seller is the Sole Provider of a Good or a Service

The primary condition of a monopoly is that there is one seller of the product and that is usually caused by barriers to entry and exit. The word "monopoly" itself is derived from two Greek words "monos" (alone or single) and "polein" (to sell).

No close substitutes/unique product

In a monopolistic market, the single seller offers a unique product that has no close substitutes. This is the opposite of the perfectly competitive market where zero differentiation is a condition. This may exist in the medical field: suppose a new medication is discovered to treat a previously untreatable illness; the company with this new-found medication has a monopoly on the market. Although other treatments may exist, they are imperfect substitutes and cannot fully treat the illness.

Lack of information

With the lack of proper dissemination of information, sellers have better information about the product than the buyer does. Because of this the buyer's demand price is reflective of a different value placed on the product; the seller may very well be willing to accept less for the price of the good but because of lack of information, the buyer ends up paying more.

High barriers to entry

Monopolies occur largely due to the existence of barriers to entry in a given industry. As mentioned, a barrier to entry is any obstacle that makes it unprofitable or impossible for new firms to enter an industry. However, there is some disagreement as to which barriers are significant. Economists agree that government can and does erect barriers that seriously harm the economy. The common government prohibition of cabotage, for example, obviously eliminates foreign competition and allows domestic carriers to keep prices somewhat higher; incumbent carriers benefit at the expense of consumers and the industry is prevented from being as large and efficient as it would otherwise be. The issue, discussed more fully in the next chapter, is whether a barrier to entry outside of government can seriously harm the economy. Some commonly mentioned barriers are listed below and explained in the following paragraphs:

- legal/government barriers;
- capital requirements;
- technology;
- natural barriers;
- labor unions;
- project risks;
- development costs.

Legal/government barriers   A major barrier to entry, particularly in international markets, is legal or government restrictions. Government regulation can help prevent market access, creating situations where an artificial monopoly may be created. One of the most common legal barriers to entry is patents and copyrights. When a patent or copyright is held, the holder of the patent has exclusive legal right to sell the product. At first glance it might seem that these are inherently anti-competitive. When a company obtains a patent on say, an aircraft design, this prevents anyone else from manufacturing that same aircraft, increasing the likelihood that the firm will enjoy some degree of monopoly power and be able to keep prices higher. In the absence of the patent another firm might well start producing the same aircraft and drive prices down substantially. However, this simplistic, static analysis is misleading. From a dynamic, long-run prospective, this monopoly power serves to motivate far more research and development of new aircraft, thereby creating more competition and a wider variety of products.

For example, suppose there had been no patent laws back when Boeing was planning the development of the 787. Boeing would have known that once the 787 was developed and had met necessary regulatory approval, a competitor could have bought one and by simply copying the design could have eventually created copies while incurring only a fraction of the development costs that Boeing itself incurred. Knowing this, Boeing might well have abandoned the risky project, and the 787 wouldn't exist. Thus, in this case, the patent prevents too much competition in the short run, so that more new products can be developed over time; more monopoly power in the short run increases output and competition in the long run. The pharmaceutical industry also has extensive experience with monopolies, as new drugs receive legal protection, and enable the company to have a monopoly over the drug. As in the example above, however, an argument can be made that patent and copyright monopolies are justified as an incentive for research and development; that is, without patents, pharmaceutical companies would have much less incentive to spend significant resources on the development of new drugs.

The aviation industry also has extensive experience with government restrictions creating barriers to entry. During US regulation of air transportation, the Civil Aeronautics Board (CAB) dictated which airlines were to fly specific routes and in many cases created monopolies on many individual routes. Since deregulation, domestic markets no longer have extensive government barriers to entry; however, international markets are still heavily regulated, and this creates a situation where government restrictions still provide substantial barriers to entry. For example, a bilateral air service agreement may restrict access into a particular market creating a situation where a monopoly is created on a specific route. The topic of aviation monopolies will be covered in more detail later in the chapter.

Capital requirements   Another possible barrier to entry in any industry is the capital required to enter the market. The capital necessary to commence production may be sufficiently large so that the potential profits do not justify the investment, the risk is too large, or the capital cannot be obtained. Firms only enter a market when they feel that they can obtain a reasonable rate of return. Because of these factors, the larger the capital required to enter a new market, the smaller the number of firms. For example, the commercial aircraft manufacturing industry requires very large capital requirements for new entrants. A new aircraft manufacturing company would call for sizeable capital for production facilities, research and development, and general overhead expenses. The capital requirements are the major reason why extremely few firms enter aircraft manufacturing. Conversely, the capital requirements for a restaurant or small retail store are considerably less than aircraft manufacturing; restaurants are therefore far more numerous. An aviation application of capital requirements as a barrier to entry are project costs; several examples are given in Table 8.3.

Airlines require tremendous capital and physical assets to enter the commercial aviation market. While start-up airlines and start-up general aviation manufacturers often complain about the problems they have raising capital, it should be emphasized that this does not necessarily reflect any inefficiency in capital markets. Profits rarely come easy in any industry and seem to be particularly elusive for most airlines. Sometimes "no" is the efficient answer to entrepreneurs long on enthusiasm but short on viable business plans. On the other hand, capital is clearly accessible to those who do have a persuasive business plan. For instance, JetBlue obtained $130 million in start-up investment, making the airline the most heavily financed start up in US airline history (Kjelgaard, 2000).

Related to capital requirements are the prospects of profitability in the industry. If the industry has narrow profit margins in addition to large capital requirements, then it is even less likely that new firms will enter that industry. Of course, if profits are not high there is no social need for new entry.

Table 8.3   Airline and Airport Development Costs

Project	Year	Costs ($ billions)
Aircraft Manufacturers:		
Boeing 707	1957	1.3
Boeing 747	1970	3.7
Boeing 777	1994	7
Airbus A380	2005	14
Airbus A350	2012	15
Airport Industry:		
JFK Terminal 8 (American)	2007	1.3
JFK Terminal 5 (jetBlue)	2008	0.75
JFK Terminal 4 (Delta)	2013	1.2
Miami International Capital Improvement Project	2011	6.5

Sources: Adopted by the authors from Miami Airport International, Delta, JetBlue and American Airlines's press releases

Technology   Depending on the industry, technology can be a substantial barrier to entry. Without a certain required level of technology, firms may be unable to compete effectively in a market. This is especially true in high technology markets, where new technology drives sales. For example, in the microprocessor industry an entering firm requires a substantial level of technology to provide a product that might compete with Intel and AMD. Additionally, the required technology is not a one-time occurrence, but must be continually upgraded in order to keep up with the industry.

While technology gains do not create sustainable monopolies, they can create monopolies for some period of time. For over 30 years Boeing held a monopoly in the very large commercial aircraft industry with its 747 aircraft; although, as discussed in the following chapter, Boeing's profits seem to be about normal. This monopoly was finally broken when Airbus launched its double-decker A380 aircraft. While Boeing's 747 monopoly was a result of several factors, the required level of technology played a significant role in preventing other companies (until Airbus) from creating a very large commercial aircraft. Another aviation example is Aerospatiale's Concorde, which remains the only supersonic passenger aircraft to undergo commercial production.[3] Before its demise, the Concorde had a monopoly, though not a very profitable one, over supersonic commercial aircraft through a technological advantage that other companies could either not replicate, or replicate efficiently. In this case, the technology for supersonic commercial aircraft represented a significant barrier to entry.[4]

---

3   Tupolev did manufacture the TU-144, a similar supersonic commercial aircraft, but it did not go into widespread production.
4   On the other hand, many economists would argue there is no problem in any of this. The Concorde did not succeed financially—the time savings were apparently not great enough to motivate enough travelers to pay for the higher costs; regular jet service across the Atlantic turned out to be a very viable substitute. Perhaps, Boeing and others could have readily mastered the technology but had the good business sense to choose to stay out of this market.

Labor unions   Another barrier to entry, particularly in the aviation industry, can be labor unions. Labor unions essentially band workers together to bargain as a monopolist of labor supply and can thereby raise members' wages above the competitive level.[5] This monopoly power stems mainly from supportive government regulation, sometimes supplemented by direct government subsidy. For example, government prohibits employers from requiring new hires to contractually agree to not join a union, and generally limits the efforts firms can make to avoid or expel unions. Labor unions can have significant power in bargaining relationships, and this can increase the barriers to entry or completely restrict entry into certain markets. Depending on the contract negotiations, conditions may be imposed that make it difficult or unprofitable for an airline to enter a specific market.

Labor unions can also increase the barriers to entry in airline markets through scope agreements. Scope agreements are contracts with labor groups that dictate various requirements, such as the size of aircraft that regional airlines can fly. For example, the scope agreement may require all aircraft with more than 51 seats to be flown by mainline pilots, as opposed to cheaper regional affiliates. In this scenario, the barriers to enter the 70-seat regional jet market are increased as a result of the scope agreement. Scope agreements can also exert monopoly power in terms of the number of aircraft that an affiliate can fly. For example, US Airways launched discount carrier MetroJet in 1998; however, the pilot contract limited MetroJet's operation to 25 percent of US Airways, thereby restricting the number of aircraft that MetroJet could use to roughly 100 aircraft (Daly, 1998). This agreement restricted MetroJet's operations creating an artificial barrier to entry.

More broadly, unions may restrict entry into new markets by indirectly restricting capital inflows. Though one can argue that unions produce some benefits, from the viewpoint of investors, unions tend to depress rates of return and are viewed negatively. Imagine, for example, the likely surge in stock prices and borrowing prospects for legacy carriers if laws were changed making airline strikes and other union work actions illegal. The fact that such a law does not exist often makes it much harder for unionized firms to raise capital.

Project risks

> The United States, at the beginning of 1960, had 12 commercial aircraft manufacturers. By 1980, following the exit of several firms from commercial aircraft production, only three remained: McDonnell Douglas, Boeing, and Lockheed. By 1981, Lockheed's L-1011 TriStar project had been a dismal failure, costing the company $2.5 billion over 13 years and forcing it to exit the market as well. Lockheed's exit left Boeing and McDonnell Douglas as the only two US commercial aircraft manufacturers (Harrison, 2003).

Lockheed's L-1011 TriStar project is a prime example of the risk involved with aircraft manufacturing. The TriStar project experienced delays that reduced its competitiveness versus the similar McDonnell Douglas DC10; these delays ultimately doomed the project. While Lockheed still remains as a defense contractor, the financial failure of the L-1011

---

5   Note that when unions succeed, at least temporarily, in forcing wages above the completive level, this more expensive labor cost will tend to ultimately reduce employment in the union sector. In turn, workers who can't get jobs in the union sector will enter non-union markets and depress wages there. Thus, unions push wages down in some sectors even as they raise them in others; the net impact on wages tends to be about zero. Thus, it is a myth that unions tend to be a significant cause of inflation.

caused Lockheed to exit the commercial aircraft manufacturing market, leaving only two (at that time) large US commercial aircraft manufacturers.

The remaining two companies, Boeing and McDonnell Douglas, pursued separate strategic business paths. Boeing developed new aircraft types while McDonnell Douglas focused on redesigning existing aircraft types. The McDonnell Douglas strategy was not successful in the commercial aircraft market, and their market share of the US market fell below 20 percent in 1993 (Harrison, 2003). In 1996, Boeing announced a $13 billion merger with McDonnell Douglas leaving only one US commercial aircraft manufacturing firm; this created a duopoly between Boeing and Airbus in the global market for large commercial aircraft (Harrison, 2003). A similar trend occurred in Europe where BAE and Fokker exited the commercial aircraft manufacturing business. This global trend is largely a result of the enormous cost of doing business in the market.

Development costs   The primary reason for Boeing's long-held monopoly of the very large aircraft market was the very high barriers to entry in this market. Not only were the financial requirements immense, but the technological and manufacturing requirements were also huge. In 1965 when Boeing decided to develop the 747, the projected launch costs were $1.5 billion (Esty, 2001). The project was widely viewed as a "daring, bet-the-company gamble on an untested product" (Esty, 2001). In addition to the large capital requirements, Boeing needed new technological breakthroughs, and they also had to construct an entirely new manufacturing complex in Everett, Washington. And, there were times when the project appeared to be a failure.

The tremendous project risk, coupled with the other extensive technical and capital requirements, make it difficult to enter the very large aircraft market, especially since a new entrant will not have a monopoly and must compete with Boeing. However in 2000, Airbus announced its intention to develop a very large double-decker aircraft to compete with the Boeing 747 in the very large aircraft market. The project was estimated to cost Airbus $13 billion, yet with subsequent delays and problems that figure climbed to $14–15 billion (Esty, 2001). Additionally, the original breakeven forecast of 270 aircraft has climbed to 420 (BBC News, 2006). While a risky project, the new competition in the very large aircraft market will undoubtedly cause greater competition between Boeing and Airbus. Examples of competitive actions already utilized in other commercial aircraft market segments include discount prices, attractive financing, and purchase of older aircraft.

> "Boeing's problems with the 747 sounds like a litany of the damned ... (and almost) threatened the company's survival ... Boeing not only had to pay penalty fees for late deliveries, but, far worse, didn't receive the large last installments until the deliveries were made. Deprived of an adequate ... cash flow, Boeing found itself seriously short of funds yet obliged to finance a huge inventory of partly build 747s."
>
> Newhouse, 1982

Natural monopoly   Natural monopolies can occur in the market due to economies of scale. In industries that have very high fixed costs, significant economies of scale can be achieved as production increases. For example, there usually is an extremely high fixed cost in constructing a hydroelectric power plant; however, once the plant is constructed, the cost of generating extra electrical power is very low. Therefore, if there is a competing hydroelectric power plant, then each of the plants can lower their prices down to the MC of producing electricity. Neither plant can recover their fixed costs under this sort of a pricing arrangement, so that one or both of the plants must go out of business in the longer term.

In the late nineteenth century, railroads were subject to this sort of situation (sometimes called "ruinous competition"). That is, there were extremely high fixed costs in acquiring the land and constructing the railroad, but the costs of adding the extra cars and engines to carry more freight were relatively low. Therefore, competing railroads could lower their prices to just cover their variable costs (so they could still operate) but could not cover the fixed costs that they had incurred in building the line. The result was a predictable bankruptcy for one of the railroads. In this situation the bankrupt railroad was usually acquired by the competing line and a monopoly was the final result. Any prospective new entrant would be faced with the prospect of requiring extensive capital costs and faced with the prospects of losing money so the monopoly was likely to last for some time.

> In a pure monopoly a single firm controls the total supply of the whole industry and is able to exert a significant degree of control over the price. In this sense, it can be referred to as a *price maker*.

PRICE/OUTPUT DECISION FOR MONOPOLIES

The first thing to note about a demand curve for a monopolist is that the market demand equals the firm's demand. This makes sense since, with only one firm competing in the market, the firm faces the entire market. The profit-maximizing output is the point where marginal revenue equals MC. The reason for this is the fact that profit increases as long as the incremental revenue achieved from producing one additional unit is greater than the incremental production cost incurred to produce the additional unit. Figure 8.8 displays the profit maximization point for a monopolist.

In Figure 8.8, the monopolist would produce at Q^*, which is the point where marginal revenue equals MC. At a production level of Q^*, the firm would have profits equal to the shaded area between the demand curve and the average cost curve. Since new market entry is not possible (by definition) in a monopoly, these profits can be enjoyed in the long run, *ceteris paribus*. However, simply having a monopoly does not guarantee long-run profits, as a combination of shifts in the demand curve and changes in average cost could put the firm in a loss-making situation. This scenario is presented in Figure 8.9.

It should be noted that the profit-maximizing (or loss-minimizing) quantity for the firm represents the total supply to the industry. Without competition, the monopolist dominates the market, therefore the monopolistic firm effectively defines the supply side of the market and this determines the price for the market. In order to see the relationship between pricing under monopolies and other market structures, consider Figure 8.10.

As mentioned previously, monopolists will set their price where marginal revenue equals MC. Therefore the corresponding price would be P_m in Figure 8.10. Once a monopoly is no longer available and competition in the market increases, the firm will move to a price/output decision where MC equals demand. This point corresponds to P_c in Figure which lies below P_m.[6] Additionally, the quantity increases from Q_m to Q_c. Because of the

---

6   All this assumes that the production costs are the same for the tiny competitor and the huge monopolist, that no economies of scale exist. Alternatively, if there were major economies of scale then it is very possible that the lower costs of the monopolist would translate into a price below the competitive level.

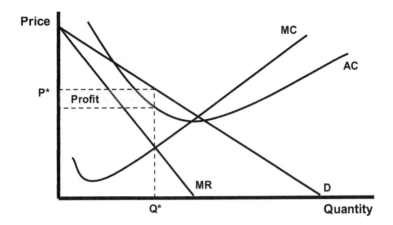

Figure 8.8   Profit Maximization for a Monopolist

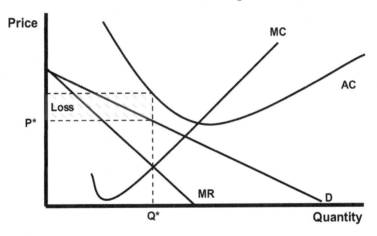

Figure 8.9   Loss Scenario for a Monopolist

increased competition, the price is lowered which increases the quantity demanded. Finally, P_r corresponds to a price where government regulation requires cost-based pricing. Under cost-based pricing, the optimal output for the firm would be where average cost equals the demand curve. This point represents a further decrease in price and a further increase in the quantity demanded. Figure 8.10 graphically portrays why prices drop when new entrants enter a formerly held protective market. For a monopoly, once an optimal output is determined, optimal price can be formulated as a function of MC and price elasticity (E_p). The formula below provides the optimal price for a monopolist to maximize total profit. Here we simply state the formula which will be derived in the next section:

$$P = \frac{MC}{1 + \dfrac{1}{E_p}}$$

However, since price elasticity is rarely constant for all levels of output, the profit-maximizing profit will also vary by price elasticity. To illustrate this point, consider Table

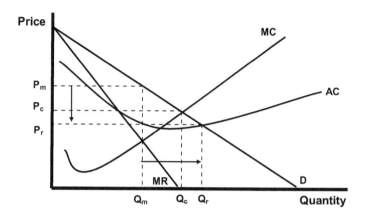

Figure 8.10   Equilibrium Price under Monopoly and Perfect Competition

8.4 which finds the profit-maximizing price for different price elasticities, assuming MC is held constant at $1,000.

As Table 8.4 depicts, the profit-maximizing price decreases as consumers become more price elastic. Eventually, consumers may become so price-elastic that the monopolist's profit-maximizing price would be equal to the MC. Based on this, cost plays a significant role in shaping the price/output decision for a monopolist and that decision varies because the elasticity of demand changes. This creates a situation where monopolists must still be reactive to the market and understand their consumers.

The following section explains the derivation of the optimal pricing formula for a monopolist as shown before. Since profit is total revenues minus total costs, the optimal price for a monopolist can be determined mathematically. Total profit (π) can be stated as:

$$\pi = TR - TC$$
$$\pi = P(Q)*Q - C(Q)$$

Since price is not constant for all levels of demand, the formula above represents price as a function of quantity [P(Q)]. The price is then multiplied by Q to take into consideration the number of units sold by the firm. Cost is also represented as a function of quantity, however it is not multiplied by Q since the function would be total cost. In order to determine an optimum price, the first order derivative with respect to quantity leads to:

$$\frac{d\pi}{dQ} = P(Q) + \frac{dP(Q)}{dQ}Q - \frac{dC(Q)}{d(Q)}$$

Setting the above equation equal to zero yields:

$$P(Q) + \frac{dP(Q)}{dQ}Q - \frac{dC(Q)}{d(Q)} = 0$$

$$P(Q) + \frac{dP(Q)}{dQ}Q = \frac{dC(Q)}{d(Q)}$$

Table 8.4   Profit-maximizing Price for Various Price Elasticities

Marginal Cost	Price Elasticity	P
$1,000	-1.1	$11,000
$1,000	-1.2	$6,000
$1,000	-1.3	$4,333
$1,000	-1.4	$3,500
$1,000	-1.5	$3,000
$1,000	-2	$2,000
$1,000	-3	$1,500
$1,000	-4	$1,333
$1,000	-5	$1,250
$1,000	-6	$1,200
$1,000	-7	$1,167
$1,000	-8	$1,143
$1,000	-9	$1,125
$1,000	-10	$1,111
$1,000	-100	$1,010
$1,000	∞	$1,000

Since price and cost per additional unit is the definition of marginal revenue and MC, the equation could be restated in those terms. However, in order to explain things further, the cost function will only be rewritten in terms of MC. The rewritten formula is:

$$P(Q) + \frac{dP(Q)}{dQ} Q = MC$$

To help solve the equation, all terms are divided by $P(Q)$:

$$\frac{P(Q)}{P(Q)} + \frac{dP(Q)}{dQ} \frac{Q}{P(Q)} = \frac{MC}{P(Q)}$$

The term $\dfrac{dP(Q)}{dQ}\dfrac{Q}{P(Q)}$ is the mathematical definition of the inverse of price elasticity; therefore the formula can be simplified into:

$$1+\frac{1}{E_p}=\frac{MC}{P}$$

Solving for the optimal price at which a monopolist should price at provides:

$$P=\frac{MC}{1+\dfrac{1}{E_p}}=MC\times\left(\frac{E_p}{1+E_p}\right)$$

If, for example, an airline is faced with a price elasticity of demand equaling (-2) and the MC of $100 per seat, the airline can markup by 200 percent and charge each additional passenger up to $200.

$$P=100\times\left(\frac{-2}{1-2}\right)$$
$$P=\$200$$

MONOPOLY PRICING AND CONSUMER WELL-BEING

To many, the pricing decision for a monopolist might seem simple: charge the market as much as possible. However while such a policy may provide extraordinary short-run profits, in the long run this practice will tend to reduce profitability. Maximum pricing will cause demand for the product to fall as consumers find ways to adapt to substitute goods or otherwise change their habits to buy less of the monopolized product. Though there may be no perfect substitute for airline travel over long distances, it is easy to see how a monopoly airline could shift travelers over to corporate jets, chartered airlines, auto travel, train travel, or utilization of communication techniques such as video conferencing. Aggressive monopoly pricing will also cause other firms to invest more in overcoming barriers and entering the market. Monopoly pricing basically pits one firm against the rest of the world and the world tends to ultimately win in that sort of contest. Therefore, even a true monopolist protected by an iron-clad barrier, such as an effective patent, tends to moderate price somewhat in an attempt to discourage the consumer adjustments and market innovations that will eventually slash monopoly profits.

Economists point out another factor, overlooked by most people, that greatly mitigates the net harm of monopoly. If a monopolist restricts output and raises price in one market this will simultaneously increase output and reduce price in other markets. If, for instance, a firm were to somehow monopolize the grapefruit market and raise prices this would result, of course, in fewer grapefruit being sold and therefore fewer resources employed in producing grapefruit. More land, fertilizer, orchard workers, and so forth would now be available for orange production so there would be a surge in production there. The harm done by monopoly in the grapefruit market would be substantially, though not completely, offset by cheaper and more abundant oranges. Likewise, any monopoly pricing in some air travel markets would shift aircraft and employees to other markets and reduce prices for consumers there. There would still be some net harm since this reallocation of resources is triggered by monopoly manipulation rather than consumer preference, but the point is that the net cost of any monopoly is far less than most people realize.

MONOPOLY MARKET POWER IN AVIATION

The three industries that were to be discussed below, aircraft manufacturing, jet engine manufacturing, and airports are all industries with firms that are capable of exerting substantial market power and, depending on the level of analysis, can easily be classified as monopolies. The airline industry is somewhat unique in that its major suppliers all have substantial market power and, depending on circumstances, can exert near monopoly power. However, airlines complete fiercely in a market that more often resembles an oligopoly, which will be discussed in more detail in the next chapter.

Airports

Another aspect of the aviation industry where market structure plays a significant role in shaping the industry is airports. Once again, the major issues concerning airports are the significant barriers to entry. The lack of available gates and slot controls at some airports act to protect incumbent carriers and make it difficult for new entry. And at hub airports, it can allow an airline to exert monopoly power.

Gate and facility availability   For an airline looking to serve a new airport, one of the first requirements is the availability of gates and ticket counters. Without the availability of both, the airline will not be able to service the airport. While these requirements are universal, the ease of obtaining them varies considerably worldwide.

In the US, the majority of gates and ticket counters are closely held by the individual airlines either through lease agreements with the airport or under full ownership. Leases with the airport provide the airline with exclusive control over a set number of gates, enabling the airline to utilize the facilities as they see fit. In a practical sense, the airline acts as the owner of the gates so that the airline can update their facilities as required. In return for this control over airport assets, airlines must make lease payments that are prescribed in a lease agreement with the airport authority. Depending on the term of the lease agreement, airlines may be stuck with the facilities unless they can find another user for them. In some instances, airlines have found it profitable to sub-let these facilities to competing airlines. And in other instances airlines have found success in terminating the lease agreements under chapter 11 bankruptcy, although this is a rather drastic remedy.

At hub airports, airlines may own facilities. For example, Northwest Airlines invested heavily in their World Gateway hub at Detroit International Airport, while Continental Airlines has ownership over Terminal E at Houston Intercontinental Airport. Under these scenarios, airlines may invest a large amount of capital into the airport to construct or revamp an entire terminal.

Regardless of the technical structure, airlines in the US have significant control over airport facilities. This control represents a significant barrier to entry as any new airline will have to acquire airport facilities in order to commence service. At airports where there are idle facilities, this usually does not pose a significant problem. However, at airports with scarce resources, this can pose a significant problem to the new entrant. While the airport authority will attempt to work with all the airlines to accommodate the new entrant airline, this may not always work if the resources are all committed. The other option for the new entrant would be to enter an agreement with an incumbent airline to either use their facilities or sub-lease facilities. This option is usually costly, and there may be some instances where, in order to suppress competition, incumbent airlines may not be willing to allow a new entrant to the airport. This is particularly true at hub airports where the hub carrier may go to extensive means to obtain gates in order to block new entrants, especially low-cost competitors.

Moreover, such a structure may not be economically efficient since airlines may sign gate leases to block competition, even if the airline does not require the gate. Even if leases contain minimum usage requirements, airlines can adjust their schedule or add extra flights to meet the minimum requirements. This creates a situation where the assets may not be used in the most efficient manner.

Another solution to airport facility usage is implemented by the majority of airports outside of the US and even by a few airports within the US. Common use facilities are facilities that all airlines can use and are assigned by the airport on a daily basis depending on demand. By doing so, facility utilization is maximized. Under a common use system, airlines pay a fee per use for using the facilities. In theory, since the facilities are being utilized more intensely, the cost per use should be diminished. However, airlines dislike common use because they have less control over their facilities and their operation. Common use facilities generally do not provide significant barriers to entry, since if facilities are available, an airline can use them. While facilities may be full at peak periods, it is unlikely that an airline could not enter an airport using common use facilities at some time of the day. Therefore gates and other airport facilities do not generally provide a significant barrier to entry outside of the US.

Slot controls   However, in Europe and other parts of the world, barriers to entry remain in the form of airport slots. A slot is the right to land or takeoff from an airport at a given time. More formally, article 1 of the European Council Regulation No. 95/93, the key regulation concerning airport slot allocation in Europe, define a slot as:

> ... the entitlement established under this Regulation, of an air carrier to use the full range of airport infrastructure necessary to operate an air service at a coordinated airport on a specific date and time for the purpose of landing and take-off as allocated by a coordinator in accordance with this Regulation (Commission of the European Communities, 2001).

Slots are allocated through a variety of mechanisms; however there are two primary rules to slot allocation according to EEC No. 95/93. The first rule, the "grandfather right," entitles an airline to the same slot in the future (if they are currently using it). While this rule was enacted to provide stability, in practice it provides the airline with quasi-ownership of the slot. Slot usage is determined by the "use-it or lose-it" rule that states that the airline must use the slot for at least 80 percent of the time during the scheduled period. If the airline fails to meet this requirement then the grandfather right does not apply and the slot is lost. This provides an incentive for the airline to continue using the slot, even if it is not economically efficient. The remaining slots are then dispersed to applicants, with only 50 percent of new slots allocated to new entrants (Matthews and Menaz, 2003). Based on the grandfather right, the use-it or lose-it rule, and the new entrant slot limit, very few slots become available for new entrants. For example, in the summer of 2000, 97 percent of London Heathrow's slots were grandfathered, leaving only 3 percent of the total slots available, with only 1.5 percent available to new entrants (DotEcon, 2001). Moreover, in all likelihood the available slots were at inconvenient times. These slot controls represent a significant barrier to entry and the grandfather right enables an airline to accumulate a significant number of slots over time.

According to EC 95/93, the allocation of slots through the formal slot allocation process is the only legal method to obtain a slot in Europe. While swapping of slots is permitted, a 1999 UK court ruling opened the door to the "grey market," where slot swapping was permitted with monetary compensation (Matthews and Menaz, 2003). This ruling created a situation in the UK where slots could be traded, leased, and sold to other airlines; however, such practices are considered illegal in the rest of Europe. The grey market is opposed throughout the rest of Europe under the rationale that a private firm should not benefit financially from a public good (Mackay, 2006). However, this rationale certainly favors incumbent airlines over new entrants into the market. While the grey market does reduce the barriers to entry, it shifts the entry requirements from a legal/structural requirement to a financial requirement. From an economic point of view financial barriers are more desirable than legal barriers since, under financial barriers, the slot is allocated to the firm that is willing to pay the highest price and therefore values the slot the most. With the recent open skies agreement signed between the EU and the US, the grey market will probably become more active, especially for slots at London Heathrow.

The US has limited experience with slot controls largely due to the fact that the US has been successful at constructing additional runways (Mackay, 2006). However, in 1969 the Federal Aviation Administration (FAA) implemented a slot system at four airports: New York LaGuardia, New York Kennedy, Chicago O'Hare, and Washington National. In 1985, the FAA permitted a full secondary market for slots, similar to the UK's grey market, where slots could be sold, traded, or leased (Department of Justice (DOJ), 2005). However secondary slot trading was not as successful as planned due to two major issues: market power, and uncertainty of duration and value.

Airlines decided to retain slots instead of sell/lease them in order to prevent new carriers from entering the market. Market power also caused airlines to increase their slot ownership in order to become more dominant. This was especially true at Chicago O'Hare where United and American both increased their slot ownership, thereby making the airport less competitive. Finally, slots were only deemed temporary by the FAA, creating uncertainty over the lifetime and true value of the slot (DOJ, 2005).

A lack of slot trading made the system relatively ineffective and the slot mechanisms were eventually abolished at all four airports. As a result of the slot removal, airlines immediately increased service to the airports, especially at LaGuardia and O'Hare where tremendous delays were experienced (Mackay, 2006). This caused the FAA to mediate with the airlines to reduce the number of flights to acceptable levels. The immediate increase in service to these airports illustrates that a slot system represents a barrier to entry, even with a buy/sell market for slots.

As a result of either slots in Europe or airport facilities (gates, ticket counters) in the US, new entrants face significant barriers to entry at the airport level. Additionally, Dresner,

Windle, and Yao (2002) determined that high gate utilization during peak periods was the major airport barrier to entry in the US. All of these factors have allowed incumbent carriers to increase their market power at airports, and in particular, at hub airports. The near monopoly power that some airlines have acquired at hub airports has been a significant barrier to entry at these airports.

Hub airlines Many studies have investigated the hypothesis that airlines have exerted their strong market power at the hub airport in the form of hub premiums. The hub premium hypothesis states that originating or terminating passengers are charged a higher fare than other passengers traveling throughout the carrier's system (Gordon and Jenkins, 1999).

Opinions on whether hub premiums actually exist are mixed. Gordon and Jenkins (1999) used proprietary Northwest Airlines data to show that there was actually a hub discount at Minneapolis-St. Paul airport. Additionally, other studies have found that the hub premiums do not exist. On the other hand, some studies have shown that hub premiums exist. Borenstein (1989) and Lijesen, Rietveld and Nijkamp (2004) claimed that a few carriers in Europe, specifically Lufthansa, Swiss, and Air France charged hub premiums. However, even if hub airlines do charge higher prices, this may simply reflect higher quality not adjusted for in such studies. For example, the dominant carrier may have more amenable airport facilities, may benefit from goodwill in the community where they are perceived as the "home-town company," may be perceived as having established greater credibility on safety, or have higher-valued frequent flier awards. The fact that most businesses and communities seem to like having a hub airline nearby supports this view.

Since the evidence for a hub premium is not conclusive and since hub airlines are frequently in bankruptcy, it stands to reason that if any actual premium exists it is likely to be small. Furthermore, even if there is some slight monopoly power at that airport the positive network effects from hub-and-spoke systems also produce benefits that, as Economides (2004) shows, may be substantially greater than the damage of higher prices at the hub. That is, the efficiency gained from economies of scale, scope, and density in the hub network may help reduce prices and increase product availability in the overall network.

While specific studies appear to have differing conclusions on hub premiums, the actual case of Pittsburgh International Airport provides a good real-world example of what may happen when a hub airport is open to competition. In fall 2004, US Airways announced that it was going to significantly scale back its Pittsburgh hub. The total number of flights at Pittsburgh was dramatically reduced and a greater number of gates became available. As the barriers to entry were lowered, low-cost carriers (LCCs) Southwest, JetBlue, and Independence Air all entered the market. While the total number of passengers using the airport decreased, originating passengers increased by 12 percent, (McCartney, 2005). Moreover, airfares dropped significantly, especially on a few dominated routes. For example, the average airfare between Pittsburgh and Philadelphia (a route between two US Airways hubs) fell from $680 to $180 (McCartney, 2005).

Commercial aircraft manufacturing

Today there are four major aircraft manufacturers, with two competing fairly evenly in each market. Boeing and Airbus compete in the large commercial aircraft market, comprising aircraft over 100 seats, while Bombardier and Embraer compete in the regional jet market.

A new entrant to the regional aircraft market in the 75- to 95-seat category, the Sukhoi Superjet made its maiden flight in May of 2008.[7] Unfortunately, as a result of a deadly crash during a demonstration flight in Indonesia in May of 2012, its future is in jeopardy.

Boeing and Airbus compete aggressively with each other in almost all segments of their individual markets. Until recently, the biggest exception to this rule was the very large aircraft market, where Boeing had a monopoly with its 747. However, with the introduction of the Airbus A380, Boeing will no longer continue to enjoy a monopoly in that market.

While the broad market for commercial aircraft is largely a duopoly, for some airlines the market is more like a monopoly. Thus, while most airlines are able to play one manufacturer against the other in negotiations, some are more constrained by operational limitations. For example, consider Southwest Airlines, an all-Boeing 737 operator (before the acquisition of AirTran's 717s). Part of Southwest's success has been a common fleet type; this has increased crew flexibility, reduced crew training cost, and reduced spare part inventories, to name just a few benefits. In this situation, switching to an Airbus aircraft would entail substantial costs. This may put Boeing in a position to exert a degree of monopoly power over Southwest Airlines. However, this is greatly tempered by the resale market for aircraft; Southwest could even buy new aircraft indirectly through another airline that was better positioned to strike bargains with Boeing, then immediately resell to Southwest.

Furthermore, Boeing must also consider the possibly severe damage to their reputation if they ever were somehow able to exploit such situations. If they are perceived to have betrayed one of their best and most loyal customers, then other airlines would be careful to avoid repeating Southwest's mistake. It is likely that any possible gain from exploiting Southwest would be far less than the loss from declining sales as the world learned that it was a grave mistake to become too dependent on Boeing. Indeed, since anyone is likely to be hesitant about relying too much on a single supplier, it may be vital for Boeing to be able to point to a very satisfied, successful "dependent" in order to encourage other airlines to, as much as possible, follow Southwest's example.[8]

Modern aircraft manufacturing is a heavily capital-intensive industry requiring immense expenditures in research, development, and manufacturing. As technology has increased and economies of scale benefits have become more important, the cost of designing and marketing an aircraft have become substantial, strengthening the barriers to entry into the industry. This trend has also seen the number of firms competing in the commercial aircraft industry drastically reduced. Therefore, there are now very few firms competing in the industry, creating a market structure that resembles an oligopoly, which will be discussed in greater detail in the next chapter.

Jet engine manufacturing

An industry with a market structure similar to the commercial aircraft manufacturing industry is the commercial aircraft jet engine manufacturing industry. While the market broadly has three major firms, General Electric (GE), Rolls-Royce, and Pratt & Whitney

---

7   ANTARA News, May 11, 2012.
8   Also, if Boeing bargained too aggressively then Southwest could ultimately switch over. This scenario may have already occurred when easyJet placed a large order for Airbus aircraft after the airline had previously been a consistent Boeing customer.

(P&W), plus two major joint ventures (CFM International (CMFI) and IAE), just as in the aircraft manufacturing industry there may be certain situations where the market is in reality a duopoly or monopoly. For example, on the new Boeing 787 only two engines are being offered to customers: GE and Rolls-Royce. Therefore, the competitive actions of the firms will largely be based on duopoly competition theory. One step further, all Boeing 737NGs are powered by CFMI engines, meaning that CFMI has a monopoly on the supply of engines for 737 aircraft. This creates an interesting dynamic as an engine's success is highly tied to an aircraft's success. Moreover, if CFMI wants to enforce monopoly power by attempting to raise the engine's price, this would hurt the competitiveness of the 737. However, with CFMI also being one of two suppliers for the rival Airbus 32X family, CFMI holds an extremely strong position in the narrow body jet engine manufacturing industry.

The capital requirements for engine manufacturing are quite large and, for the existing firms, have generally come mainly from earlier military contracts. The "Big Three" all have roots tracing back to military applications. Rolls-Royce and P&W were heavily involved with engine manufacturing during the Second World War, while GE made the jump from military applications in the late 1960s (Smith, 1997). Without military applications and grants, the required technology would be extremely expensive to acquire for a new entrant. If a new aircraft project looked profitable, then undoubtedly one of the "Big Three" would become involved, leaving new entrants the less desirable aircraft designs.

While some market power can be held by the engine manufacturers, airlines routinely play one engine manufacturer against the other in order to get the best deal. This means that engine manufacturers may have to provide deep discounts on engine purchases, but are able to recover through maintenance agreements or "power by the hour" contracts. Therefore, whenever possible, engine manufacturers capitalize on economies of scope benefits. This is especially true of GE which leases aircraft with GE engines through GE Commercial Aircraft Services (GECAS). Economies of scope represent another barrier to entry into the industry as it provides a competitive advantage to the incumbent firms.

MONOPSONY

A monopsony is a market condition in which only one buyer faces many sellers. The defense industry in the US may be a monopsony in which there is only one buyer, the US government, and there are several sellers. For example, the Boeing B-1 Lancer bomber is a prime example of a monopsony structure, where the bomber is used exclusively by the United States Air Force (USAF). A single-payer health care system, in which the government is the only buyer of health care, is another example. A market condition consisting of only one buyer and only one seller is called a bilateral monopoly. For example, in some countries the national airline is the only employer of pilots, but there is only one supplier of pilot from the pilot union members.

SUMMARY

This chapter introduces the reader to the concept of market structures. The chapter covers the more theoretical economic models of a competitive market and also those of a more monopolistic market. The competitive market model covers the conditions necessary for a competitive market and the overall determination of price within such a market—to include the price faced by an individual competitor or firm. It also outlines the long-term equilibrium solution within such a market. The chapter covers a monopolistic market in a similar fashion to include the price output decision for the monopoly firm. The chapter also discusses some of the reasons that a monopoly might arise and some of the factors that might act to reduce monopoly power. And, although very few markets are perfect examples of competition or monopoly, the models are useful as benchmarks against which one can measure real-world market structures. The last part of the chapter contains a discussion of the market structure of the various parts of the aviation industry and how they may approximate monopolistic conditions under certain circumstances.

DISCUSSION QUESTIONS

1. What are the four basic market structures?
2. What is a real-life example of a monopsony?
3. What is normal profit?
4. State the four basic conditions that characterize a competitive market.
5. Does perfect competition exist in the real world?
6. Are perfectly competitive firms "price makers" or "price takers"? How is the profit-maximizing price and quantity decided upon by the perfectly competitive firm?
7. What happens when markets do not have enough competition?
8. What are the barriers to entry in the airline industry?
9. Is the aircraft industry an example of perfect competition?
10. Is it true that a monopoly can charge any price and customers will still have to buy the product? Do you agree or disagree? Why?
11. You are the manager of a firm that sells its product in a competitive market at a price of $50. Your firm's cost function is $TC = 40 + 5Q^2$.
    a   Calculate the profit-maximizing output.
    b   Calculate the total profit
12. Explain how the price set by a perfectly competitive firm compares with the price under monopoly competition.
13. Assume a firm is operating under perfect competition with the following total cost and demand curves:

    $$TC = 100 + 0.50 \times Q^2$$
    $$P = 50$$

    a   Find equilibrium price, and quantity.
    b   Is the firm making a profit or loss?
    c   Is this a long-run or short-run equilibrium?
14. GE is the only producer of new jet engines for general aviation aircraft. Demand for a single engine is $P = 2,000,000 - Q$ while the MCs of producing an engine are: MC = 1,999 Q.
    a   What would be the monopoly price and quantity of these engines?
    b   What economic profit would GE earn on the sale of these engines?
    c   What would happen to price and quantity if the market were competitive (assuming the same costs)?

15. Suppose you are the manager of NavGas, a major FBO, exclusively serving general aviation aircraft. Based on the estimates provided by a consultant, you know that the relevant demand and cost functions for your product are:

    $Q = 25 - 0.5P$

    $TC = 50 + 2Q$

    a   What is the FBO's inverse demand function?
    b   What is the FBO's marginal revenue when producing four units of output?
    c   What are the levels of output and price when you are maximizing profits?
    d   What will be the level of profits?
16. Explain what will happen in the long run in a competitive market when some of the firms are not covering their average total costs.
17. Is it true that a monopoly firm will always make a profit? Explain why or why not.

REFERENCES

BBC News. (2006). *Airbus Hikes A380 Break-even Mark*. Retrieved on March 28, 2007 from http://news.bbc.co.uk/go/pr/fr/-/1/hi/business/6067540.stm.

Borenstein, S. (1989). Hubs and High Fares: Dominance and Market Power in the US Airline Industry. *Rand Journal of Economics*, Vol. 20, No. 3, 344–365.

Commission of the European Communities. (2001). *Proposal for a Regulation of the European Parliament and of the Council amending Council Regulation (EEC) No. 95/93 of 18 January 1993 on Common Rule for the Allocation of Slots at Community Airports*, June 20. Retrieved on April 4, 2007 from http://www.ectaa.org/ECTAA%20English/Areas_dealt_with/en_501PC0335.pdf.

Daly, K. (1998). Winds Rise in the East. *Airline Business*, September. Retrieved on March 12, 2007 from Air Transport Intelligence.

Department of Justice (DOJ). (2005). *Congestion and Delay Reduction at Chicago O'Hare International Airport*. Comments of the United States Department of Justice. Docket No.

FAA-2005-20704. Retrieved on April 5, 2007 from http://www.usdoj.gov/atr/public/comments/209455.pdf.

DotEcon. (2001). *Auctioning Airport Slots*. A Report for the HM Treasury and the Department of the Environment, Transport and the Regions. Retrieved on April 4, 2007 from http://www.dotecon.com/publications/slotauctr.pdf.

Dresner, M., Windle, R. and Yao, Y. (2002). Airport Barriers to Entry in the US. *Journal of Transport Economics and Policy*, Vol. 36, No. 3, 389–405.

Economides, N. (2004). *Competition Policy in Network Industries: An Introduction*. NET Institute Working Paper No. 04-24. Retrieved on May 2, 2007 from http://ssrn.com/abstract=386626

Esty, B. (rev. 2001). *Airbus A3XX: Developing the World's Largest Commercial Jet (A)*. Harvard Case, 9-201-028, August 24.

Gollish, D., Clausen, H., Koggersvol, N., Christey, P. and Bruner, R. (rev. 1997). *The Boeing 777*. Darden Case, UVA-F-1017.

Gordon, R. and Jenkins, D. (1999). *Hub and Network Pricing in the Northwest Airlines Domestic System*. Washington, DC: The George Washington University.

Harrison, M. (2003). *US versus EU Competition Policy: The Boeing-McDonnell Douglas Merger*. American Consortium on European Union Studies on Transatlantic Relations Cases, 2.

Kjelgaard, C. (2000). JetBlue Holds $50m-plus Cash Despite Cost Over-run. *Air Transport Intelligence News*, July 2. Retrieved on March 9, 2007 from Air Transport Intelligence.

Lijesen, M., Rietveld, P. and Nijkamp, P. (2004). Do European Carriers Charge Hub Premiums? *Networks and Spatial Economics*, Vol. 4, No. 4, 347–360.

Mackay, L. (2006). *Overview of Mechanisms to Deal with Airport Congestion*. Unpublished work, Embry-Riddle Aeronautical University.

Matthews, B. and Menaz, B. (2003). *Airport Capacity: The Problem of Slot Allocation*. Retrieved on April 4, 2007 from http://www.garsonline.de/Downloads/Slot%20Market/031107-matthews.pdf.

McCartney, S. (2005). The Middle Seat: Why Travelers Benefit When an Airline Hub Closes. *Wall Street Journal*, November 1. Retrieved on April 2, 2007 from Proquest.

Newhouse, J. (1982). *The Sporty Game*. New York: Alfred A. Knopf.

Smith, D. (1997). Strategic Alliances in the Aerospace Industry: A Case of Europe Emerging or Converging? *European Business Review*, Vol. 97, No. 4, 171–178. Retrieved on March 27, 2007 from Emerald.

9

Hybrid Market Structure and the Aviation Industry

United has little to fear from numerous small competitors. We should be able to compete effectively by advertising our size, dependability, and experience, and by matching or beating their promotional tactics... In a free environment, we would be able to flex our marketing muscles a bit and should not fear the treat of being nibbled to death by little operators.

Richard Ferris, CEO United Airlines, 1976

While the previous chapter introduced the two extremes of the market structure continuum (Table 9.1), this chapter will analyze the two middle hybrid market structures: monopolistic competition and oligopolies. Unlike perfect competition which rarely exists in actuality, both oligopolies and monopolistic competition are prevalent in modern industry, with the airline industry heavily influenced by the characteristics of oligopolies. The mergers between US Airways–America West, Delta–Northwest, and Continental–United have made this oligopolistic industry more concentrated. The acquisition of AirTran by Southwest has promoted Southwest as the largest US carrier by domestic passenger volume and brought it within striking distance of the new domestic revenue passenger mile (RPM) leader, United–Continental. This chapter will analyze both hybrid market structures and show how the market structure impacts companies operating in this environment. The general outline for the chapter is:

- Monopolistic Competition
  - Price-output decision
- Oligopolies
  - Differing views of oligopolies
  - High cost of capital
  - High exit barriers
- Examples of Oligopoly
  - Airlines
  - Commerical aircraft manufacturing
  - Jet engine manufacturing
- Contestability Theory
- Kinked Demand Curve Theory
- Cournot Theory

- Price-Output Determination under Oligopolistic Market Structure
- Profitability Issues
- Competition and Anti-trust Issues
  - Predatory pricing
  - Cartels and collusion
- Industry Consolidation
  - Four-firm concentration ratio
  - Herfindahl-Hirschman Index
- Beyond Market Concentration Considerations
- Anti-trust, Market Evolution and Cooperation
- Summary
- Discussion Questions

MONOPOLISTIC COMPETITION

Monopolistic competition is, perhaps, the most common market structure of the four types displayed in the market continuum in Table 9.1. Monopolistically competitive markets contain many sellers, but not quite to the degree of a perfectly competitive market. Therefore, each firm produces a small fraction of industry output. In other words, there exists little market consolidation. Like perfect competition, it is relatively easy to enter the market. There are of course still costs of entry. The key difference is that the ease of obtaining capital is considerably less in a monopolistically competitive market than in an oligopoly market. Other barriers to entry may include customer loyalty or regulatory restrictions. A monopolistically competitive industry has the following characteristics:

- a large number of sellers;
- low barriers to entry;
- product differentiation;
- full dissemination of information.

Some examples of monopolistic competition include:

- accounting firms;
- books;
- convenience stores;
- garment industry;
- jewelry shops;
- law firms;
- radio stations;
- restaurant industry.

The key difference between perfect competition and monopolistic competition is that in perfect competition companies mainly sell homogeneous products while monopolistically competitive firms sell heterogeneous products with many close substitutes. Firms in a monopolistic competitive environment sell products that are somewhat unique, and have

Table 9.1   Market Continuum

	Perfect Competition	Monopolisitic Competition	Oligopoly	Monopoly
Number of Sellers	Large	Many	Few	One
Type of Product	Homogenous	Unique	Homogenous or differentiated	Unique
Control over Price	None	Very little	Strong	Very strong
Entry Condition	Very easy	Easy	Difficult	Impossible
Example	Agriculture	Retail	Airlines	Public utilities

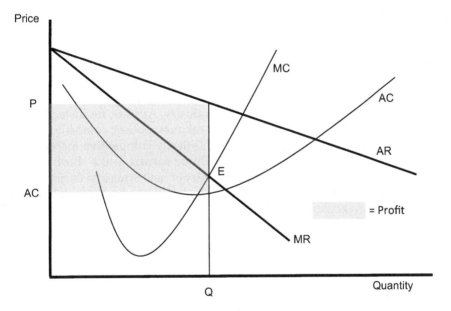

Figure 9.1   Short-run Equilibrium: Monopolistic Competitive Market

some power to set their own prices. From a firm's point of view this is a more desirable situation, since the more a product can be differentiated, the more control a firm has on its price. Since all the firms have the same incentive, there is still a good deal of price competition in monopolistically competitive markets.

Monopolistic competition is extremely common in today's business environment including bookstores, grocery stores, and pharmacies. Restaurants are a prime example of monopolistic competition since they provide somewhat unique services, are fairly common in most markets, and exist in a market with relative ease of entry.

Price–output decision

Demand for a firm's product in perfectly competitive markets is essentially horizontal since the firm is a price taker and has virtually no power in setting prices. In this situation, the elasticity of demand is perfectly elastic. As the amount of product differentiation increases, the elasticity of demand decreases, which shifts the firm away from a perfectly competitive market and toward a more monopolistically competitive or oligopolistic market. As mentioned above, this increase in product differentiation creates the potential for a firm to have more control over the price that they charge. This situation is depicted in Figure 9.1 above. That is, the demand curve for a firm shifts from a horizontal line to a downward sloping line. As the firm progresses through the market structure continuum, the slope of the demand curve will become steeper until it eventually reaches a point where it encompasses the entire market, and it is at this point that the firm becomes a monopoly. Just as in a monopoly, in a monopolistically competitive environment, firms produce the quantity (Q) where marginal cost equals marginal revenue (MR) and charge the highest price (P) that sells that quantity.

$$MC = MR$$

$$P > MR$$

The price the firm charges would be on the demand curve. As should be clear from the figure, the more elastic the demand curve facing the firm, the less control the individual firm has over its price. Figure 9.2 illustrates the typical demand curves across the various market structures. Oligopoly is an especially complex case and the nature of demand can vary significantly given the particulars of each industry. Airlines, for instance, often seem to have no more control over price than the typical monopolistic competitor.

Since it is relatively easy to enter a monopolistically competitive market, we would expect that any above–normal profits that might be earned in the short run would be competed away in the longer term through the entry of new firms into the industry. Figure 9.3 shows the longer-term equilibrium in a monopolistic competitive industry where a typical firm earns zero economic profit.

OLIGOPOLIES

The next step along the market continuum from monopolistic competition is oligopoly, which is characterized by the market dominance of a few firms. Oligopoly is the most relevant market structure with regard to aviation and its relevance will be covered throughout this chapter. Despite the large number of airlines that operate globally, individual routes are typically served by one or two dominant carriers and a handful of lesser competitors. Major airlines like Singapore Airlines, Cathay Pacific, British Airways, Air France, and United Airlines operate their routes with only a few close competitors, but they are also influenced by competition from smaller low-cost airlines. Similarly, the aircraft manufacturing industry is dominated by a handful of manufacturers (Boeing, Airbus, Embraer, Bombardier, and Sukhoi Superjet[1]).

---

1   The Sukhoi objective is to compete effectively with its Embraer and Bombardier counterparts by offering substantially lower operating costs.

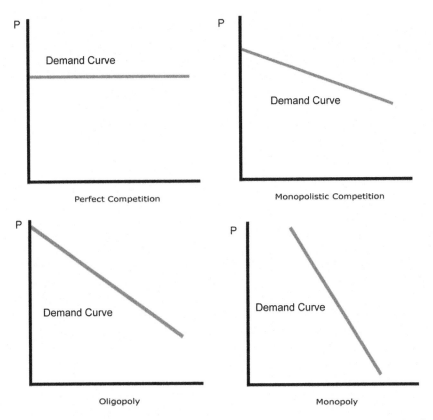

Figure 9.2   Demand Curves as the Firm Progresses through the Market Continuum

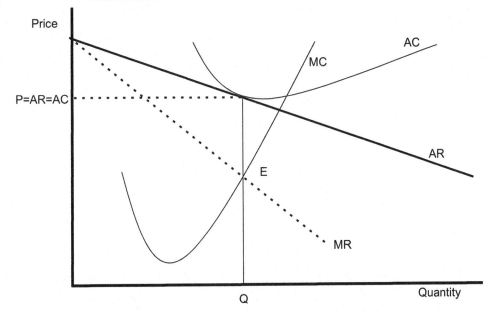

Figure 9.3   Longer Run Equilibrium: Monopolistic Competitive Market

Unlike perfect competition or monopolistic competition, the actions of one firm substantially affect the market. These actions often alter the actions of competitors. This creates a complex interdependence amongst the firms; that is, each firm's actions will be conditioned on how they believe the competition will react. For example, an airline might be more likely to reduce fares if it thought competitors would leave their fares unchanged, but would prefer to leave fares constant if it believed competitors would instantly match price cuts. Thus, each airline's pricing is based, in part, on what it believes its competitors' actions and reactions will be. Clearly this is a complex problem and one that often presents no clear, optimal strategy.

Therefore, almost any short-run outcome is theoretically possible in oligopoly. It might be possible, for instance, that firms would sometimes practice "tacit collusion" where they keep prices relatively high and "go along to get along" by avoiding any aggressive competitive act that would lead to price wars. On the other hand, oligopoly can, as seems too often the case for airlines, produce aggressive "cut throat" competition where the average firm is routinely operating in the red.

Though normal long-run profits are a given in monopolistic competition, it is theoretically possible for long-run profits to be above normal in oligopoly if there is a sufficiently high barrier to entry. Many oligopolies including such former paragons as General Motors struggle just to earn normal long-run profits. Indeed, most legacy airlines' long-run profits have been well below normal. The barriers to entry in oligopoly markets keep the number of competitors relatively small. These barriers include high start-up/ fixed costs, existence of sizeable economies of scale, control over scarce resources, or exclusive patent/legal rights.

Oligopoly is often viewed as inherently undesirable—better than monopoly but not nearly as good as perfect competition. While most economists would probably agree that there is a certain amount of truth to this perspective, there are some complications. A few large firms can often enjoy economies of scale and produce at far lower costs and sell at far lower prices than could an industry composed of smaller, more numerous firms. It is far cheaper for Boeing or Airbus to develop and produce 1,000 aircraft than it would be to have 100 small manufacturers develop and produce ten comparable aircraft each. Boeing or Airbus can spread research and development costs over more units, thus pricing them lower, and can benefit from the experience gained, becoming increasingly efficient with each additional aircraft produced. There is no doubt that airlines and air travelers are better served by having two manufacturers of large aircraft rather than having 200. Economies of scale, scope, and density are also important in the airline industry so, again, we are probably better off with an airline oligopoly than any feasible alternative.

This does not mean, of course, that a movement to fewer, larger firms is automatically more efficient and better for consumers. Such industry consolidation might be beneficial but it might also artificially suppress competition. So, how can we decide, say, when two competing airlines should be allowed to merge? Let us sketch two different views.

Differing views of oligopoly

According to a common school of thought there is a presumption that any substantial increase in market concentration is generally undesirable and therefore oligopolistic competitors should generally not be allowed to merge. For the airline industry, the assumption is that as the industry consolidates through mergers and acquisitions, airfares will increase as competition on routes is eliminated. This theory is further examined later in the chapter. The main exception is the situation where denying a merger cannot prevent increased market concentration because the weaker firm will simply liquidate if the merger is not approved. This was the rationale when Boeing acquired McDonnell Douglas in 1996. Underlying this staunch anti-merger view is the perception that efficiency gains from economies of scale are likely to be less significant than the increased danger of oligopoly abuse through increased pricing power and artificially high profits.

An important foundation for this view is the belief that new entry is often exceedingly difficult because many barriers to entry tend to be naturally powerful. The staunchest proponents of this view would even argue, for example, that advertising and brand name recognition are potentially enough in themselves to seriously impede new entry and thereby allow oligopolies to enjoy high profits even in the long run (see, for example, Galbraith, 1979). An alternative view called the market process view, is that very few, if any, barriers to entry (other than legal barriers erected by government) are significant. According to this theory, it is best to let the market evolve in whatever way firms choose. Efficiency gains are likely and will generally be impossible for government regulators to estimate and predict so it is best for them to merely stand aside and let the market process work. High profits, as in the perfect competition model, will always be short lived as new firms will enter and existing firms will increase capacity, thus driving down prices and profits. Government can improve efficiency only by getting its own house in order — eliminating international trade barriers and other government policies that seriously limit competition and harm the economy.

High cost of capital

Let us compare and contrast these differing views by considering the difficulties posed by the very high capital requirements for a new entrant into, for example, aircraft manufacturing. Any entrepreneur intending to start a company that would compete with Boeing and Airbus in the production of large aircraft would face quite a challenge. Most likely, he or she would need billions of dollars' worth of specialized capital equipment to begin production. Since raising billions of dollars of capital isn't easy this would, according to the traditional view, constitute a serious barrier to entry.

Very few firms, or countries, have the resources to design and manufacture commercial aircraft, and this is a primary reason that manufacturing of aircraft has become something of a global enterprise. For example, Boeing's 787 Dreamliner has development costs between $8–10 billion and utilizes partners from multiple countries, especially Japan (Kotha et al., 2005). Another example is Boeing's 777, which was launched in 1990 with an estimated development cost of $4–5 billion (Gollish et al., 1997).

However, the market process proponents would point out that modern capital markets have many trillions of dollars' worth of assets; billions can be readily raised if one offers a persuasive business plan to investors. Many of the Internet start-ups of the 1990s demonstrated just how easy it can be to quickly raise billions of dollars if investors are excited about your prospects. At this point, it seems that two producers, Airbus and Boeing, are enough. If, somehow the industry grows sufficiently or Airbus and Boeing somehow otherwise manage to enjoy high prices and profits then Lockheed Martin, or some other firm, will enter their market. This will compete with Boeing's 737-600/700 and the Airbus A318/A319. On 13 July, 2008, in a press conference on the opening of the Farnborough Airshow, Bombardier Aerospace announced the launch of the CSeries, a family of narrow-body, twin-engine, medium-range jet airliners. New aircraft would seem to be strong competition to either the A320 or 737.

High exit barriers

Another potentially serious entry barrier is high exit barriers. Investors must consider worst case scenarios—what happens to their investment in a company that performs so poorly that it must be liquidated, its assets sold off to the highest bidder? If our hypothetical aircraft manufacturer liquidates it will face the problem of very limited resale markets. That is, much of its equipment might have only two possible buyers, Airbus and Boeing, and end up being sold as scrap metal if neither of those two is interested. Thus, any major investment in such illiquid assets may face unusually high risk. Theoretically, this risk might inhibit new entry and thereby allow Boeing and Airbus to enjoy higher than normal returns.

However, market process proponents reply that it is a standard principle of finance that riskier investments must offer higher expected profits. In other words, if Boeing ever does earn unusually high long-run profits, it can be reasonably argued that such profits reflect merely the greater risk; that is, the risk-adjusted rate of return would still be a normal rate of return. After all, their investors also face the risk of illiquid resale markets should Boeing ever fail. More broadly, since economies of scale are so important in this industry and since the physical capital may be so specialized, it makes sense in terms of social welfare to be cautious before capital is plunged into aircraft production. In other words, because it is so hard to exit that industry it is perfectly appropriate, in this view, to hesitate before plunging in. From society's perspective, returns should, arguably, be quite high, before new entry occurs.

EXAMPLES OF OLIGOPOLY

Oligopolies are very common in modern economies. When you go to rent a car, you find that almost all of the cars are provided by a relatively small number of rental companies: Alamo, Avis, Budget, and a few other firms produce a vast majority of rental cars in the US. The rental market is dominated by relatively few car rental companies and relatively high barriers to entry. Common examples of oligopolistic industries are:

- airports in close proximity;
- automobile industry;
- breakfast cereal;
- cigarettes;
- long-distance telephone companies;
- film and camera;
- soft drinks;
- supermarkets;
- television cable companies.

Oligopolistic markets are of particular interest since most major aviation-related industries are oligopolies. The following sections will take a closer examination into three major forms of oligopolies that make up the greater part of the entire aviation industry:

- airlines;
- aircraft manufacturers;
- jet engine manufacturers.

Airlines

The airline industry is clearly an oligopoly market as it only has a few firms participating in the typical city-pair market. While oligopoly market theory suggests that firms should compete on service, since price cuts can be so readily matched, the US domestic airline industry has totally reversed this trend as the airlines have cut costs and service amenities. Part of this stems from the fact that many non-price competition aspects can be easily copied by competing airlines, such as frequent flier programs, but the key point is probably that consumers are often driven mainly by price concerns. Airlines might prefer less price competition but customer preferences seem to consistently force vigorous price competition. However, even an airline like Southwest Airlines, which has traditionally had a price leadership strategy, is also known for its friendly service and provides a frequent flight schedule on many of its city-pairs. Southwest's awareness of service quality is one of the reasons it has been successful. Other airlines, such as Emirates and Singapore, which pride themselves on their service quality, have been very successful by adopting this strategy.

Commercial aircraft manufacturing

The commercial aircraft manufacturing industry is largely a duopoly (an oligopoly with two firms) with Airbus and Boeing competing in the 100-seat plus aircraft category and Bombardier and Embraer competing in the regional aircraft market. Consolidation in aircraft manufacturing occurred with the acquisition of McDonnell-Douglas by Boeing and the exit of Lockheed from commercial aircraft manufacturing. Since commercial aircraft manufacturing is extremely capital intensive, it is unlikely that another manufacturer will enter the market, except for possibly a Chinese manufacturer, in the next ten to 15 years.

In 2007, China announced its intention to start making large commercial aircraft by 2020, providing a possible new entrant to the market (Associated Press (AP), 2007). While the barriers to entry are immense, China may be able to overcome them by having a robust economy with a positive trade balance, a strong domestic economy, and an abundance of labor. China is expected to buy 2,230 new aircraft before 2025 so the demand is clearly there; however, the required technological skills will still be a major barrier for the Chinese (AP, 2007). In an attempt to overcome this barrier, China reached an agreement with Airbus to open an A320 final assembly line (AP, 2007). China hopes the technological skills gained from this venture will translate into success for its large aircraft program. So while the barriers to enter the market are extremely high, they are not impenetrable.

In both aircraft manufacturing duopolies, the manufacturers offer similar products (that is, 737 vs. A320; CRJ vs. ERJ) and have prices that are very close to each other. The only major difference has been Airbus's insistence on a super-jumbo aircraft (A380) while Boeing has decided to focus on smaller super-long range aircraft (B-787). The B-787 aircraft is very fuel efficient with a cruising speed of Mach 0.85 and it is also smaller than the A380 and can access regional airports without a problem. The aircraft also has a range that can extend to 8,500 nautical miles carrying about 280 passengers. Since both manufacturers price their aircraft almost identically, competition is in the area of additional services such as financing agreements or agreed buyback of older aircraft. The competition between aircraft manufacturers is also characterized by a nearly even market share for recent aircraft deliveries, as shown in Table 9.2.

Table 9.2   Large Aircraft Manufacturers' Market Share

Large Aircraft Manufacturers				Regional Aircraft Manufacturers			
Year	2009	2010	2011	Year	2009	2010	2011
Boeing deliveries	481	462	477	Embraer—commercial only	125	101	73
Market share	49%	48%	47%	Market share	51%	51%	48%
Airbus deliveries	498	510	534	Bombardier — commercial	121	97	78
Market share	51%	52%	53%	Market share	49%	49%	52%

Source: Compiled by the authors from Boeing, Airbus, Embraer and Bombardier

Jet engine manufacturing

Oligopolistic market characteristics also apply to the commercial aircraft jet engine manufacturing industry. Not only are the products of the industries highly related, but they exhibit similar barriers to entry: high capital requirements, economies of scale, and advanced technological knowledge and skills. Unlike commercial aircraft manufacturing where there are four major firms competing in two distinct market segments, engine manufacturing is an oligopoly of three, or the "Big Three": General Electric (GE), Rolls-Royce, and Pratt & Whitney (P&W).

In addition to the "Big Three," two consortiums were formed to add additional players to the engine market. CFM International (CFMI) was a partnership between GE and Snecma, the French state-owned engine manufacturer. International Aero Engines (IAE) was formed in 1983 between P&W, Rolls-Royce, Daimler-Benz, Fiat, and Japan Aero Engines. These consortiums pooled technological knowledge, reduced risk, and lowered production and development costs for individual manufacturers. The consortiums created an interesting situation where manufacturers could be both partners and competitors concurrently. In addition to the two consortiums, P&W formed an alliance with GE for the development of the GP7200 platform, an engine designed for very large commercial aircraft such as the A380 (Bowen and Purrington, 2006).

Figure 9.4 displays the worldwide market share for commercial jet engines in terms of deliveries for the past 45 years. Up until the early 1980s, P&W was the dominant jet engine manufacturer. However, in the early 1980s P&W made a strategic decision that shaped the market for many years. P&W, a principal McDonnell-Douglas supplier, made the decision to focus on supplying engines for Boeing's new 757 instead of the 737, believing that the

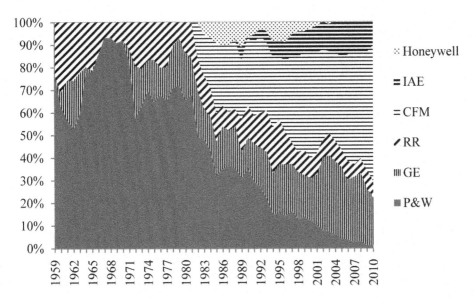

**Figure 9.4   Commercial Jet Engine Manufacturing Market Share by Engine
Deliveries**

Source: Compiled by the author using the Airline Monitor (2011)

757 was going to be the aircraft of the future (Bowen and Purrington, 2006). This created
a situation where CFMI became the sole jet supplier for 737s while P&W and Rolls-Royce
split orders for the 757s. While the 757s sold well, the 737s became the most successful
commercial aircraft in history. P&W lost even more market share with the demise of
McDonnell Douglas.

The main benefactor of P&W's declining market share has been GE. GE, a long-time
military engine manufacturer, entered the commercial market with the backing of its large
parent company (which includes aircraft lesser GECAS). GE strengthened its position in
commercial engine manufacturing with its CFMI joint venture with Snecma. CFMI in
2010 had a 58 percent market share while GE had a 20 percent market share compared to
P&W's 1.3 percent. P&W, however, is involved with two major joint ventures, the Engine
Alliance with GE which manufactures engines for the Airbus A380, and International
Aero Engines company with Rolls-Royce, MTU Aero Engines, and the Japanese Aero
Engines Corporation which manufactures engines for the Airbus A320 and the McDonnell
Douglas MD-90 aircraft.

Figure 9.5 provides another comparison between the major commercial engine
manufacturers in terms of in-service engines. Figure 9.5 is lagged, by roughly an engine's
life, from Figure 9.4. This lag is the major reason why P&W still remains a market leader,
as many of their older products, such as the JT8D remain in service on aircraft such
as the Boeing 727 and MD-80 (Bowen and Purrington, 2006). Therefore, P&W can still
turn substantial revenues through maintenance agreements and selling of spare part
inventories. However, as the older aircraft are retired, P&W's market share has declined
while GE and CFMI have gained.

Engine manufacturers also compete in service categories by creating the most fuel-
efficient engines and by providing the most attractive "power-by-the-hour" contracts.
Under a power by the hour arrangement, the engine manufacturers provide fixed-cost

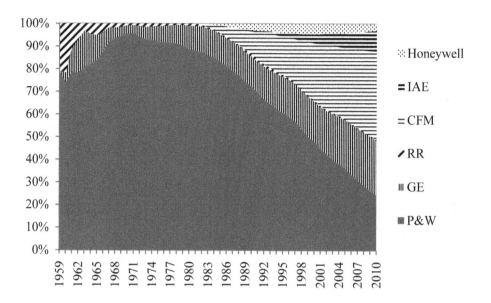

**Figure 9.5   Commercial Jet Engine Manufacturing Market Share by In-
Service Engines**

Source: Compiled by the author using The Airline Monitor (2011)

maintenance based on the number of hours flown each year. The airlines provide a fixed level of funding and expect to receive a given level of support by the engine manufacturers. The contractor expects to be provided a fixed level of funding up front and anticipates a long-term support arrangement. Similar situations occur for avionics, aircraft interiors, and in-flight entertainment systems.

The technological knowledge and skills required in the commercial aircraft engine manufacturing industry are substantial, and are major barriers to entry. Moreover, this knowledge requirement is a primary motivator for the number of alliances and agreements generated in the industry, as resources and risk are spread across multiple firms.

CONTESTABILITY THEORY

Thinking back to the perfect competition model presented in the previous chapter, note that it is not the large number of competitors that reliably drives profits to normal level but rather the expansion of output and entry of new firms. Profits can temporarily be quite high regardless of how many multitudes of competitors there are in wheat farming or anything else. Likewise, having only a few competitors in a market does not at all guarantee any prospect for high profits. If new competitors can readily enter a market then even a single firm may be driven to behave in a basically competitive manner, earning only normal profits on average in order to discourage new entry.

"That a market is vulnerable to competitive forces even when it is currently occupied by an oligopoly or a monopoly. That is, if any incumbent is inefficient or charges excessive prices or exploits consumers in any other way, successful entry must be possible and profitable."

Bailey, 1981

Pure "contestability theory" takes this idea a step further and posits that, in the absence of significant entry barriers, the number of firms in an industry is completely irrelevant. The key element of contestability theory is that successful entry must be possible and profitable as pointed out in the quotation above.

Before deregulation many economists speculated that pure contestability theory might well apply to the airline industry. Since aircraft are inherently mobile the thinking was that they could, under certain conditions, readily be reallocated to whatever routes were commanding higher prices, thus driving those prices down. Since each airline knew this they would refrain from significantly raising prices; potential competition would have the same effect as actual competition. However, several studies examined the airline industry and found a positive relationship between airfares and market concentration levels; that is, the fewer the airlines in a given market the higher the fares on average, suggesting that airline markets are not perfectly contestable (Strassmann, 1990; Whinston & Collins, 1992; Oum, Zhang and Zhang, 1993). One possible explanation is that economists underestimated the cost of entering a new market. Suppose, for example, that an airline had service to airport A and to airport B but no non-stop service connecting A and B. It might seem that an aircraft could fly out of one market and be reassigned to a new A-B route almost instantly in pursuit of the greatest profit. But, in reality a new route must be planned and announced to consumers well ahead of time, normally at least three months, and probably some special spending on advertising will be necessary. These are not massive costs but perhaps create enough friction to the entry process to prevent pure contestability results.[2]

However, there is considerable doubt that these slight entry costs can fully explain observed price variances. Network effects in the context of intense competition may offer a better explanation. An airline network is more than the sum of its separate routes, particularly for the legacy carriers that aspire to offer seamless travel to "almost anywhere." Suppose, for instance, that such a carrier found it necessary to operate a non-stop route to Las Vegas because it is such a popular vacation destination and, among other things, many key customers preferred to redeem their frequent flier awards for a Las Vegas trip. In this case the value of the Vegas route might far exceed the actual revenue garnered from paying customers on that particular flight. Thus, the airline would sensibly keep that route rather than reallocating the aircraft to another route, say non-stop to Minneapolis, that would produce more direct revenue but less total value and revenue for the network as a whole. Thus, the price of a flight to Minneapolis could remain higher than the price to Las Vegas even if the market were purely contestable. Roughly the same thing happens in grocery stores when a particular item, a "loss leader," is sold at an especially low price

---

2   One of the key elements of contestability theory is that entry and exit from markets must be free and easy (Bailey, 1981). The complete absence of barriers of entry would satisfy pure contestability theory but in the airline industry there can be sizeable barriers of entry into a given airport. This is certainly the case at airports such as London Heathrow which is slot controlled, restricting competition. However, there are competing airports which makes contestability theory apply to a certain extent even in this market. Ultimately, in oligopoly markets where there are low barriers to entry, any market power of carriers will be much less than in markets where there are high barriers to entry.

in order to bring customers into the store who will then, hopefully, buy other items with higher markups while they are in the store.

In this case we need to look at profits for the entire airline rather than the prices of particular city-pairs to judge the contestability of the industry. Since, as we shall see in more detail later, the long-run profits and rates of return for the airline industry are exceedingly low, it may well be that the industry is basically contestable in this broader sense.

Of course, it is possible to drown in a deep spot within a pond where *average* depth is only knee deep. Likewise, in this network setup there can still be considerable pain for those particular cities and city-pairs that face relatively high prices even as the system-wide average fare is an extraordinary consumer bargain, actually below the cost associated with a normal profit.

Fortunately for consumers in those 'high end" markets, low-cost carriers (LCCs) with simpler point-to-point networks are entering these markets with increasing frequency. There have been numerous examples where carriers who abuse their market power have generated competition from other carriers. LCCs AirTran and Frontier created their own hubs in the Atlanta and Denver markets that were formerly dominated by oligopolistic legacy carriers. Virgin Atlantic evolved to provide British Airways with legitimate competition on long-haul flights. While these examples provide an application of contestability theory on a widespread scale, contestability theory is probably most relevant in small markets where there may only be one or two airlines serving the market. In these situations air fares may remain somewhat high, but not exorbitantly so, since this would encourage competition to enter that market.

Figure 9.6 displays a hypothetical demand curve for a firm in an oligopoly market. Under contestability theory, there are two components to the firm's demand curve. The first occurs when the firm wants to increase the price. In a non-contestable market (dashed line above P), the firm can increase prices and not lose a substantial amount of demand, thus suggesting a gain in total revenue. However, in the presence of contestability (solid

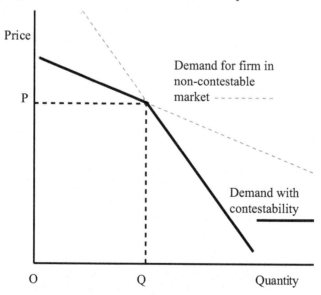

Figure 9.6   **Price Competition and Market Reaction for Contestable Market**

line above P), this increase in price will result in an even larger decrease in quantity demanded.[3] Here demand is relatively elastic since the competition is reluctant to match the price increase.[4] The second component of the kinked demand curve is when a firm is pondering a price decrease in the hope of increasing quantity demanded and thus increasing total revenue. This is the case in a non-contestable market as demonstrated by the dashed demand curve below P. Alternatively, in a contestable market, competitors will match the price decrease, creating the situation where the increase in the firm's quantity demanded from the decrease in price will not be as large as the firm had hoped. Here, demand is relatively inelastic since the competition is willing to follow the price decrease. This is demonstrated by the solid demand curve below P.

KINKED DEMAND CURVE THEORY

The kinked demand curve is another theory that predicts price stability in oligopolistic markets. The theory states that in an oligopolistic market, there exists a band where price stability exists. This band is the kinked portion of the demand curve.

Consider a duopoly with two firms who have slightly different demand curves for their product. Figure 9.7 displays the two demand curves for firms, D(1) and D(2). These two demand curves intersect at point A on Figure 9.7; this point lies roughly above the $800 price level and the 320 demand level. Both companies want to operate at a point where marginal revenue equals marginal cost. From this point, the optimum price can be found from the corresponding point along the demand curve. In a duopoly market, the firm whose prices are the lowest will provide the market price which the other firm will match in order to remain competitive.

Based on this information, the market demand curve can be constructed. Up until point A in Figure 9.7, firm 1's demand curve is less than firm 2's, therefore based on simple supply and demand, firm 1 would charge a lower price than firm 2. This situation is reversed after point A where firm 2's demand curve is significantly less than firm 1's. Therefore, the market demand curve will be the point $1000:0 to point A and from point A to $0:750.

In order to construct the market marginal revenue curve, firm 1's marginal revenue curve should be used up to a demand level of 320 units, and firm 2's marginal revenue curve should be used thereafter. This creates a situation where the market marginal revenue curve contains a vertical portion, line B-C. Since the intersection of the marginal revenue and marginal cost curve will yield the optimum price for the industry, when the marginal cost curve intersects between points B and C, the market clearing price would be roughly $820, or the price at point A. Therefore, if the marginal cost curve's intersection shifts anywhere in between points B and C, the market clearing price will remain the same. This creates a situation of long-run price stability in the market since there is a relatively wide range over which marginal cost can change (B to C on Figure 9.7) without changing the profit-maximizing price (A on Figure 9.7).

At times, the airline industry seems to provide some indication that this might be occurring. Only major shifts in the demand curves will create significant fluctuations in

---

3   In October of 2006, United raised fares in several markets, but when it became clear that airlines such as JetBlue and Northwest Airlines would not raise their fares, United rescinded fare increases.
4   Other airlines may not follow a price increase by one airline; therefore demand will remain relatively elastic. Furthermore, an increase in price would not lead to an increase in the total revenue of the airline.

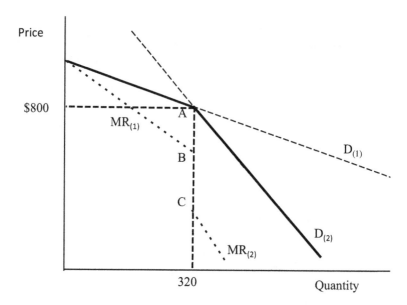

Figure 9.7   Kinked Demand Curve

the market marginal revenue curves, thus changing the market clearing price. Such a major shift in demand occurred shortly after the terrorist attacks of September 11, 2001. In this case demand shifted down disrupting and altering previously stable equilibriums.

COURNOT THEORY

Cournot theory helps explain competition and market equilibrium based on firms competing through output decisions. The theory assumes that products are homogenous, market entry is difficult, firms have market power, cost structures are similar, and each firm assumes that the other will not respond to changes. To clarify this last point, the model assumes that, for instance, Boeing believes that Airbus will not respond to any changes in price and output initiated by Boeing. Likewise, Airbus assumes Boeing will also be completely unresponsive. Though this assumption is probably not realistic, the model still offers some insights and, given that Airbus and Boeing cannot be sure how the other will respond, the theory may sometimes approximate reality.

Consider a duopoly market where the firm's marginal costs are zero. The demand curve for the entire market is described in Figure 9.8, with the total output in the industry being Q. The first firm's marginal revenue curve is also described in Figure 9.8, and is exactly half the demand curve, since the demand curve is linear. The optimal output decision for the first firm is where marginal revenue equals marginal cost. Since marginal costs are assumed to be zero, the firm would want to produce at the point where the marginal revenue curve crosses the x-axis. This point is exactly half of the total demand, or point $Q/2$. The corresponding price for this level of output from the demand curve is P_1.

Since firm 1 takes half of the market for itself, the second firm's maximum demand would be $Q/2$. Therefore, the demand curve for the second firm is shifted to the left and intersects the x-axis at the point $Q/2$. This is displayed in Figure 9.9. Using this new

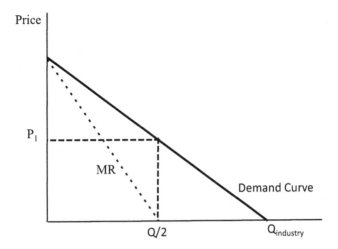

Figure 9.8   **Initial Output Decision for First Firm**

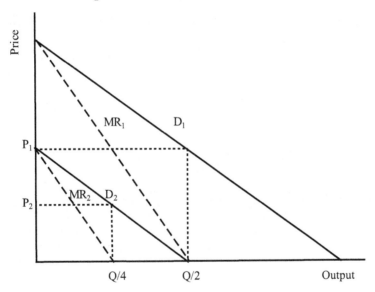

Figure 9.9   **First Round Cournot Theory**

demand curve, the second firm's optimal output level is where its marginal revenue curve intersects the x-axis. This occurs exactly at $Q/4$ and the corresponding price point is P_2. Based on this, the first firm would take half the market and the second firm would take a quarter of the market, meaning that the firms' total market share is $3/4Q$.

However, these firms have not reached equilibrium since the second firm's price is dramatically below the first firm's price. With this disparity in prices, most consumers would opt for firm 2's product over firm 1's, since the products are homogenous and the price is significantly lower. Based on this, the firm's will readjust their output in an effort to obtain equilibrium. These readjustments do not occur all at once, but occur over various rounds. Table 9.3 provides the various rounds and corresponding market output for the firms as they progress through their readjustments.

Table 9.3   Cournot Theory Progression

	Round 1	Round 2	Round 3	Round 4
Firm 1	1/2 Q	3/8 Q	11/32 Q	1/3 Q
Firm 2	1/4 Q	5/16 Q	21/64 Q	1/3 Q
Total Market	3/4 Q	11/16 Q	43/64 Q	2/3 Q

The Cournot solution is where both firms are in equilibrium with the same price level and the same level of output. In a duopoly, the Cournot solution would have both firms obtaining 1/3Q; therefore the total market share obtained by both firms would be 2/3Q. Note that the total market share declines as firm's readjust to equilibrium and that firm 1's market share declined while firm 2's market share increased. Cournot theory predicts that firms will continue to readjust the level of output until they have achieved market equilibrium at the same price level.

The aircraft manufacturing industry seems to resemble the characteristics of the Cournot theory. While Boeing and Airbus's products are not identical, they are fairly close in technical performance and requirements. The barriers to entry in the aircraft manufacturing industry are high and each firm has tremendous market power. Through government aid, both firms have similar cost structures, but not identical ones. Both Airbus and Boeing have roughly equal market share and the readjusting progression has been evident in the past years as Boeing's market share has slowly declined and Airbus's market share has increased (although this has turned around in 2006 and 2007 as Boeing has regained market share). Differences between the two companies' output can be attributed to cost structure differences and/or product differentiation.

On the other hand, the Cournot solution is rarely achieved in the airline industry as airlines are able to differentiate their product (mainly through route structure and frequency of flights), every airline has a different cost structure, and each individual carrier has little market power. Because of this, we rarely see markets where the market share is evenly distributed across multiple carriers. In fact, the development of the hub system has led to situations where one airline tends to dominate the market share at a hub airport.

PRICE–OUTPUT DETERMINATION UNDER HYBRID MARKET STRUCTURE

In order to understand the mathematics involved with determining the optimal price–output level for a firm in an hybrid market, consider the case of luxury airline DirectJet. DirectJet's management has asked its financial managers to study short-run pricing and production policy. DirectJet's financial managers were able to determine the airline's price, fixed cost (FC), and variable cost functions. They were:

$P = 10,000 - 8Q$

$FC = \$200,000$

$VC = 2{,}200Q + 5Q^2$

Since the ultimate goal is to determine the optimal price–output decision for DirectJet, both the marginal revenue and marginal cost curves need to be found. On the revenue side, the total revenue function can be found by simply multiplying the price function by quantity (Q). This results in:

$TR = 10{,}000Q - 8Q^2$

The first-order derivative of the total revenue function will produce the marginal revenue function of:

$MR = 10{,}000 - 16Q$

On the cost side, the two cost components (fixed and variable) need to be combined to create the total cost function. From there, the first-order derivative of the total cost function will yield the marginal cost function. The two functions are:

$TC = 200{,}000 + 2{,}200Q + 5Q^2$

$MC = 2{,}200 + 10Q$

The final step in determining the optimal price–output combination for DirectJet is to set the marginal revenue and marginal cost curves equal to each other in order to determine the optimal quantity. This is computed below and also displayed graphically in Figure 9.10.

$MR = MC$

$10{,}000 - 16Q = 2{,}200 + 10Q$

$7{,}800 = 26Q \rightarrow Q = 300$

Based on this calculation, DirectJet's optimal output is 300 seats per day. At this optimal level, the average ticket price for DirectJet's flights and the total revenue are:

$P = 10{,}000 - 8Q$

$P = 10{,}000 - 8(300)$

$P = \$7{,}600$

$TR = 10{,}000Q - 8Q^2$

$TR = 10{,}000(300) - 8(300)^2$

$TR = \$2{,}280{,}000$

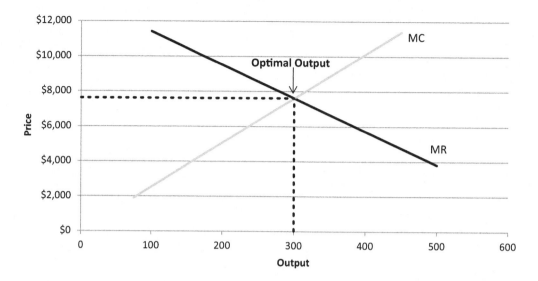

Figure 9.10   Marginal Revenue and Marginal Cost for DirectJet

Or:

TR = P*Q → $7,600 * 300 = $2,280,000

While total revenue is not maximized at this point, total firm profit is maximized at this value. In order to determine total profit at the optimal level, the total cost at the optimal output is:

$TC = 200,000 + 2,200Q + 5Q^2$

$TC = 200,000 + 2,200(300) + 5(300)^2$

TC = $1,310,000

This yields a total profit at the optimal output level for DirectJet of:

TP = TR − TC

TP = $2,280,000 − $1,310,000 = $970,000

No other combination of price and output based on the cost functions provided by DirectJet's financial managers will yield the airline with a greater profit. This is displayed graphically in Figure 9.11, which displays the total revenue, total cost, and total profit functions. Since the total cost curve is an upwards "u-shape," the output level of 300 is a maximum point and is where total profit is maximized.

While the above scenario applies in the short term, it is very much different in the long run. Similar to perfectly competitive markets, the super normal profits earned by monopolistic competitive firms in the short run will attract new firms to the market. These

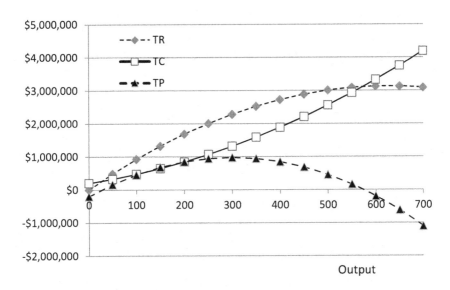

Figure 9.11   Total Revenue, Cost, and Profit for DirectJet

new firms will offer similar competing products, but with the increase in new firms and products entering the market, the degree of differentiation between products diminishes. As a result of the decreasing degree of product differentiation, the elasticity of the firm's demand decreases, causing the firm's demand curve to become more horizontal. This creates a situation similar to perfectly competitive markets as monopolistic competitive firms tend to become price takers in the long run. This causes the super normal profits to diminish and zero economic profits are earned. In essence, monopolistic competition acts like perfect competition in the long term; therefore the differentiating characteristic between the two is the length of the "short-run" period where super normal profits are realized.

PROFITABILITY ISSUES

Normal long-run profit levels might be explained by the industry being contestable in the manner discussed above, but contestability in itself should not produce the below-normal returns observed in the airline industry. The explanation may relate to the industry's oligopoly nature combined with very high fixed costs and very low marginal costs. Recall, since an airline's schedule is usually approximately fixed three months in advance, most costs, even labor and fuel, are essentially fixed for that period. The marginal cost of placing a passenger in an otherwise empty seat on an aircraft that will be flying in any case (with or without that passenger) is extremely low—consisting mainly of the cost of processing their ticket or, in some cases, of paying a travel agent commission. Thus, each individual airline is in a position where even a very low price is better than nothing for an otherwise empty seat. As discussed in the chapter on revenue management (Chapter 11), each airline strives to make this low fare available only to those passengers who would not have flown at a higher price on *their own airline*. However, each airline is naturally happy to draw passengers away from a competitor. Suppose for example that Joe would have paid $200 to fly airline A but is lured into flying airline B for $150. Likewise, Jane would have flown airline B for $200 but is lured over to airline A for $150. Each airline acts independently but the collective result is that each receives $50 less, perhaps incurring a loss rather than making a profit.

If each could refrain from "stealing" the other's customers they might both enjoy a normal profit rather than being in bankruptcy. This is a classic "prisoner's dilemma."[5] Since each airline cannot "trust" the other to refrain from this "cutthroat pricing" they both do it, even though both would prefer to cooperate and stop such aggressive price competition. That is, if only one airline stops offering the $150 deal that airline will lose both Jane and Joe to the competition. Of course, any attempt at cooperating to reign in this cutthroat competition is complicated by the fact that government anti-trust policy typically prohibits arranging such cooperation via a formal contract and makes it difficult, probably prohibitively risky, to make an illegal arrangement in secret. Another complication is that other airlines will tend to enter the market with aggressive price cuts even if the two airlines do somehow manage to cooperate. US airlines also argue that the bankruptcy laws exacerbate the problem by providing a subsidy[6] that keeps failed airlines from actually leaving the market so that even in the long run it is difficult for the industry to decrease capacity enough to keep prices high enough to support a normal profit.[7]

At first glance, the misery experienced by airline investors from pricing below costs appears to be a joyous gain for air travelers — as if most carriers were perpetually selling at below cost, "going out of business" bargain rates. And that may in fact be the case. Conventionally, most economists have viewed excessively low long-run returns as a problem that will eventually take care of itself as needed. That is, if many investors and lenders don't think it is worth investing in the airlines, then they can stop financing them until capacity decreases enough to raise prices, normalize profits, and warrant future investment. Moreover, the strong performance by airlines like Ryanair and Southwest suggests that the business can be profitable if it's "done right." So, most economists are probably content to let the industry evolve as it will, even with rampant bankruptcies, until or unless there is clear evidence of a problem for consumers.

However, some economists, proponents of the "empty core" theory, suggest there may already be a serious problem.[8] It is possible for the aforementioned cut-throat competition to be so severe that it actually does harm consumers by preventing airlines from offering some higher priced products that consumers would prefer to have. For example, suppose there is no available non-stop service on a given route and that two competing airlines offer service, via a stop and transfer at their respective hubs, for a price of $200. Suppose that consumer demand is such that *one* airline could offer non-stop service on this route for $280 and, if the other competing airline maintained its same $200 service, both airlines would be financially viable in the market. In other words, the market is capable of

---

5   The term follows from the idea that two criminals might both go free if each lies to protect the other. But, with police questioning them separately, each knows that he will face a very stiff sentence if he lies, protecting his partner, but his partner then tells the truth. Unless each can somehow be pretty certain that the other will lie, they have incentive to implicate each other in exchange for a lighter sentence.

6   For example, many airlines have effectively 'dumped" the cost of their pension programs onto taxpayers via government assumption of these pension obligations, though with some cuts for wealthier pensioners.

7   Although this may be changing of late, the much higher cost structure of the legacy carriers also lead to increased capacity from LCCs more than off-setting the capacity cuts by the legacies. Unusally powerful unions at the legacy carriers are also often cited as contributing to the legacy arilines' ongoing struggles.

8   See Raghavan and Raghavan (2005) for a discussion of an "empty core problem."

supporting both airlines but only one with a non-stop flight. However, if competition leads either to duplication of the non-stop service and/or a significant price cut then the non-stop service becomes financially impossible to maintain. So, it is possible to offer consumers a chance to pay a premium and obtain the more desirable non-stop service only if there is no strong competitive response.

Suppose, though, that every time one airline adds non-stop service, the other either duplicates that service or substantially slashes prices on its stop and transfer service so that the non-stop service becomes a financial loser and is abandoned.[9] Moreover, since the two airlines compete in numerous markets they each learn of this tendency and therefore choose never to start such non-stop service in similar markets. This problem might explain why airline customers complain so much about declining quality while at the same time making choices that drive airlines to cut quality.[10] There is no practical way for an airline to contract with customers to get them to keep flying the new non-stop route after competitors respond, even though customers might be willing to do so if they understood that booking a bargain today would eventually result in poorer service/higher prices in the future.[11]

In such a case, consumers would be better served if the two airlines could freely negotiate a solution: perhaps one airline could be induced to not respond by receiving a small side payment, or the two airlines might make a trade where they take turns adding non-stop service in various more marginal markets. We have the counter-intuitive situation where consumers might actually be better served by an alliance that seems to closely resemble a cartel! Actually, though, this is a less unique situation than it might at first appear. It has, for example, long been common practice, especially in manufacturing, for firms that are normally competitors to occasionally team up on particular projects. Similarly, one airline will conduct maintenance or baggage handling for a competitor. One reason for this is that economies of scale for certain products may be such that only a single producer can be efficient enough to viably deliver the product to market. This relates to the fact we mentioned earlier that reducing the number of competitors may sometimes increase efficiency and increase consumer well-being. If the airlines were allowed to act in concert it is easy to see how, for instance, in a given market, two large aircraft with 95 percent load factors might have much lower costs per passenger and therefore lower prices than three smaller aircraft with 75 percent load factors operated by three separate airlines. Just as aircraft prices might be lower with only the two current producers than if we had several more it may be that, at least in some cases, reducing competition on some routes could benefit consumers.

It is probably fair to say that most economists in the traditional vein would view allowing competing airlines to cooperate and form alliances as too radical a step, seeing the risk of price collusion conspiracy as outweighing potential gains. However, many other economists would probably be willing to consider it, especially on some limited experimental basis. Some would argue that, given the financial plight of legacy carriers,

---

9   This situation explains, incidentally, the puzzling fact that a flight from A to B to C is sometimes cheaper than a flight from A to B.
10   Of course, it is also very possible that consumers really do want lower prices primarily and just enjoy complaining about quality because they unrealistically want extremely low prices *and* high quality!
11   To illustrate, suppose many customers' first choice is $150 with stop and transfer, second choice is $280 non-stop and third choice is $200 with stop and transfer. Suppose airline D starts the non-stop $280 service but then airline U offers the $150 service in response. This renders the non-stop service non-viable and it is abandoned, the market returns to $200 service by both airlines, leaving these customers with their least preferred option.

the risk of them colluding to make "too much profit" is not significant. Also, strong proponents of market process would argue that new entry, and perhaps potential new entry (contestability), could effectively restrain any harmful anti-competitive impulses. In other words, price conspiracies are unlikely and, even if the airlines in a given market attempted collusion, a new entrant would undercut them; that is, the market is contestable, at least in a basic, practical sense.

Ironically, in this situation staunch anti-trust regulation may ultimately decrease the number of competitors. Since current regulation may prevent airlines from cooperating to improve efficiency in particular markets the end result may be more airline failures, eventual liquidation and/or desperation mergers that anti-trust enthusiasts have no choice but to accept. Although speculative, it is possible, for instance, that Lockheed or McDonnell-Douglas might have remained as competitors to Boeing and Airbus had they been allowed to cooperate on some projects.

When regulatory preferences conflict with the economic reality of substantial economies of scale, scope, or density, economic reality will ultimately win. If more cooperative efforts between airlines are needed they will eventually emerge, if not through approved alliances then through bankruptcy, liquidation, and mergers that reduce the number of independent airlines. Just as regulators could not stop the aircraft manufacturing market from evolving into a duopoly, with each firm able to enjoy considerable economies of scale, it may be that a similar process is unfolding for airlines. Of course, there is tremendous uncertainty about all of this. However, it does seem likely, given the financial disarray of many legacy carriers, that major changes of some sort will occur in the industry.

COMPETITION AND ANTI-TRUST ISSUES

Anti-trust regulation has become very controversial. Virtually all economists see substantial problems with at least some aspects of anti-trust regulation and some even maintain that the costs of these regulations clearly exceed benefits and that they should be completely abolished.[12]

One difficulty is that the most problematic anti-competitive behavior is completely exempt from anti-trust oversight. Economists would generally be thrilled if it were possible, for instance, to challenge international trade barriers under anti-trust law, but all government policies are exempt from these laws. Since, as shown above, there is also some question about the real power of any barrier to entry outside of government, it follows that there is some debate as to whether there is enough of a private monopoly problem to justify a government regulatory program. Even if there are some imperfections in market competition, as most economists would probably agree, it is not easy for regulators to make things better for consumers by imposing fines and escalating legal costs on firms; after all, such costs tend ultimately to be passed on to consumers in the form of higher prices. There is also likely to be substantial bias on the part of regulators. If, for instance, regulators publicly admit that there are few problems in the marketplace, then they will likely experience major budget cuts and may soon find themselves out of a job. If, on the other hand, they claim that there are monopoly problems and potential problems lurking everywhere, then they tend to receive budget increases and a greater chance for job security, promotion, and higher pay.

---

12   See for example, Crandall and Winston (2003) and Armentano (1986).

The problem is exacerbated by the fact that, as we shall see, truly anti-competitive behavior is usually very difficult to distinguish from healthy, competitive business practices. Indeed, the purpose of any business is to attract consumers and thereby *harm their competitors*. It is theoretically possible for a firm to harm its competitors too much, or in a certain way, such that it is ultimately harmful to the overall economy, but is difficult, some would say impossible, for regulators to draw a clear line to divide appropriate competitive acts from inappropriate competitive acts. Let us consider some examples.

Predatory pricing

Predatory pricing theoretically occurs if a firm: 1) cuts price below cost intentionally low in order to driving competitors out of the market; then 2) raises prices to a monopoly level once competition is gone. Note that aggressive price cuts, even if they drive competitors out of business, are not in themselves predatory. Airlines often find themselves losing money, fighting over a market that isn't big enough to sustain all existing firms. In this case they may fight it out until some firms leave the market and raise prices high enough to support normal profits. Therefore, aggressive price cuts and less-efficient firms going out of business are the routine results of healthy competition. Only if prices go up to monopoly levels can the action be said to be predatory. In many countries, including the US, predatory pricing is considered an anti-competitive practice and is illegal under anti-trust laws.

Also, there is some question as to whether predatory pricing is likely to occur at all. It is certainly a high-risk strategy that would likely fail in many cases, even in the absence of any regulation. Assuming a firm survives stage 1, that its competitors go bankrupt before it does, it is likely to face new entrants once price is jacked up at stage 2. Indeed, new entrants would know that the predator could not easily afford another round of predatory cuts after already enduring losses in driving out the first group of competitors. If the predator did successfully repeat his predation than a third round of new entrants would likely be drawn by the knowledge that the predator would struggle to survive a third duel to the death and so on. In other words, a successful predator must be so fierce that he completely frightens off the rest of the world. This is theoretically not impossible, and so could constitute a barrier to entry, but not likely to be common. A popular example of a failed alleged predatory pricing strategy comes from Delta and ValuJet. ValuJet began service hubbed out of Atlanta in 1994. By December of 1994, Delta had matched Valujet's cut-rate fares—as low as $29 one way—on flights between Atlanta and the 11 cities that both carriers served. On January 7, 1997, ValuJet suspended its low-fare service between Mobile and Atlanta. The next day Delta raised its lowest fare for the route from $58 to $404—an increase of about 600 percent.

However, normal price competition is often mistaken for predatory behavior. If, for example, a competitor enters a market with a close substitute offered at a lower price, then clearly the incumbent firm must either match that price cut, at least approximately, or leave the market altogether. Airline A can't charge a price much above Airline B if B offers essentially the same product. If A was previously charging a much higher price before it then matches the much lower price of B, then A will also likely increase output. In effect, low-cost B forces A to abandon its higher-price/lower-volume strategy and embrace a low-price/high-volume strategy; the only alternative for A is to abandon the market completely. If it should happen that B eventually pulls out of the market then A may find it optimal to return to the high-price/low-volume strategy. The standard business procedure of matching a competitor's price cuts when necessary is indistinguishable from predatory pricing. Does it make any sense to forbid legacy carriers from matching the lower prices of low-cost entrants?

US courts considering predatory pricing charges in recent decades have focused not on the price cuts but on the feasibility of a "predator" raising prices to monopoly levels. In, for example, *Frontier Airlines vs. American Airlines*, the judge summarily dismissed the case because, he maintained, the government really had no case at all; that is, no credible evidence of predation. American merely matched the prices of Frontier but, in the court's view, had no hope of gaining any monopoly power even if it destroyed Frontier since there were numerous other competitors in the market. The fact that American, like all legacy airlines, struggled to earn even normal profits over the long run is supportive of the court's decision. Though some economists may disagree with this approach, the courts' deep skepticism combined with the fact that the legacy carriers have been, in recent years, struggling just to survive, seems to have dampened regulators' enthusiasm for bringing predatory pricing charges.

In 2000, Spirit Airlines launched a lawsuit alleging that Northwest Airlines engaged in predatory pricing and other predatory tactics in the leisure passenger airline markets for the Detroit–Boston and Detroit–Philadelphia routes beginning in 1996. In 1996, the two carriers entered a price war over these two routes. Spirit claimed that Northwest's fares were so low that these prices would force Spirit to exit the markets and consequently, Northwest would raise fares to monopoly levels and consumers would be harmed. Northwest responded that the low fares in these two markets reflected head-to-head competition between the airlines. The district court awarded summary judgment to Northwest, but a panel of the Sixth Circuit unanimously reversed the district court's decision.[13]

Cartels and collusion

Firms that fix prices and output in a formal agreement are called a cartel. Firms that fix prices and output in covert informal agreements are said to be in collusion. Cartels and collusion may enable firms to exert monopoly-like power in their pricing policies (Hirschey, 2003). While cartels and collusion are generally illegal in the US, they are allowed in many foreign markets. In the US under the Sherman Antitrust Act of 1890, the Clayton Antitrust Act of 1914, and the Federal Trade Commission Act of 1914 such collusive agreements are illegal. Nonetheless, there are several examples where they can be found, particularly in the sports professions. The National Football League, Inc. (NFL), Major League Baseball, Inc. (MLB), the National Basketball Association (NBA), and the National Collegiate Athletic Association (NCAA) are often cited as examples of cartels.[14] Around the world, there have been famous cartels in oil and diamonds. Probably the most famous and most important cartel in the world economy is the Organization of the Petroleum Exporting Countries (OPEC). While OPEC cannot directly set the price of a

---

13   *Spirit Airlines, Inc. v. Northwest Airlines, Inc.*, 431 F.3d 917 (6th Cir. 2005), cert. denied, 166 L. Ed. 12 (US 2006).

14   However, it might also be argued that these sports leagues are not really cartels at all since their product, entertainment, faces many substitutes. Cooperation among sports franchises might be of the same sort that exists among different franchises of a given restaurant chain.

barrel of oil, its control over much of the supply enables the cartel to dramatically impact the price by either increasing or decreasing output.

The international nature and often extreme competition in the airline industry possess a strong potential for collusion. For example, in June 2006 British Airways became involved in a price-fixing scandal involving fuel surcharges on long-haul flights (Simpkins, 2006). Investigations by both British and American authorities uncovered the fact that calls were made to Virgin Atlantic concerning the timing and level of increases in fuel surcharges (Simpkins, 2006). British Airways admitted fixing cargo surcharges from 2002 to 2006 and passenger fuel surcharges from 2004 to 2006 (typical surcharges rose from £5 to £60 per ticket). British Airways was charged $300 million in August 2007 by the Office of Fair Trading (OFT) and US Justice Department but Virgin Atlantic was not fined as it was given immunity after reporting British Airways's actions.

There is a distinct problem when analyzing cartel and collusion issues in the airline industry, and this is the fact that the prevalence of information in the industry makes it fairly simple for airlines to match the prices and output of competitors. It is argued that this is indeed a form of collusion termed "tacit collusion." Tacit collusion is coordination without express communication usually through price signaling. For example, one airline raises its prices with the hope that the other airlines interpret this move as an invitation to collude and respond by matching the price increase. However, the fact that two airlines have price fluctuations that match exactly does not mean they are in collusion, but more likely that they are competing fiercely.

Airline alliances create interesting issues related to airline collusion, which ultimately deal with coordinating schedules and prices of flights. In order for alliances to be allowed, they must receive regulatory approval from the necessary bodies; however limits may be placed on their coordination. For instance, American Airlines and British Airways have several stringent restrictions placed on them, while the KLM/Northwest relationship was given extensive anti-trust immunity by US regulators. Anti-trust immunity is given to potential alliances based on a variety of factors, including the level of consolidation that would exist in the industry.

Again, however, there is fierce debate as to whether government anti-trust actions against alleged collusion have been appropriate and beneficial to society. The abysmal rate of return for legacy airline investment, perhaps the lowest for all industries, suggests that there is no shortage of competition. Even if collusion were attempted, it is difficult to orchestrate high prices for any length of time. For one thing, such high prices invite new competitors to enter the market and undercut the cartel. Even before new entry there is always strong incentive for an individual firm to violate the collusive agreement since one can potentially earn far greater profits by slightly undercutting one's partners.

In fact, the behavior of US airlines in the age of regulation illustrates the strong tendency to break a cartel agreement. Regulators severely limited price cuts, much as a conventional cartel would do, and even prevented any new entry for some 40 years. Yet airlines struggled to earn even normal profits. Airlines competed by improving quality—improving the food, giving away liquor, utilizing larger aircraft, providing roomier seating, and so on. Since even this quasi-cartel, run with the aid of government, still failed to suppress competition, this illustrates how difficult it is for firms to secretly suppress competition on their own.

Using regulation to prevent price fixing is problematic because there is frequently, as in the case of alleged predatory pricing, no clear way to distinguish innocent behavior from illegal collusion. For example, two airlines may constantly raise and lower their fares in tandem not because they are colluding but because of logical, independent reactions to market conditions. One airline's product is normally a very close substitute for another's—in the leisure market it may approach being a perfect substitute. Inherently, it is not possible for the prices of two close substitutes to vary greatly; most consumers would flock to the one that is substantially cheaper. Thus, it is necessary to generally match any price cut by a competitor. The prevalence of online information in the industry also makes it particularly easy for airlines to monitor and quickly match the prices and output of competitors. Likewise, any airline attempting to increase price because of, say, higher fuel costs, will retreat from the price increase unless competitors follow suit. So, one observes either a general increase in airline prices or no lasting price increase at all.

In the 1990s US regulators took note of the fact that airlines seemed to be constantly signaling each other to raise prices. For instance, an airline would normally announce its intention to raise prices several weeks in advance. If, following the announcement but before the scheduled price increase, the competition eventually announced that they too would increase fares by similar amounts then the announced price hikes would in fact materialize. On the other hand, if competitors left prices unchanged then the airline would cancel the previously announced fare increase.

In their defense, airlines pointed out that most announced increases were in fact cancelled, that the overall long-run trend in inflation-adjusted fares was downward, and that the lack of industry profitability indicated that prices were not "fixed" but rather "broken." Moreover, as already explained, it is understandable that prices for close substitutes will naturally either move in tandem or not move at all. It is also understandable that firms sustaining substantial losses would make every effort to raise prices.

Nevertheless, anti-trust regulators insisted on changes and, among other things, forced airlines to agree to no longer announce fare increases in advance. Many travel agents and consumer groups complained about the change since forbidding airlines from announcing planned price hikes seemed to make fare increases harder to predict and plan for. Airlines responded to the regulatory constraint by implementing fare increases on Saturdays, when relatively few people book flights. Thus, if competitors refused to join in the price increase then the new prices could be cancelled by the following Monday before they had significant impact.

Critics of anti-trust regulation argued that the whole episode seemed absurd and that it was particularly ironic for regulators to force firms to actually raise prices as opposed to just announcing an intention to eventually raise prices. Still, it does seem at least theoretically possible that preventing airlines from signaling a desire to increase prices might ultimately benefit consumers, at least slightly. Of course, regulators claimed that their actions and general vigilance would make sure that airline profits and prices would never rise too much.

INDUSTRY CONSOLIDATION

The level of concentration in the industry helps determine the market's structure. Industries that are highly concentrated may be more prone to exhibit characteristics of monopolies and oligopolies, while industries with multiple players may tend to exhibit characteristics of perfect and monopolistic competition. Table 9.4 presents the market share of the US car rental companies over the past few years. It is clear from the table that Enterprise Holdings is the primary provider in this industry, with their market share continuing to increase over time. Despite the relatively large number of car rental companies, the industry is highly consolidated with the top three car rental companies holding a combined 94.5 percent market share in 2011.

Table 9.4   Market Share of US Rental Car Companies

	2009	2010	2011
Enterprise Holdings	47.40%	47.60%	53.90%
Hertz	19.70%	19.80%	20.60%
Avis Budget Group	20.00%	18.90%	20.00%
Dollar Thrifty Automotive Group	7.30%	7.90%	8.00%
U-Save Auto Rental System Inc.	0.50%	0.50%	0.60%
Fox Rent A Car	0.50%	0.50%	0.70%
Payless Car Rental System Inc.	0.50%	0.60%	0.70%
ACE Rent A Car	0.50%	0.50%	0.50%
Zipcar	0.60%	0.70%	0.90%
Rent-A-Wreck of America	0.20%	0.20%	0.20%
Triangle Rent-A-Car	0.20%	0.20%	0.20%
Affordable/Sensible	0.20%	0.20%	0.20%
Independents	2.50%	2.40%	2.50%

Source: Compiled by the authors from company information

There are two widely used methods for evaluating industry consolidation: the four-firm concentration ratio and the Herfindahl index.

Four-firm concentration ratio

The concentration ratio is a measure of the total market share held by a certain number of firms in the industry. The general form of the concentration ratio formula can be described as:

$$CR_m = \frac{\sum_{m=1}^{n} Q_m}{n} \times 100$$

Where "n" is the number of firms measured and Q is output.

The concentration ratio is simply the summed output of *n airline* companies divided by the total industry output. The most commonly used concentration ratio is the four-firm concentration ratio, which measures the output of the four largest firms in the industry. The market can then be classified according to a continuum of the percentage share of the top four.

In this example, American, Delta, Southwest, and United would be grouped together as indicated using 2010 data from Figure 9.5. To find the four-firm concentration ratio of

Table 9.5   Total US Domestic Industry Output

	Domestic	
	2009	*2010*
Delta	67,384	110,095[1]
Southwest	98,171	98,558
American	92,966	93,173
United	69,334	67,520
US Airways	53,250	52,700
Continental	49,528	48,796
Industry total	683,844	689,907

Source: Compiled by the authors from US DOT Form 41

[1] As a result of the merger with Northwest in 2010

the US domestic airline industry in 2010, the four largest airlines in terms of output need to be grouped together and compared to the industry total. In this example, American, Delta, Southwest, and United would be grouped together. These four firm's combined output is 369,346 available seat miles (ASMs). When divided by the total industry output, this produces a four-firm concentration ratio of 53.5 percent. The four largest airlines produce 54 percent of the industry's total output.

Markets are separated into categories along the market continuum based on the four-firm concentration ratio. Markets where the four-firm concentration ratio is less than 20 percent are assumed to mimic perfectly competitive industries. On the other end of the spectrum, markets where the concentration ratio is above 80 percent are highly concentrated and are assumed to be closer to the monopoly end of the market continuum spectrum. In the middle of the spectrum, where the majority of industries lie, monopolistic competition markets would have a four-firm concentration ratio between 20 and 50 percent, while oligopolies would have a 50 to 801 percent four-firm concentration ratio.

- Perfect Competition: less than 20 percent.
- Monopolistic competition: between 20–50 percent.
- Oligopoly: between 50–80 percent.
- Monopoly: above 80 percent four-firm measurement.

Earlier in the chapter it was claimed that the airline industry is an example of an oligopoly market, and it appears that the four-firm concentration ratio that was calculated above does in fact support this claim with a 53.5 percent four-firm concentration ratio in 2010. The reason for this is that, for the airline industry, the four-firm concentration ratios should be calculated on an airport-by-airport basis. The fact is, for most airports, there are only a few major carriers and this produces an oligopolistic market in most of the industry. Since consumers are ultimately impacted on an individual market basis, assessing the four-firm concentration ratio on an airport-by-airport basis provides a more realistic picture of the air transport industry.

Table 9.6 provides a synopsis of the four-firm concentration ratios when calculated on an airport-by-airport basis for six major airports in the US. While the industry's concentration ratio was only 54 percent, all six of the airports analyzed had concentration ratios substantially above that. In the case of Atlanta and Dallas, one dominant hub carrier receives almost monopoly power as they effectively control the market. Chicago, with both United and American operating hubs out of the airport, is essentially a duopoly. Finally, Los Angeles and Las Vegas both exhibit strong oligopolistic market tendencies with their four-firm concentration ratios equaling roughly 65 percent. Similar statistics would be found if the analysis were applied to other airports.

Using an airport-by-airport analysis of the domestic aviation market, it is clear that the industry resembles a strong oligopoly instead of the monopolistically competitive market that the industry-wide four-firm concentration ratio might suggest. While every market is unique, with varying levels of concentration, it is unlikely that any one airport would have a low four-firm concentration value. The four-firm concentration ratio enables the analyst to get a quick look at the amount of concentration in the industry, but it usually requires a more in-depth analysis to fully understand the specific market situation.

Table 9.6   Four-firm Concentration Ratio

Airport	2010	2011
Atlanta (ATL)	92.37%	93.77%
Chicago (ORD)	71.88%	68.52%
Dallas (DFW)	89.97%	88.42%
Los Angeles (LAX)	65.62%	62.45%
Las Vegas (LAS)	67.81%	66.13%
Phoenix (PHX)	84.95%	84.56%

Source: Compiled by the authors

Herfindahl-Hirschman Index

Another method used to analyze the amount of concentration in an industry is the Herfindahl index (also known as the Herfindahl-Hirschman index, HHI). This is a popular measure that was used in the first chapter to analyze the amount of consolidation that exists in the industry. HHI is obtained by squaring the market-share of each of the players, and then adding up those squares:

$$HHI = \sum_{m=1}^{n} S_m^2$$

Where:

- S is m-firms' market share
- N in number of firms

The measure is simply the cumulative squared value of the market share for every firm in the industry. Therefore, the higher the index, the more concentration and (within limits) the less open the market. By squaring the market share values, firms with a large market share receive more weight in the calculation than firms which have a smaller market share. The US Department of Justice considers a market with a result of less than 1,000 to be a competitive marketplace. A result of 1,000–1,800 is a moderately concentrated marketplace and a result of 1,800 or greater is a highly concentrated marketplace. It should also be noted that market share can be calculated in terms of different products; therefore, unique HHI indices could potentially be created for the same market.

For example, in a duopoly market if each of the two firms in the market has a market share of 50 per cent, the HHI index would be:

$$HHI = (50)^2 + (50)^2 = 2500 + 2500 = 5000$$

With two firms that have of 80 percent and 20 percent respectively, then:

$$HHI = (80)^2 + (20)^2 = 6,800$$

Based on data obtained through Form41, the industry HHI index for the domestic US airline industry was calculated for the past several years. The respective market share for each airline was based on the number of passengers enplaned. The general trend in the US airline industry was deconcentration until 2010, as the HHI value has dropped over 200 points since 1998. In 2010, the HHI returned to its pre-1998 levels, a further indication of the cyclicality of the aviation industry. The current HHI value of around 800 is fairly typical of an oligopolistic market as values greater than 1,000 generally indicate a high degree of concentration in the market. However, just as in the four-firm concentration ratio, a much higher degree of concentration exists at individual airports. The HHI for the top ten US airlines is contained in Table 9.7.

The HHI is frequently used by the Department of Justice to determine if a proposed merger is acceptable for anti-trust reasons. If we consider the mergers of Delta/Northwest, AirTran/Southwest and United/Continental, Table 9.8 presents the market share for the major carriers pre-merger and Table 9.9 provides the new market share post-merger as well as the industry HHI.

Based on the proposed mergers, the HHI would increase by slightly more than 500 points. The Department of Justice usually does not like mergers that raise the industry HII above 1,000, as any point above that is deemed too monopolistic. In addition to these industry-wide HHI measures, certain markets, particularly in the northeast, would experience far greater increases in the HHI index. Since the HHI is only one of many factors employed by the Department of Justice when evaluating mergers, mergers are often approved on other grounds.

Markets in which the HHI is:[15]

- Less than 1,000; competitive marketplace.
- 1,000 and 1,800; moderately concentrated.
- Higher than 1,800; concentrated.

---

15   US Department of Justice and the Federal Trade Commission, Merger Guidelines § 1.51.

Table 9.7   **US Domestic Airlines Herfindahl Index, 2011**

Year	Herfindahl Indices: 1998–2010
1998	1,335.81
1999	1,309.30
2000	1,281.84
2001	1,246.10
2002	1,251.70
2003	1,206.97
2004	1,187.27
2005	1,172.30
2006	1,195.86
2007	1,170.08
2008	1,152.02
2009	1,151.66
2010	1,349.22

Source: Compiled by the author using Form41 data
* Based on enplaned passengers

Table 9.8   **US Airline, Pre-merger Market Share (2009)**

Carrier	Market Share
Delta	12.7%
Northwest	7.7%
American	16.0%
United	10.5%
Continental	8.2%
US Airways	9.5%
jetBlue	4.2%
AirTran	4.5%
Southwest	18.9%
Frontier	1.8%
Virgin America	0.7%
Alaska	2.9%
Hawaiian	1.6%
Allegiant	1.0%
HHI	1,153

Source: Compiled by the author using Back Aviation Form41 data*

* The HHI values calculated included data from carriers not listed in the tables. Table 9.8 and 9.9 merely provide an overview of the major carriers' market share

Table 9.9   US Airline, Post-merger Market Share (2012)

Airline	Market share
American	16.0%
Delta	20.4%
jetBlue	4.4%
United	17.9%
US Airways	9.5%
Southwest	24.0%
Frontier	1.7%
Virgin America	0.7%
Alaska	3.0%
Hawaiian	1.5%
Allegiant	1.1%
HHI	1,685

Source: Compiled by the authors from Form 41

BEYOND MARKET CONCENTRATION CONSIDERATIONS

Although the above calculations of market concentration lend a certain aura of rigor, it remains a very arbitrary decision criterion, partly because it completely ignores how mergers may increase efficiency and/or reduce prices through positive impacts of economies of scale, scope, and density. Of course, it is not possible for anyone, including the merging airlines themselves, to know with certainty exactly how efficient the newly combined airline will be. The mix of corporate cultures, merging of separate labor unions, and/or other factors creates uncertainties that become clear only long after the merger actually occurs. Proponents of strong anti-merger regulation argue that, with no guaranteed gains in efficiency, it makes sense to keep the number of competitors as high as possible for as long as possible.

On the other hand, opponents of such vigorous regulation maintain that the dismal rate of return for the airline industry shows that there is more than enough competition, implying that firms should generally be free to combine as they choose. The struggling industry should be allowed to repair itself. For instance, Ben-Yosef (2005: 265–266) suggests that, had government regulators allowed them to merge, it is quite possible that United Airlines and US Airways might have avoided bankruptcy and been in a better position to keep fares low enough to profitably compete with the low-cost airlines. Thus, rather than focusing on concentration ratios, regulators might better serve the public interest by generally allowing troubled firms to merge as they see fit, and allowing the industry to evolve as it will as long as there is no indication of higher than normal long-run profits being earned.

So far, regulators have allowed such free choice in mergers only if it becomes obvious that one firm is on its way to shutting down anyway. However, airline alliances, which might be viewed as a sort of partial merger, have often been allowed considerably more freedom. The KLM/Northwest alliance, for example, is quite extensive, having received anti-trust immunity from US regulators. Sometimes governments severely restrict the action of alliance partners, such as in the case of American Airlines and British Airways, but at least some cooperation is allowed. The greater degree of freedom allowed in alliances seems to represent some compromise; regulators may be implicitly admitting that rigid focus on concentration ratios is not appropriate in an industry largely floundering in bankruptcy. At the same time, from a pro-regulatory viewpoint, if alliances should prove to be somehow anti-competitive, they can more readily be altered or even completely undone.

ANTI-TRUST, MARKET EVOLUTION, AND COOPERATION

As mentioned above some economists believe there is a an "empty core" problem where the high fixed cost/low marginal costs of the airline industry require more cooperation between firms to avoid competition that destroys profits and prevents some efficient production and product innovation. The rivalry between Airbus and Boeing may also illustrate how some cooperation can benefit consumers. Consider, for example, the problematic production of the "jumbo aircraft," the A380. Suppose Boeing had decided to make its own version of the A380 and had then begun to encounter problems similar to those of Airbus. It might have been, in this situation, that both companies would have decided to simply abandon production. With each of them having to share demand with the other the costs might have been prohibitive. Of course, it never came to this because Boeing chose not to enter the A380 market. Whether intentional or not, there was a sort of implicit cooperation when Boeing stepped aside to make the project viable for Airbus. Governments may unwittingly facilitate such cooperation through patent laws, which can have the effect of segmenting the market for different producers.

Such implicit cooperation is less likely to take place, though, where the number of firms is greater. The airline industry may need explicit contracts to coordinate an efficient allocation of resources for consumers. Airlines might be able to offer more non-stop service or move to larger, more efficient and comfortable jets on more routes if they were able to explicitly cooperate. Some of this might be arranged through mergers and some through more limited alliances that might sometimes resemble cartels but could be aimed at arranging efficient production rather than suppressing competition. Removal of the restrictions on cabotage and international mergers would facilitate this and would also reduce entry barriers to help reduce the possibility of the cooperation taking an anti-competitive turn.

Of course, current regulators, and many economists, would oppose such a move. It might be for instance, that the lack of industry profitability reflects simple overcapacity rather than the need for complex cooperation. Eventually, bankruptcy, capacity cuts, and liquidation may decrease supply, increase price, and return the industry to normal profitability. In any case, many economists argue that airline consolidation in some form, both in Europe and the US, is inevitable. According to this view, the record shows that the industry cannot be profitable in its present state, whether because there is simply too much capacity from too many airlines or a more complex lack of coordination. Since investors will ultimately require a reasonable rate of return to keep capital in the industry, some capital will be withdrawn so the number of large airlines is probably bound to decline somewhat. Regulators might slow this decline in numbers but cannot prevent it and may, as outlined above, cause the adjustment to be less orderly and more severe than it would be if they simply got out of the way. Only time will tell.

SUMMARY

This chapter examined the models of monopolistic competition and oligopoly, the so-called hybrid markets. Most aviation industries fit the oligopoly model but there are different views of what this implies. Some economists see oligopoly as inherently problematic while others point to the lack of high profits in many oligopolies, particularly the airlines, and conclude that entry barriers are not so significant after all. The chapter then provided an overview of various theories of oligopoly; these included contestability, kinked demand curve, and Cournot models. Empirical evidence indicates that the airline industry may be contestable when viewed as a network. From this viewpoint, it may even be that increased concentration and cooperation through more alliances can benefit consumers, primarily through economies of scale, and return the airline industry to reasonable long-run profits level. Finally, the chapter introduced various indices to measure the amount of concentration in the industry; these included the four-firm concentration ratios and the HHI.

DISCUSSION QUESTIONS

1. What are the best examples of monopolistic competition in the real world?
2. How is price established in an oligopoly market?
3. The anti-trust laws do not allow firms in the same industry to agree on what prices they will charge. Is that correct for the airline industry?
4. How is price established in a monopolistic competitive market?
5. If monopolies are socially undesirable why does a government actually support having some?
6. Provide examples illustrating how markets change from one structure to another when technology or other market conditions change.
7. What are the implications for the regulatory authorities of the existence of contestable markets?
8. You are the manager for DirectJet and unable to determine whether any given passenger is a business or leisure traveler. Can you think of a self-correction mechanism that would permit you to identify business and night leisure?
9. Graphically depict a shut-down case for a monopolistic competitive firm. When should any firm shut down in the short run?

10. Focus on the airline industry: why is the upper portion of the kinked demand curve elastic, and the lower portion inelastic?
11. What are four distinguishing characteristics of monopolistic competition?
12. South Charleroi Airport is a regional airport serving leisure travel market. The inverse demand curve for this airport is $P = 150 - Q$. Assume that there are only two airlines serving this airport, each with the identical marginal cost (MC) of $30.
    a Supposing they perform as Cournot oligopolists, determine the price and total firm productivity.
    b Compare with the result under pure monopoly and perfect competition.
13. Assume that there are only two airlines serving an airport each with the MC of $40. The market demand curve at this airport is $MR = 160 - Q$.
    a Supposing they perform as Cournot oligopolists, determine the price and total airlines outputs.
    b Compare with the result under pure monopoly and perfect competition.
14. Suppose that a typical monopolistically competitive firm faces the following demand and total cost equations for its product:

    $Q = 50 - P$

    $TC = 375 - 25Q + 1.5Q^2$

    a Where P is the price of the product and Q is the number of units produced.
    b What is the firm's profit-maximizing price and output level?
    c What is the relationship between P and average total cost (ATC) at the profit-maximizing output level?
    d Is this firm earning an economic profit? Is this firm in short-run or long-run monopolistically competitive equilibrium? Will new firms enter into or exit from this industry?
15. Calculate the change in the HHI for the period of 2006–2012:
    a US airline industry.
    b US major airports.
    c Have these industries become more or less concentrated over time?

REFERENCES

Armentano, D. (1986), *Antitrust Policy: The Case for Repeal*. Washington, DC: Cato Institute, Associated Press (AP). (2007). China to Develop Large Commercial Aircraft by 2020.

International Herald Tribune, March 12. Retrieved on March 28, 2007 from http://www.iht.com/articles/ap/2007/03/12/business/AS-FIN-China-Homegrown-Jet.php.

Bailey, E. (1981). Contestablility and the Design of Regulatory and Antitrust Policy. *American Economic Review*, Vol. 71, No. 2, 178–183.

Ben-Yosef, E. (2005). *The Evolution of the US Airline Industry Theory, Strategy and Policy Series: Studies in Industrial Organization, Vol. 25*. Springer.

Bowen, K. and Purrington, C. (rev. 2006). Pratt & Whitney: Engineering Standard Work. Harvard Case, 9-604-084, March 27.

Crandall, R. and Winston, C. (2003). Does Antitrust Policy Improve Consumer Welfare? *Journal of Economic Perspectives,* October 19, 3–26.

Crandall, R. and Winston, C. (2006). The Breakdown of "Breakup". *The Wall Street Journal,* Vol. 6. March 9.

Galbraith, J. K. (1979). *Age of Uncertainty.* Boston, MA: Houghton Mifflin.

Gollish, D., Clausen, H., Koggersvol, N., Christey, P. and Bruner, R. (rev. 1997) The Boeing 777. Darden Case, UVA-F-1017, November.

Hirschey, M. (2003). *Managerial Economics* (10th ed.). Ohio: South-Western.

Joesch, J. and Zick, C. (1994). Evidence of Changing Contestability in Commercial Airline Markets during the 1980s. *Journal of Consumer Affairs,* Vol. 28, No. 1, 1–24.

Kotha, O., Nolan, D. and Condit, M. (rev. 2005). Boeing 787: The Dreamliner. Harvard Case, 9-305-101, June 21.

Oum, T., Zhang A. and Zhang, Y. (1993), Inter-firm Rivalry and Firm-specific Price Elasticities in Deregulated Airline Markets. *Journal of Transport Economics and Policy,* Vol. 27, No. 2, 171–192.

Raghavan, S. and Raghavan, J. (2005). Application of Core Theory to the US Airline Industry. *Journal of the Academy of Business and Economics,* Vol. 5, No. 3, 116–125.

Simpkins, E. (2006). BA in the Dock—The Airline is Being Investigated for Potential Collusion on Fuel Surcharges. *The Daily Telegraph,* June 26, 6.

Strassmann, D. (1990), Potential Competition in the Deregulated Airlines. *Review of Economics and Statistics,* Vol. 72, No. 4, 696–702.

Whinston, M. and Collins, S. (1992), Entry and Competitive Structure in Deregulated Airline Markets: An Event Study Analysis of People Express. *RAND Journal of Economics,* Vol. 23, No. 4, 445–462.

10

Aviation Forecasting and Regression Analysis

So we went to Atari and said, "Hey, we've got this amazing thing, even built with some of your parts, and what do you think about funding us? Or we'll give it to you. We just want to do it. Pay our salary, we'll come work for you." And they said, "No." So then we went to Hewlett-Packard, and they said, "Hey, we don't need you. You haven't got through college yet."

Steve Jobs, Apple Computer Inc. Founder

Forecasting is one method for reducing the uncertainty in the business world and whether we realize it or not, we all forecast in one way or another. For example, if I take an umbrella out with me, I am obviously forecasting rain with some degree of probability. And, of all industries, the aviation industry stands out as one of the most uncertain and unpredictable. Since the variability of such inputs as jet fuel prices and passenger demand can dramatically affect airlines, forecasting is extremely useful in reducing the uncertainty associated with this volatility.

This chapter will look at the applications of forecasting in the aviation industry and discuss many of the major forecasting methods used. Greater emphasis will be placed on quantitative tools, such as regression analysis since it is the most powerful forecasting method to be discussed. The outline for this chapter is as follows:

- Aviation Forecasting Applications
- Qualitative Forecasting Methods
  - Focus group
  - Market survey
  - Market experiments
  - Barometric forecasting
  - Leading economic indicators
  - Lagging economic indicators
  - Coincident economic indicators
  - Historical analogy
  - Delphi method
- Quantitative Forecasting Methods
  - Time-series statistics
  - Cross-section statistics
- Descriptive Statistics

- Mean
- Variance
- Standard deviation
- Time-series analysis
- Trend analysis
- Seasonal variations
- Cyclical variation
- Random effect
- Time-series Forecasting
  - Moving average
  - Weighted moving average
  - Exponential smoothing
  - Trend analysis
- Forecast Accuracy
- Regression Analysis
  - How to estimate a demand function
  - Goodness of fit
  - Performing regression analysis
  - Dummy or binary variables
  - Autocorrelation
  - Multicollinearity
- Data Sources
  - US Department of Transportation (DOT)/Bureau of Transportation Statistics (BTS)
  - Federal Aviation Administration (FAA)
  - International Air Transport Association (IATA)
  - International Civil Aviation Organization (ICAO)
  - Official Airline Guide (OAG)
  - Airports Council International (ACI)
  - Flightglobal (formerly Air Transport Intelligence)
  - Airline Monitor
  - UK Civil Aviation Authority (CAA)
  - Transport Canada (TC)
  - Eurocontrol
  - The Aircraft Owners and Pilots Association (AOPA)
  - Bureau of Economic Analysis (BEA)
  - Bureau of Labor Statistics (BLS)
  - Organization for Economic Cooperation and Development (OECD)
- Summary
- Discussion Questions

AVIATION FORECASTING APPLICATIONS

Forecasting has many applications in the aviation industry, and probably the chief among these is the forecasting of demand. Since demand is not monolithic and varies for every flight, sophisticated forecasting tools need to be applied to help forecast the size and nature of demand. Forecasting the nature of demand could include the mix between price-sensitive (leisure) and time-sensitive (business) travelers and the expected booking rate for the flight; when these forecasting methods are applied to every flight, the forecasting operation becomes quite large. Moreover, since strategic planning and yield management are dependent on demand forecasts, it is probably one of the most important applications of forecasting in the airline industry.

However, forecasting is not just limited to demand forecasting. Planning of human resources, financial resources and needs, route developments, aircraft fleet, and infrastructure expansion are all based on some expectation of future events. While the forecasting methods employed can range from rudimentary to sophisticated, some type of forecasting is still applied. In 2010, Airbus forecasted that over the 2011–2030 periods, world passenger traffic is set to increase by 4.8 percent per year. The world's passenger aircraft fleet (above 100 seats) will grow from 15,000 at the beginning of 2011 to just over 31,000 by 2030. At the same time, some 10,500 aircraft from the existing fleet will be replaced by more eco-efficient models.[1]

Forecasts concerning the amount of flying, crew requirements, training schedules, absenteeism, and employee turnover ratios are all important for the airlines. Additionally, project viability and profit projections are all based on the expectation of future events. Since projects are analyzed over their lifespan, forecasts need to be created concerning future expected cash flows. Based on these forecasts, multi-million dollar projects are either approved or rejected. For example, analyzing the installation of audio-visual on-demand (AVOD) in-flight entertainment systems across the fleet could be based on forecasts concerning the installation schedule, future maintenance costs, and the passengers' opinions of the new in-flight entertainment system.

As the examples above highlight, forecasting spans multiple functional areas. Therefore, it is critical to understand the many aspects of forecasting before attempting to apply it to aviation. To do this, four critical skills are needed:

- knowledge of the airline industry;
- knowledge of economic principles and statistics;
- computer applications;
- communication.

While all four skills are required to forecast, this text will only cover the first two skills in detail. The text itself provides knowledge of the airline industry and basic economic principles, while this chapter will discuss the basic statistics used in forecasting. Of course, the chapter cannot replace a complete statistics textbook, but is meant to provide an introductory and applied overview. A variety of computer applications can be used to help forecast, including Microsoft Excel and SPSS, that can help perform basic regression and statistical analysis.

QUALITATIVE FORECASTING METHODS

No flying machine will ever fly from New York to Paris … [because] no known motor can run at the requisite speed for four days without stopping.

Orville Wright, 1908

---

1   Airbus, Global Market Forecast, 2011–2030.

Forecasting methods can be broken down into two categories: qualitative and quantitative. Qualitative forecasting methods use subjective techniques to help forecast. Qualitative forecasts do not use statistical databases or provide measures of forecast accuracy since they are based on opinions, surveys, and beliefs. On the other hand, quantitative forecasts use statistical relationships to help forecast future events and, while they are more mathematical in nature, they may or may not be any more accurate than qualitative forecasts. Table 10.1 displays a few of the main advantages and limitations of qualitative forecasting.

One of the main advantages of qualitative forecasts is that they are flexible and can be easily altered to reflect any changes in the economy or environment. The flexibility of qualitative forecasting also enables early signals of changes and anomalies in data to be recognized. On the other hand, one of the limitations of qualitative forecasting is that it can be difficult to track and isolate the primary variable that is causing changes in the dependent variable. Additionally, the lack of tests for accuracy always creates the situation where there is no way to know how good the forecast is.

While there are numerous types of qualitative forecasts, six major methods will be discussed. They are:

- focus groups;
- market survey;
- market experiment;
- barometric forecasting;
- historical analogy;
- Delphi method.

Table 10.1   Advantages and Limitations of Qualitative Forecasting

Qualitative Forecasting	
Advantages	Limitations
Flexibility:	Complex:
Easily altered as the economy changes	Hard to keep track of interactions in the primary variables
Early signals:	Lack of tests of accuracy:
Can catch changes and anomalies in data	Cannot easily test the accuracy in prior periods

Focus group

A focus group is a relatively informal information gathering procedure in marketing research. It typically brings together eight to 12 individuals to discuss a given subject. Usually focus group participants are brought into a room where a moderator asks questions to help move the discussion forward. Researchers observe the participants and their responses, providing a quick and relatively inexpensive insight into their research problem. Focus groups can be quite effective in evaluating new product options, such as new aircraft seats or in-flight entertainment systems. However, researchers need to be concerned that participants may not produce completely honest responses and may feel pressured into accepting what everyone else believes. Additionally, if the focus group is not a representative sample of the target population, then the responses are likely to be inaccurate. An airline, by forming a focus group, can reach out to its potential passengers for feedback and comment. Airlines generally use focus groups in planning, marketing, or serving a new destination domestically or internationally. Finally, a focus group can help airlines managers to identify the right level of in-flight and cabin crew services.

Market survey

A survey is simply a method of acquiring information by asking people what they think will happen. The most common method of a market survey is a questionnaire, but there are many other methods. Depending on the nature of the questions asked, questionnaires can provide the researcher with both quantitative and qualitative results. One of the major benefits of market surveys is that they are easy to use and do not require advanced theory or econometric analysis to interpret the results. A potential flaw with market surveys is that the accuracy of the survey depends on the size and responsiveness of the sample. In the aviation industry, market surveys are often used to find out what improvements an airport can make, and airlines use them in the form of customer comment cards to improve service.

Market experiments

A more expensive method of qualitative forecasting is a market experiment. Market experiments involve testing new product factors, such as prices or packaging, in a few test markets (Hirschey, 2006). Market experiments use real-life markets and this can be risky if the change is not accepted by consumers; that is, the change may permanently alienate them from the product. (Allen et al., 2005). Due to the costs involved, market experiments are rarely used in forecasting demand in the aviation industry. However, market experiments have been conducted with in-flight food and beverages. For example, an airline may wish to test the acceptance of new buy-on-board meals on a few flights to help forecast demand for the product.

Barometric forecasting

Barometric forecasting involves using current values of certain variables, called indicators, to help predict future values of other variables (Truett and Truett, 1992). A leading indicator is a variable whose current changes give an indication of future changes in other variables. A lagging indicator is a variable whose changes typically follow changes in other economic variables. Depending on one's point of view, the relationship between any two variables can either be a leading or lagging indicator.

Leading economic indictors

For forecasting, a leading indicator is the most useful, since it provides an early signal of what may come. Leading economic indicators are believed to change before the economy as a whole changes and allow a prediction to be made. One of the common leading indicators is the stock market; stocks generally increase or decline before the rest of economy.

In aviation, gross domestic product (GDP) is a leading indicator of airline demand as changes in GDP are usually followed by changes in demand for air transportation. Therefore, if the GDP growth rates decline sharply over a period of time, airlines can expect a decline in demand shortly afterwards. Part of the reason for this lagging relationship is that consumers are generally slow to adjust their spending patterns to reductions in income.

Lagging economic indicators

A lagging economic indicator is one that follows the event. The nature of it does not allow for the prediction of events but rather allows for confirmation that the forecasted event or trend is occurring or has occurred. The unemployment rate, for example, serves as a lagging indicator to the general economic condition. For an airline, profits and costs per unit of output are good lagging indicators of performance; these metrics are not useful for predictions but rather confirm the airline's performance over a period of time.

Coincident economic indicators

A final type of indicator, a coincident indicator, is not tremendously useful for aviation forecasting, but does have other applications. Coincident indicators are variables whose changes roughly coincide with changes in other economic variables. In aviation, crude oil prices are a coincident indicator of jet fuel prices. Since changes in the price of crude oil roughly occur at the same time as changes in jet fuel prices, crude oil is not a useful indicator at predicting future jet fuel prices. However, crude oil could be used as a proxy variable in jet fuel hedging since crude oil is more heavily traded than jet fuel.

Historical analogy

Historical analogy is a simple forecasting technique where the future is forecasted based on historical events. While many quantitative forecasting methods use historical data to help predict the future, historical analogy is, by definition, on a qualitative level. The success of historical analogy is largely dependent on the depth of knowledge and history that the forecaster has. A forecaster who has seen all facets of an industry over a long period probably has a better prediction of the future than a more recent employee. Therefore, historical analogy is only as good as the person making the forecast.

Delphi method

The Delphi method is related to historical analogy in that the forecast is largely based on opinion; however, the Delphi method collects forecasts and opinions from an independent panel of experts. Each expert provides their analysis and opinion independently, and then a consensus forecast is created based on the analysis provided by each member of the panel. By having members independently submit their opinions, the Delphi method benefits from not having steamroller or bandwagon problems (Hirschey, 2006). Additionally, the Delphi method benefits from having multiple experts analyze the issue instead of just one or a few people in historical analogy. In theory the accuracy of the forecast is based on the collective knowledge of the expert panel; however, because every opinion is equally weighted, the collective knowledge may not be as reliable as just a few experts.

QUANTITATIVE FORECASTING METHODS

In contrast to qualitative methods, the quantitative methods use statistical data to analyze and forecast future behavior of specific variables. Statistical information is divided into time-series and cross-sectional data.

Time-series statistics

Time-series data represent observations of particular variables over a number of time periods. For example, the number of passengers at various past points in time at a given airport, daily oil prices, and monthly profits are all useful time-series data used by the aviation industry. Time-series analysis looks for patterns in data while regression analysis assumes a casual relationship between two or more variables.

Cross-section statistics

Cross-sectional data are compiled for different variables at a single point in time; for example, the number of passengers over different geographically located airports, or the number of aviation accidents over different countries for one time period. While there are many methods of quantitative forecasting, we shall cover only two broad categories: time-series analysis and regression analysis. Both methods will be analyzed in greater detail.

There are several advantages and limitations of quantitative forecasting that are summarized in Table 10.2.

Table 10.2   Advantages and Limitations of Quantitative Forecasting

Quantitative Forecasting	
Advantages	*Limitations*
Organic relationships	Economy changes may distort results
Behavioral relationships	Extensive data mining of information
Tests of reliability determine forecast accuracy	Only a crude approximation of actuality

One of the major advantages of quantitative forecasting is that tests of reliability can easily be performed to determine the accuracy of the forecast. In time-series analysis, the most accurate forecasting method can be chosen based on the test of reliability. In regression analysis, not only are forecasters able to provide a probability of how accurate the overall forecast is, but also the reliability of the individual variables in the forecast

(Spirtes, Glymour and Scheines, 2001). One major drawback of using only quantitative forecasting is that history is not always a correct predictor of the future. In addition, historically, another chief drawback of quantitative forecasting is that forecasts require extensive data collection and processing. However, with the introduction of advanced statistical computer software, data collection and processing is a much simpler task, making quantitative forecasting much easier. Another limitation exists with quantitative forecasting in that it is dependent on the quality of the data that is used. Depending on how good the data are, quantitative forecasts may distort reality or model it perfectly. The garbage in–garbage out cautionary statement would apply here.

DESCRIPTIVE STATISTICS

Prior to analyzing various forecasting methods in detail, a fundamental understanding of elementary statistics is required and for most forecasting that amounts to three basic statistics: the mean, the variance, and the standard deviation. Descriptive statistics are numerical estimates that organize, sum up, or present the data and provide simple summaries about the sample and the measures. To meet that objective, a full range of indicators has been developed, and we will provide definitions and applications of the most important ones. To illustrate descriptive statistics, we will utilize the following scheduled revenue passenger miles (RPMs) data for the time period 2005–2011.

Table 10.3   Scheduled Revenue Passenger Miles, 2005–2011

Year	RPM (billions)	Year	RPM (billions)
2005 Q1	186.05	2008 Q3	218.10
2005 Q2	206.38	2008 Q4	187.03
2005 Q3	213.26	2009 Q1	177.90
2005 Q4	189.44	2009 Q2	202.38
2006 Q1	188.57	2009 Q3	213.19
2006 Q2	211.56	2009 Q4	186.51
2006 Q3	214.37	2010 Q1	181.20
2006 Q4	195.61	2010 Q2	208.26
2007 Q1	193.90	2010 Q3	221.39
2007 Q2	219.12	2010 Q4	198.20
2007 Q3	226.53	2011 Q1	187.17
2007 Q4	203.47	2011 Q2	215.44
2008 Q1	200.24	2011 Q3	225.00
2008 Q2	217.65	2011 Q4	193.37

Source: Compiled by the authors from OAG Form 41 data

Mean

The mean, probably the most common indicator of a data set, is simply the average of the data. For the above data, the mean can be calculated by taking the sum of the RPMs and dividing by the number of years. In statistics, the mean is usually denoted with the Greek letter, μ. In Microsoft Excel, the mean can simply be calculated by using the average function. From this, the general form for the mean is:

Mean = μ

$$\mu = \frac{\sum_{i=1}^{n} X_i}{n}$$

Based on this information, the average, or mean, RPMs between 2005 and 2011 can be calculated in a variety of ways (quarterly, yearly, or over a number of years). Over the time period shown in Table 10.3, the average RPM is 202.90 billion.

Variance

The variance of a sample measures how the observations are spread around the mean. A large variance means the observations are widely scattered around the mean. The variance of variable x is simply the summed squared difference between the actual values, x, and the mean of x. The variance shows the dispersion of the data from the mean. In statistics, the variance is denoted as σ^2, with the general form being:

$$\sigma^2 = \frac{\sum_{i=1}^{n} (x_i - \mu)^2}{n-1}$$

The variance of a data set is important since it gives some idea of the accuracy of the mean. For example, the variance in the previous class average example is 204.[2] The variance is computed in Table 10.4.

Standard deviation

Standard deviation is directly related to variance since it is the positive square root of the variance. The standard deviation is a statistic that shows how tightly all the observations are clustered around the mean in a set of data. When the observations are spread around the mean, it indicates that we have a relatively large standard deviation. In order to avoid problems with the negative signs of some deviations from the mean (note that the sum of the values of the deviations from the mean in the above example is zero and this will

---

2   The population standard deviation is represented by the Greek letter sigma "σ" and the sample standard deviation represented by S.

Table 10.4   Variance Calculation of Revenue Passenger Miles, 2005–2011

Year	RPM (billions)	$x - \mu$	$(x - \mu)^2$
2005 Q1	186.05	-16.85	284.03
2005 Q2	206.38	3.48	12.09
2005 Q3	213.26	10.36	107.26
2005 Q4	189.44	-13.46	181.26
2006 Q1	188.57	-14.33	205.44
2006 Q2	211.56	8.66	74.94
2006 Q3	214.37	11.47	131.49
2006 Q4	195.61	-7.29	53.19
2007 Q1	193.9	-9.00	81.06
2007 Q2	219.12	16.22	262.98
2007 Q3	226.53	23.63	558.23
2007 Q4	203.47	0.57	0.32
2008 Q1	200.24	-2.66	7.09
2008 Q2	217.65	14.75	217.47
2008 Q3	218.1	15.2	230.94
2008 Q4	187.03	-15.87	251.96
2009 Q1	177.9	-25.00	625.16
2009 Q2	202.38	-0.52	0.27
2009 Q3	213.19	10.29	105.82
2009 Q4	186.51	-16.39	268.74
2010 Q1	181.2	-21.7	471.03
2010 Q2	208.26	5.36	28.70
2010 Q3	221.39	18.49	341.76
2010 Q4	198.2	-4.70	22.12
2011 Q1	187.17	-15.73	247.53
2011 Q2	215.44	12.54	157.17
2011 Q3	225.00	22.10	488.27
2011 Q4	193.37	-9.53	90.88
μ	202.9	σ	203.97

always be true by definition), the values of the deviations are displayed in squared terms. By taking the square root of the variance, the standard deviation returns the variance to a more easily interpretable number. However, it should be noted that while the squaring procedure eliminates the problem of negative deviations canceling out positive ones, it also gives much greater weight to outlying observations. That is, the further away from the mean the observation is, the greater the difference between the observation and the mean and therefore the greater the squared value of this observation.

In the first example, the standard deviation is simply 13.52. Standard deviation is usually denoted as, σ, with the general form being:

$$\sigma = \sqrt{\frac{\sum_{i=1}^{n}(x_i - \mu)^2}{n-1}}$$

The concept of standard deviation is not difficult to understand. Assume we have collected one month of ticket prices, for example from New York to London, about 1,250 observations, and entered them into a spreadsheet to calculate the average. Suppose the average price is calculated as $870. This number, by itself, is of limited value. By measuring the standard deviation of the ticket price, however, we can gain an idea of how volatile the ticket price really was (the larger the standard deviation the more volatile the ticket price). For a perfectly normal distribution, 68.4 percent of all the observations fall within plus or minus one standard deviation of the average, 95.4 percent fall within plus or minus two standard deviations of the average and 99.7 percent of the observations fall within plus or minus three standard deviations of the average.[3] To summarize:

About 68.4% of the data will be within: $X \pm 1\sigma$

About 95.4% of the data will be within: $X \pm 2\sigma$

About 99.7% of the data will be within: $X \pm 3\sigma$

TIME-SERIES ANALYSIS

As mentioned, time-series analysis measures the status of some activity, such as aviation accidents, number of aircraft operations, or number of enplanements over a period of time. A time-series analysis records the activity with measurements taken at equally spaced intervals with a consistency in the activity and the method of measurement. Observations may be carried out annually, quarterly, monthly, weekly, daily, or every hour. All the time-series data contain four components:

- trend analysis;
- seasonal variations;
- cyclical variations;
- random effect.

---

3   A normal distribution is a distribution where the area to each side of the mean under the distribution curve is equal to 0.5.

Trend analysis

The trend component accounts for the movements of time series over a long period of time. Any regular patterns of values above and below the trend component are likely attributable to the cyclical component of a time series. For the air transport industry, this shifting or trend is usually attributed to factors such as liberalization, deregulation, change in disposable income, introduction of new technology, population growth, and/or privatization. The overall trend of demand has been consistently increasing (Figure 10.1). The volume of US traffic during the 1969–2011 period grew at a healthy rate of 6.15 percent per year.[4]

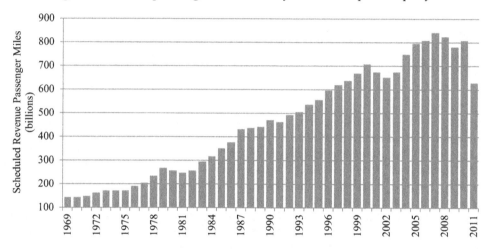

Figure 10.1   Scheduled Revenue Passenger Miles, US 1969–2011

Source: Compiled by the authors from OAG Form 41 data

Seasonal variations

The seasonal component accounts for regular patterns of variability within certain time periods, such as over a year. For an airline, the number of passengers may be very high in certain months or seasons and low in others. Figure 10.2 shows the seasonal variation.

Cyclical variations

The cyclical component refers to long-term fluctuation of time-series statistics over time. The changes in traffic from 1969 to 2011 indicate a cyclical trend (Figure 10.2). We can see that the cyclical variation in the number of passengers during this time period is more irregular because of the business cycle. In general, it is harder to predict the cyclical components of time-series data than trend and seasonal variations. The most recent cycle was in 2000 when air traffic started to decline and this cycle continued until late 2003.

---

4   The growth rate in each period is the ratio of the absolute change in RPM to an earlier value.

$$g_{RPM} = \frac{RPM_t - RPM_{t-1}}{RPM_{t-1}}$$

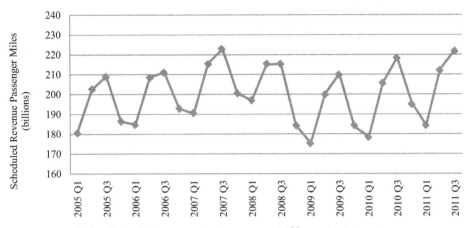

Figure 10.2   Scheduled Revenue Passenger Miles, 2005–2011

Source: Compiled by the authors from OAG Form 41 data

Random effect

Finally, random factors of a series are short-term, unanticipated, and non-recurring factors that affect the values of the series; these factors are part of the natural variability present in all measurements. Or, we can argue that the random component is that which is left over when all the other components of the series (trend, seasonal, and cyclical) have been accounted for. For example, events such as September 11, 2001 are impossible to predict. As such, even if an airline could perfectly predict cyclical operating costs and revenues, the costs (and benefits) of random effects can never be fully taken into account. Other random effects may stem from aviation accidents, airline mergers, and bankruptcies. The trend component analyzes the data over a long period of time, the cyclical component occurs over a medium term, seasonal variations occur in the short term (that is, a year), while random events are unique incidents.

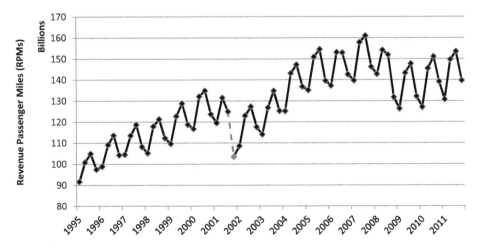

Figure 10.3   Random Effect of September 11, 2001 on US Revenue Passenger Miles

Source: Compiled by the authors from OAG BackAviation Form 41 data

TIME-SERIES FORECASTING

While there are several different methods that can be used to forecast time series which take into account the various components of a time series, we shall just discuss the following four major methods:

- moving average;
- weighted moving average;
- exponential smoothing;
- trend analysis.

Moving average

Moving average is a smoothing technique that uses the average of the most recent data values to help forecast the next period. Moving average is a very simple technique that contains the underlying assumption that the most recent values are the best representation of the future. Mathematically, the formula for moving average is:

$$MA = \frac{\sum_{i=1}^{n} X_i}{n}$$

In order to understand the applicability of moving average, consider the following data, which display the historical bookings for the last ten days of a DirectJet flight.

Table 10.5   Time-series Data of Bookings for a DirectJet Flight

Day	Bookings
1	115
2	100
3	105
4	120
5	135
6	130
7	145
8	150
9	130
10	160
11	?

If the goal is to forecast the number of bookings on the eleventh day with a moving average, the first step is to determine n, which represents the number of recent data values to include in the forecast. Since every value is given equal weight in the moving average, the larger the n, the more weight historical values are given. Assuming that a three-day moving average is desired, the three most recent data values (days 8, 9, and 10) are used to help forecast DirectJet's bookings on day 11. The forecast for the eleventh day is:

$$F(11) = \frac{(150 + 130 + 160)}{3}$$

$$F(11) = \frac{440}{3}$$

$$F(11) = 146.67$$

Using the last three data values, the forecast number of bookings on the eleventh day is 146.67, or rounded up to 147. If it is believed that a three-day moving average provides too much emphasis on the most recent days, then a five-day moving average can be used instead. The five-day moving average provides a forecast for the eleventh day of:

$$F(11) = \frac{(130 + 145 + 150 + 130 + 160)}{5}$$

$$F(11) = \frac{715}{5} = 143$$

Based on the five-day forecast, the forecasted value of 143 is less than the three-day forecast value of 146.67. While results will vary depending on the data set, any moving average attempts to smooth out any distortions in the data; this can be extremely useful especially with highly variable data.

Moving average can easily be calculated in Microsoft Excel using the moving average function in the "Data Analysis Toolpack." With the aid of Microsoft Excel, the three-day and five-day moving average forecasts were created for multiple days to compare the forecasted values and the actual values. Table 10.6 presents these data.

Note the fact that the three-day moving average forecasts cannot be computed for the first three days and likewise the five-day average for the first five days since these are needed to start the series. The accuracy of the forecasts can be determined by comparing the difference between the actual and forecasted value, and these techniques will be discussed in more detail later in the chapter.

Table 10.6   Three-day and Five-day Moving Average Forecasts

Day	Bookings	Three-Day Moving Average	Five-Day Moving Average
1	115		
2	100		
3	105		
4	120	107	
5	135	108	
6	130	120	115
7	145	128	118
8	150	137	127
9	130	142	136
10	160	142	138
11	?	147	143

Weighted moving average

Weighted moving average (WMA) is similar to moving average. It still uses historical data to provide a forecasted value; however, instead of each value receiving equal weighting, as in moving average, values receive different weightings. For example, in a three-period moving average, each value receives an equal weighting of 1/3. However, weighted moving average enables the forecaster to weight the values as desired. Mathematically, the formula for WMA is presented as:

$$WMA = \sum_{i=1}^{n} W_i \times X_i$$

Using the same data contained in Table 10.5 and displayed in Table 10.7, a weighted moving average for day 11 can be created assuming that the most recent value receives a 50 percent weighting, the next most recent value receives a 30 percent weighting, and the third value receiving 20 percent. Based on this, the forecasted value is:

$$F(11) = 0.5(160) + 0.3(130) + 0.2(150)$$
$$F(11) = 80 + 39 + 30 = 149$$

Based on the designed weightings, the forecast value for day 11 is 149. However, the forecast value can change based on the assigned weightings. Assuming that the most recent value receives an 80 percent weighting, the second value a 15 percent weighting, and the final value a 5 percent weighting, the forecast value for the eleventh day would be:

$$F(11) = 0.8(160) + 0.15(130) + 0.05(150)$$
$$F(11) = 128 + 19.5 + 7.5 = 155$$

With this new weighting, the forecast value for the eleventh day is considerably higher because in this circumstance, the most recent value received a high weighting. Ultimately the weightings that are assigned are based on the forecaster's judgment. Therefore, the more experience and expertise the forecaster has, the more likely that the assigned weightings will be accurate. Weightings can be assigned for any number of periods, as long as the total sum of the weightings equal 100 per cent or 1.0.

Table 10.7   Weighted Three-day Moving Average Forecasts

Day	Bookings	Weights	
1	115		
2	100		
3	105		
4	120		
5	135		
6	130		
7	145		
8	150	50%	80%
9	130	30%	15%
10	160	20%	5%
11	?	149	155

Exponential smoothing

A third smoothing technique that can be used to forecast time-series data is exponential smoothing. Unlike a moving average which uses multiple historical values to help forecast, exponential smoothing only uses data from the previous period. Exponential smoothing indirectly takes into consideration previous periods by using the previous period's forecast value in helping determine the forecasted value. This creates a situation where the weighting for a value gets exponentially smaller as time moves on. The general formula for exponential smoothing is:

$$F_{t+1} = \alpha Y_t + (1-\alpha)F_t$$

Where:

- F_{t+1} is the forecast value in the next period
- Y_t is the actual value in the previous period
- F_t is the forecasted value in the previous period
- α is a smoothing constant with values between 0 and 1.

The smoothing constant helps determine what weighting of the forecast value should be based upon the actual value from the previous period and the forecast value. The higher the smoothing constant, the greater the weighting the actual value receives. Like the two previous forecasting methods, the forecaster must make a judgment in assigning the value for the smoothing constant. While higher smoothing constants usually provide more accurate forecasts, the overall objective of the forecast is to be as accurate as possible. Since the formula contains a term on the right-hand side that shows a previously forecast value, the question arises as to where that value will come from for the first observation. The answer to this is that value comes from the actual value of the first period. This means that no matter what value is picked for the constant, the first value of the forecast will equal the first period of the series. Subsequent values will of course differ between constants because the actual and forecast values will differ.

Using the exponential smoothing function in the "Data Analysis Toolpack" from Microsoft Excel, forecasts can be created for DirectJet. Table 10.8 provides forecasts with two different smoothing constants (α): 0.3 and 0.8.

From Table 10.8, it is clear that a smaller smoothing constant provides greater fluctuation in the forecasted value while a larger constant provides less variability in the forecasts.

Table 10.8   Exponential Smoothing Forecasts for Using Two Different Smoothing Constants

Day	Bookings	$\alpha = 0.3$	$\alpha = 0.8$
1	115		
2	100	115	115
3	105	105	112
4	120	105	111
5	135	115	112
6	130	129	117
7	145	130	120
8	150	140	125
9	130	147	130
10	160	135	130
11	?	153	136

Trend analysis

The fourth and final time-series method to be investigated is trend analysis. Scatter diagrams and line graphs provide a good first approximation in identifying the existence of a trend line between independent and dependent variables. Depending on how closely the points group together, we may be able to identify a trend in the data. Unfortunately, trends are not always easy to see graphically and there may also be a problem with units. A more quantitative method to identify a trend line is to use regression analysis. Regression analysis attempts to create a linear trend equation to describe the data (Anderson, Sweeney and Williams, 2006). Such equations can then be used to provide a forecast for a future value. The general form for these equations follows:

$$F_t = b_o + b_1 t$$

Where:

- F_t is the forecast value in period t
- b_o is the intercept of the trend line
- b_1 is the slope of the trend line.

Hence, in order to calculate the forecast value, the parameters b_0 and b_1 must first be calculated. The formulas for these values are presented here:

$$b_1 = \frac{\sum_{t=1}^{n} t Y_t - \frac{(\sum_{t=1}^{n} t \sum_{t=1}^{n} Y_t)}{n}}{\sum_{t=1}^{n} t^2 - \frac{(\sum_{t=1}^{n} t)^2}{n}}$$

$$b_o = \bar{Y} - b_1 \bar{t}$$

Where:

- Y_t is the actual value in period t
- n is the number of periods
- $\bar{Y}$ is the average value of the time series
- $\bar{t}$ is the average value of t

Based on these formulae, the linear trend line can be constructed. To complete the calculation, however, some additional information is required. Therefore, the original DirectJet problem is expanded in Table 10.9.

Table 10.9   Expanded Data Set for DirectJet

Day	Bookings	Day*Bookings	Day Squared
1	115	115	1
2	100	200	4
3	105	315	9
4	120	480	16
5	135	675	25
6	130	780	36
7	145	1,015	49
8	150	1,200	64
9	130	1,170	81
10	160	1,600	100
55	1,290	7,550	385

From the expanded data set, the values required to determine the slope and the intercept can be found as follows:

$$\sum_{t=1}^{10} tY = 7,550$$

$$\sum_{t=1}^{10} t = 55$$

$$\sum_{t=1}^{10} Y_t = 1,290$$

$$n = 10$$

$$\sum_{t=1}^{10} t^2 = 385$$

$$\bar{Y} = \frac{\sum_{t=1}^{10} Y_t}{n} = \frac{1290}{10} = 129$$

$$\bar{t} = \frac{\sum_{t=1}^{10} t}{n} = \frac{55}{10} = 5.5$$

$$b_1 = \frac{\sum_{t=1}^{n} tY_t - \frac{(\sum_{t=1}^{n} t \sum_{t=1}^{n} Y_t)}{n}}{\sum_{t=1}^{n} t^2 - \frac{(\sum_{t=1}^{n} t)^2}{n}}$$

$$b_1 = \frac{7550 - \dfrac{(55*1290)}{10}}{385 - \dfrac{(55)^2}{10}}$$

$$b_1 = \frac{7550 - 7095}{385 - 302.5} = \frac{455}{82.5} = 5.52$$

$$b_o = 129 - 5.52(5.5)$$

$$b_o = 129 - 30.33 = 98.67$$

Using the values of the slope and intercept, the trend line is:

$$F_t = 98.67 + 5.52t$$

Based on this formula, forecasts for the number of bookings for DirectJet can be created by solving the equation. Note that trend analysis only forecasts the trend portion of a time series. Cyclicality, seasonality, and random factors can cause distortions from the trend line. Using the above equation, the forecast for the number of bookings on the eleventh day is:

F(11) = 98.67 + 5.52(11)

F(11) = 98.67 + 60.72 = 159.39

Trend analysis can be performed more quickly through computer programs such as Microsoft Excel and SPSS. By graphing the time-series data, a trend line can be fitted to the data and the equation can also be provided. Figure 10.4 displays the trend line, with the computer producing the exact same formula as was calculated above.

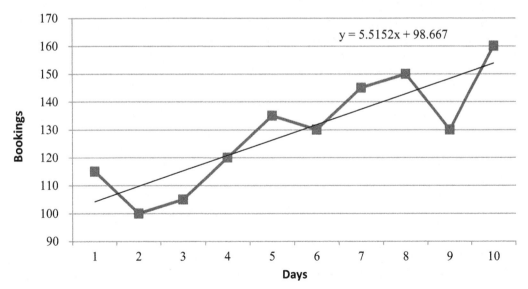

Figure 10.4   Graphical Representation of Trend Line for the Time Series

FORECAST ACCURACY

We soon saw that the helicopter had no future, and dropped it. The helicopter does with great labor only what the balloon does without labor, and is no more fitted than the balloon for rapid horizontal flight. If its engine stops, it must fall with deathly violence, for it can neither glide like the aeroplane or float like the balloon. The helicopter is much easier to design than the aeroplane, but is worthless when done.

Wilbur Wright, 1909.

Forecasting is ultimately useful only if the forecasts are reasonably accurate. While the actual accuracy of the forecast is not known until the event has occurred, historical time-series data can be analyzed to give an indication of how well the technique works. There are many examples of forecasting errors in the aviation industry which caused significant loss to the company involved. For example, in 2006, the Airbus's parent company EADS announced that the company needed to sell 420 A380s to breakeven, up from its initial announced prediction of 270 aircraft. The two major methods of analyzing forecast accuracy are mean squared error (MSE) and mean absolute deviation (MAD).

MSE averages the squared difference between the actual value and the forecasted value. The values are squared to eliminate the effect of negative errors canceling out positive errors (similar to the squaring of the deviations from the mean that was used to calculate the variance) and also to give greater weight to larger errors. The basic formula for MSE can be written as:

$$MSE = \frac{\sum_{t=1}^{n}(Y_t - F_t)^2}{n}$$

The tables below provide the MSE calculation for the different forecasting methods employed in the DirectJet example. Table 10.10 provides the MSE calculation for both the three-day and five-day moving averages. Based on the results, the three-day moving average appears more accurate since its MSE value is less than the five-day MSE.

Table 10.11 provides the MSE for the weighted average forecasts used in the DirectJet example. Both were three-day moving averages; however, the first forecast used a 50/30/20 weighting, while the second forecast used an 80/15/5 weighting. Based on the MSE, the more evenly distributed forecast provides the most accurate forecast for this particular time series.

Table 10.12 provides MSE for exponential smoothing forecasts with smoothing constants of both 0.3 and 0.8. Based on all these calculations, the exponential smoothing forecast with a smoothing constant of 0.3 provided the most accurate forecast, since it had the lowest MSE.

Another measure of forecasting accuracy is MAD. MAD finds the average of the absolute value of the deviations. Since the deviations are not squared, large deviations are not given extra weight. The general formula for MAD is:

$$MAD = \frac{\sum_{t=1}^{n}|Y_t - F_t|}{n}$$

Table 10.10   Mean Squared Error Calculation for Moving Average Forecasts

Day	Bookings	Three-day Moving Average	Forecast Error	Squared Forecast Error	Five-day Moving Average	Forecast Error	Squared Forecast Error
1	115						
2	100						
3	105						
4	120	107	13	169			
5	135	108	27	729			
6	130	120	10	100	115	15	225
7	145	128	17	289	118	27	729
8	150	137	13	169	127	23	529
9	130	142	-12	144	136	-6	36
10	160	142	18	324	138	22	484
			MSE	274		MSE	401

Table 10.11   Mean Squared Error Calculation for Weighted Moving Average Forecasts

Day	Bookings	50/30/20 Forecast	Forecast Error	Squared Forecast Error	80/15/5 Forecast	Forecast Error	Squared Forecast Error
1	115						
2	100						
3	105						
4	120	109	11	121	112	8	64
5	135	106	29	841	102	33	1,089
6	130	116	14	196	109	21	441
7	145	127	18	324	123	22	484
8	150	136	14	196	135	15	225
9	130	139	-9	81	133	-3	9
10	160	144	16	256	145	15	225
			MSE	301		MSE	369

Table 10.12   Mean Squared Error Calculation for Exponential Smoothing

Day	Bookings	α = 0.3	Forecast Error	Squared Forecast Error	α = 0.8	Forecast Error	Squared Forecast Error
1	115						
2	100	115	-15	225	115	-15	225
3	105	105	0	0	112	-7	49
4	120	105	15	225	111	9	81
5	135	115	20	400	112	23	529
6	130	129	1	1	117	13	169
7	145	130	15	225	120	25	625
8	150	140	10	100	125	25	625
9	130	147	-17	289	130	0	0
10	160	135	25	625	130	30	900
			MSE	230		MSE	369

MAD can also be calculated for the various time-series forecasts in the DirectJet example. The absolute value of the forecast error is found in Microsoft Excel by using the "abs()" function. Table 10.13 provides the MAD for the moving average forecasts, Table 10.14 for the weighted moving average forecasts, and Table 10.15 for the exponential smoothing forecasts.

Table 10.13   Mean Absolute Deviation Calculation for Moving Average Forecasts

Day	Bookings	Three-day Moving Average	Forecast Error	Squared Forecast Error	Five-day Moving Average	Forecast Error	Absolute Forecast Error
1	115						
2	100						
3	105						
4	120	107	13	13			
5	135	108	27	27			
6	130	120	10	10	115	15	15
7	145	128	17	17	118	27	27
8	150	137	13	13	127	23	23
9	130	142	-12	12	136	-6	6
10	160	142	18	18	138	22	22
			MAD	16		MAD	19

Table 10.14   Mean Absolute Deviation Calculation for Weighted Moving Average Forecasts

Day	Bookings	50/30/20 Forecast	Forecast Error	Squared Forecast Error	80/15/5	Forecast Error	Squared Forecast Error
1	115						
2	100						
3	105						
4	120	109	11	11	112	8	8
5	135	106	29	29	102	33	33
6	130	116	14	14	109	21	21
7	145	127	18	18	123	22	22
8	150	136	14	14	135	15	15
9	130	139	-9	9	133	-3	3
10	160	144	16	16	145	15	15
		MAD	16			MAD	17

Table 10.15   Mean Absolute Deviation Calculation for Exponential Smoothing Forecasts

Day	Bookings	$\alpha = 0.3$	Forecast Error	Squared Forecast Error	$\alpha = 0.8$	Forecast Error	Squared Forecast Error
1	115						
2	100	115	-15	15	115	-15	15
3	105	105	0	0	112	-7	7
4	120	105	15	15	111	9	9
5	135	115	20	20	112	23	23
6	130	129	1	1	117	13	13
7	145	130	15	15	120	25	25
8	150	140	10	10	125	25	25
9	130	147	-17	17	130	0	0
10	160	135	25	25	130	30	30
		MAD	13			MAD	16

From the exponential smoothing calculations, the forecast with a smoothing constant of 0.3 appears once again to be the most accurate forecasting method for this particular time-series data. In this situation, both MSE and MAD picked the same forecasting method as the most accurate; however this will not always hold true as it will depend on the observed data. Both are commonly used in practice. The MSE gives more weight to large errors, while the MAD is easier to interpret. Based on the measures of accuracy, an appropriate forecasting method can be chosen. Using this method, forecasts can be created for future periods. While the measures of accuracy highlight the most accurate forecasting method based on historical information, the data provide an ongoing repository of continually growing measurements that can be refined and updated. As this happens, new forecasting methods may be substituted for the original selection. Therefore, the forecaster must use some judgment in choosing which measure of accuracy to use, how much data should be incorporated, and how often to use it. Regardless of the choices, the ultimate goal of the forecaster is to provide the most accurate forecasts available.

REGRESSION ANALYSIS

The other major quantitative forecasting method is regression analysis. Regression analysis assumes a casual relationship between the dependent and independent variables. The specific two-variable linear regression model is:

$$Y_i = \beta_0 + \beta_1 \times X_i + \varepsilon_i$$

Where:

- Y_i is the dependent variable
- β_0 and β_1 are the coefficients of the regression line (the intercept and slope)
- X_i is the independent variable
- ε_i is the predictor error or so called residual.

A dependent variable is a variable that relies on other factors and variables, while an independent variable has a value which does not rely on any other factors (Ovedovitz, 2001). In order to understand the applicability of regression analysis, first consider the following dataset, Table 10.16, with consumption (C) and income (Y) values, and the corresponding Figure 10.5.

Using both the dataset and the graph, it is very clear that there is a simple linear relationship between consumption and income. In this relationship, since consumption is dependent upon income, consumption is the dependent variable and income the independent variable. Forecasting in this situation is fairly easy, since the pattern is obvious from the graph. In this case, for every 100 unit increase in income, consumption increases by 95 units. Based on the linear relationship, it is easy to forecast the level of consumption for an income level of 1,300. For this example, and using the graph above, consumption would be 1,335 for an income level of 1,300. Using the formulas from the trend analysis to calculate the slope and the intercept (the best fit formulas are the same for both techniques because they both assume a linear relationship between the variables) it can be easily determined that the slope of the function is 0.95 (95/100) and the y-intercept value is 100. This creates the function below, and this function can be used for future forecasts.

$C = 100 + 0.95Y$

Table 10.16   Dataset for Perfect Relation between Consumption and Income

C	Y
100	-
195	100
290	200
385	300
480	400
575	500
670	600
765	700
860	800
955	900
1,050	1,000
1,145	1,100
1,240	1,200
?	1,300

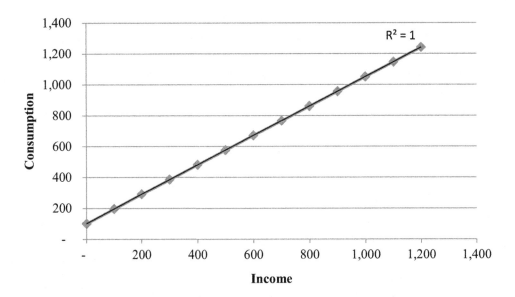

Figure 10.5   Graphical Perfect Relation between Consumption and Income

In this example, the relationship between consumption and income is very straightforward since they form a perfect linear relationship. However, real world data are never perfectly correlated, since random events and other factors cause distortions in the relationship. Consider the same relationship where the values are slightly modified.

Table 10.17   Dataset for Strong Relation between Consumption and Income

C	Y
100	-
220	100
300	200
450	300
700	400
590	500
800	600
400	700
950	800
700	900
1,020	1,000
1,200	1,100
1,240	1,200
?	1,300

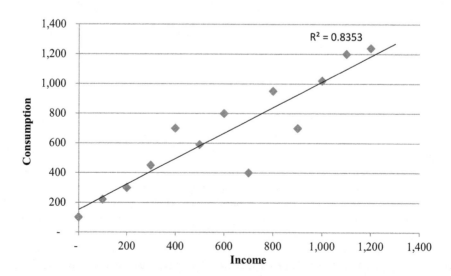

Figure 10.6   Graphical Strong Relation between Consumption and Income

In this situation there is no clear linear relationship between consumption and income. However, when the data from Table 10.17 are plotted in Figure 10.6, the points lie in a somewhat random nature but with a general upward trend. Since the previous trend analysis method would not provide an accurate forecast, regression analysis needs to be employed. As Figure 10.6 displays, regression analysis fits a trend line for the data points. In simple terms, regression analysis calculates a quantifiable linear relationship for the various data points. It should be noted that regression analysis can also estimate exponential, quadratic, or other relationships, depending on the general trend of the data points, but such estimating techniques require a functional form transformation that is linear in the parameters.

In this example, the data points have a general linear trend, therefore linear regression analysis should be chosen. However, the linear trend line is not perfectly accurate since all the values do not lie directly on the trend line. Figure 10.6 is an example of a scatter diagram, which essentially plots the points on a Cartesian plane with the dependent variable plotted on the y-axis and the independent variable plotted on the x-axis. Prior to performing regression analysis, a scatter plot should be created to help understand the nature of the data points. From this, the appropriate regression analysis (linear, exponential, and so on) can be chosen to provide the most accurate forecast of the dependent variable.

While all types of regression analysis perform the same function, there are many different methods for creating the trend line. The most common method is ordinary least squares (OLS), which minimizes the squared value of the residuals. Although it would seem more natural to minimize the sum of the errors, we run into the same problem that we observed earlier with the variance; very large positive errors (deviations above the line) would tend to cancel out very large negative errors (deviations below the line). Therefore, the errors are squared to eliminate this problem. This of course means that the larger errors carry more weight in the procedure.

Thus, the formula for OLS regression can be stated as:

$$OLS = \min \sum_{i=1}^{n}(y_i - \hat{y}_i)^2 \equiv \min \sum_{i=1}^{n} e_i^2$$

Where:

- e_i^2 is the residual
- y_i is the actual value
- $\hat{y}_i$ is the forecast value.

Therefore, the residual is simply the difference between the actual value and the forecast value along the trend line. In the first example, the residuals would equal zero since all the data points were located directly on the linear curve. However, in the second example, a residual value exists since there is a vertical gap between the data points and the trend line. Since the derivation of the OLS trend line (that is, the values of b_0 and b_1 that define the forecast line that minimizes the sum of the squared residuals) involves some elementary calculus, only the resulting values are presented here. In any event, the formulas that are derived are exactly the same as those given for the best fit trend line discussed earlier in the chapter.

For example, by using the information on Table 10.18, the short-run consumption function has been estimated as follow:

$C = 150 + 0.8615Y$

Where:

- $b_0 = 150$ and $b_1 = .8615$
- $C = $ Consumption
- $Y = $ Income.

Since the basic formula for a residual is the difference between the actual and forecasted value, we can calculate the difference for each observation as follows:

$$\text{Residual} = e^2{}_i = (C_i - \hat{C}_i)^2$$

These values are contained in Table 10.18.

Table 10.18   Residual Values for Forecasts of Consumption

C_i	Y	$\hat{C}_i$	$e\ (C_i\text{-}\hat{C})$	e^2
100	0	150	-50	2,500
220	100	236	-16	256
300	200	322	-22	484
450	300	408	42	1,764
700	400	495	205	42,025
590	500	581	9	81
800	600	667	133	17,689
400	700	753	-353	124,609
950	800	839	111	12,321
700	900	925	-225	50,625
1,020	1,000	1,012	8	64
1,200	1,100	1,098	102	10,404
1,240	1,200	1,184	56	3,136
			$\Sigma\,e^2$	266,385

Because OLS regression minimizes the sum of the residuals, no other linear trend line would produce a sum of the residuals less than 266,385. The accuracy of the forecast is ultimately determined by how large the residuals are. Forecasts with extremely large residuals imply the spread between the forecast value and the actual value is wide, and therefore the forecast is not as accurate. However, depending on the nature of the data, any trend line may be extremely accurate or not accurate at all. Measures of accuracy are discussed in the next section.

How to estimate a demand function

- collect available data about the number of tickets sold at different ticket prices;
- record these data points on a graph with price as the vertical axis and quantity sold as the horizontal axis;
- draw a line of best fit through the data points;
- measure the slope, m, (which must be negative) and the vertical intercept, b;
- create a demand function out of the parameters you've measured: price = mQ + b.

Goodness of fit

The "goodness of fit" of the regression model evaluates the strength of the proposed relationship between the dependent and independent variables and can be measured in several ways (Doane and Seward, 2007). The test that is most often used for the accuracy of any given regression is the coefficient of determination or R-squared. The coefficient of determination measures the percentage of variability in the dependent variable that the independent variable(s) explain. It is usually used to help state the degree of confidence one has in the forecast. The coefficient of determination ranges from 0 to 1, with one being a perfectly accurate forecast (similar to the first example provided) and zero being a completely inaccurate forecast. Using the following formula for the coefficient of determination, the R-square value is computed in Table 10.19.

$$R^2 = \frac{\sum_{i=1}^{n}(C_i - \overline{YY}C_\mu)^2}{\sum_{i=1}^{n}(C_i - C_\mu)^2}$$

Where:

- R^2 = the coefficient of determination
- $\hat{C}_i$ = the forecast value
- C_i = the actual value
- $\overline{Y}C_\mu$ = the mean of the actual values.

From this calculation in Table 10.19, the coefficient of determination for the linear regression model is 0.8352. This shows that 83.52 percent of the variability of the dependent variable, C, is explained by the independent variable, Y or, in simpler terms, the model is roughly 84 percent accurate. While the process of calculating the coefficient of determination is lengthy, almost all statistical programs display the R-square value in the regression output.

In order to better understand the accuracy of regression analysis, consider the same example, but with modified values of consumption. Table 10.20 provides the new dataset and Figure 10.7 displays the new scatter plot with the trend line.

Table 10.19   Coefficient of Determination Calculation

C_i	Y	$\hat{C}_i$	e $(C_i\text{-}\hat{C})$	e^2	$(\hat{C}_i\text{-}C\mu)$	$(\hat{C}_i\text{-}C\mu)^2$	$C_i\text{-}C\mu$	$(C_i\text{-}C\mu)^2$
100	0	150	-50	2,500	-517	267,289	-567	321,489
220	100	236	-16	256	-431	185,761	-447	199,809
300	200	322	-22	484	-345	119,025	-367	134,689
450	300	408	42	1,764	-258	66,564	-217	47,089
700	400	495	205	42,025	-172	29,584	33	1,089
590	500	581	9	81	-86	7,396	-77	5,929
800	600	667	133	17,689	0	0	133	17,689
400	700	753	-353	124,609	86	7,396	-267	71,289
950	800	839	111	12,321	172	29,584	283	80,089
700	900	925	-225	50,625	258	66,564	33	1,089
1,020	1,000	1,012	8	64	345	119,025	353	124,609
1,200	1,100	1,098	102	10,404	431	185,761	533	284,089
1,240	1,200	1,184	56	3,136	517	267,289	573	328,329
667	C_μ		$\Sigma\,e^2$	266,385	$\Sigma\,(\hat{C}_i\text{-}C\mu)^2$	1,351,238	$\Sigma\,(C_i\text{-}C\mu)^2$	1,617,277
							R^2	0.8352

Table 10.20   Dataset for Weak Relation between Consumption and Income

C	Y
700	0
220	100
300	200
450	300
700	400
590	500
400	600
400	700
950	800
500	900
1,020	1,000
1,300	1,100
200	1,200
?	1,300

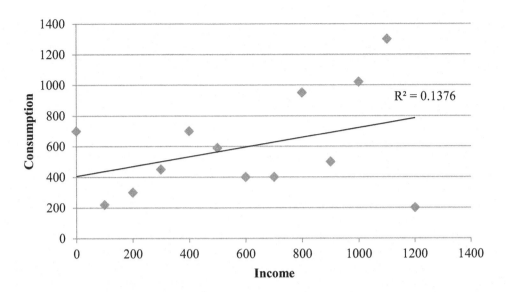

Figure 10.7   Graphical Weak Relation between Consumption and Income, with Trend Line

In this new scenario, the data points have less of a linear relationship, increasing the summation of the squares of the residual values. While OLS regression provides the best possible trend line for the data points, the increase in the size of the residuals over the previous example indicates that this trend line is not as accurate as the previous one. This is confirmed through a coefficient of determination value of 0.1376, which is significantly less than the previous value of 0.8352. The forecast for this third dataset yields an R-square value of only 0.1376, which means this forecast is not very powerful and may have little usefulness. While the coefficient of determination determines the accuracy of the overall regression, it is also possible to determine the statistical relevance of separate independent variables in the regression. This topic is discussed in the next few sections.

Performing regression analysis

For ease of exposition, the previous examples contained only one independent variable; however, most regressions will include multiple independent variables since the dependent variable can rarely be explained by one factor. For example, demand for airline services includes a host of factors such as ticket price, income, competitor's price, seasonality, and customer service. Since it would be nearly impossible performing OLS regression by hand for multiple independent variables, computer programs such as Microsoft Excel and SPSS allow regressions to be performed quickly and easily.

In order to understand applied regression analysis more completely, and to identify the important factors to analyze when interpreting regression results, we introduce a concrete example. Consider the demand for air travel between Orlando and Los Angeles. Four independent variables are used to help determine the demand. These are: average ticket prices, income, seasonality, and the presence of a random one-time event, such as September 11, 2001. Prior to any forecasting, a hypothesis should be created to help identify the expected relationship between the dependent and independent variables. This hypothesis can then be used to help determine if the results from the regression analysis are accurate. In this example, ticket price should have a negative coefficient, income a positive coefficient, seasonality could potentially have either a positive or negative (depending on the season) and 9/11 should have a negative value. The dataset used in the regression analysis is contained in Table 10.21.

In order to forecast demand, historical data on the number of passengers flying between Orlando and Los Angeles need to be found. O&D[5] data uses a 10 percent sample of total bookings to help quantify the total number of passengers flying the city-pair, regardless of whether they are flying on a non-stop flight or on connecting flights. Since it is difficult to determine which income needs to be measured, GDP is a reasonable proxy for income. Finally, the average ticket price for all travelers is determined through O&D data for each quarter. Quarterly data were used from 1998 in order to provide a sufficient number of observations. In order for regression analysis to be accurate, an appropriate number of observations are required. While controversial, a minimum of 30 observations is usually safe.

Dummy or binary variables

To capture the effects of a random event such as September 11, 2001, or the impacts of qualitative events such as seasonality, we can apply dummy variables.[6] A dummy variable is an independent variable that takes on only two values: one or zero. Dummy variables, or binary or categorical variables, require no additional economic data (Barreto and Howland 2006).

Many studies such as Anderson and Mittal (2000) and Brandt (1987) use regression analysis with dummy variables to identify the actual nature of the relationship between the dependent variable and independent (explanatory or exogenous) variables. The variable only determines if the presence of a factor exists or does not exist. In the case of seasonality, three unique independent variables can be created. The Q1 dummy variable takes on a value of one during the first quarter of every year, and the value of zero for every other quarter. Similar dummy variables were created for the second and third quarters. A fourth seasonal dummy variable is not needed since the fourth quarter is acting as the baseline for all the other quarters.[7] So that, for example, the coefficient in the regression equation for the first quarter would measure the additional (or smaller) quantity demanded over the fourth quarter. With this setup the first three quarters are being compared to the fourth quarter. Of course, we could have excluded any of the four quarters and then the regression coefficients on the remaining three would be compared to the quarter excluded.

The other dummy variable used takes into consideration the one-time shift in demand caused by the tragic events of September 11, 2001, often simply referred to as "9/11." Since

---

5   O&D is a data tool operated by OAG Aviation; www.oagaviation.com.
6   Also known as categorical variables.
7   In fact, if all the classes for a binary variable are included in a regression equation that includes a constant, the regression cannot be estimated since a linear dependence exists between the independent variables. This is the so-called dummy variable trap. See Hanushek and Jackson (1977) for a more complete description of this.

Table 10.21   Dataset for Forecasting Demand for Orlando to Los Angeles Flight

Quarter	Demand (# of Passengers)	Income (GDP in billions)	Average Ticket Price	Q1	Q2	Q3	11-Sep
2003 Q1	56,270	$ 8,726.40	$ 219.77	1	0	0	1
2003 Q2	45,640	$ 8,657.90	$ 226.55	0	1	0	1
2003 Q3	51,910	$ 8,825.40	$ 206.74	0	0	1	1
2003 Q4	47,230	$ 8,953.80	$ 214.97	0	0	0	1
2004 Q1	50,270	$ 9,253.00	$ 228.14	1	0	0	1
2004 Q2	47,650	$ 9,174.10	$ 214.77	0	1	0	1
2004 Q3	52,350	$ 9,313.50	$ 180.20	0	0	1	1
2004 Q4	60,000	$ 9,519.50	$ 195.95	0	0	0	1
2005 Q1	73,450	$ 9,715.60	$ 176.58	1	0	0	1
2005 Q2	69,190	$ 9,421.50	$ 168.25	0	1	0	0
2005 Q3	66,240	$ 9,862.10	$ 201.28	0	0	1	0
2005 Q4	64,390	$ 9,953.60	$ 198.64	0	0	0	0
2006 Q1	65,760	$ 10,356.60	$ 237.66	1	0	0	0
2006 Q2	63,090	$ 10,128.90	$ 242.60	0	1	0	0
2006 Q3	60,670	$ 10,135.10	$ 200.30	0	0	1	0
2006 Q4	51,320	$ 10,226.30	$ 180.81	0	0	0	0
2007 Q1	50,650	$ 10,576.50	$ 240.64	1	0	0	0
2007 Q2	37,703	$ 10,426.60	$ 226.70	0	1	0	0
2007 Q3	52,360	$ 10,527.40	$ 210.50	0	0	1	0
2007 Q4	53,220	$ 10,591.10	$ 208.60	0	0	0	0
2008 Q1	63,360	$ 10,705.60	$ 205.50	1	0	0	0
2008 Q2	59,590	$ 10,831.80	$ 192.99	0	1	0	0
2008 Q3	62,300	$ 11,086.10	$ 175.84	0	0	1	0
2008 Q4	64,720	$ 11,106.10	$ 173.56	0	0	0	0
2009 Q1	82,500	$ 11,920.00	$ 165.78	1	0	0	0
2009 Q2	77,650	$ 11,649.30	$ 210.84	0	1	0	0
2009 Q3	70,250	$ 11,760.50	$ 179.46	0	0	1	0
2009 Q4	76,560	$ 11,970.30	$ 175.50	0	0	0	0
2010 Q1	80,250	$ 12,173.20	$ 176.26	1	0	0	0
2010 Q2	68,260	$ 12,266.40	$ 200.15	0	1	0	0
2010 Q3	78,670	$ 12,573.50	$ 170.56	0	0	1	0
2010 Q4	69,640	$ 12,730.50	$ 195.40	0	0	0	0
2011 Q1	76,150	$ 13,008.40	$ 192.57	1	0	0	0
2011 Q2	81,080	$ 13,197.30	$ 208.06	0	1	0	0

Source: Compiled by the authors

these events impacted the demand for air travel, all quarters following and including the third quarter of 2001 received a value of 1 to identify the impact of 9/11. In this case, the excluded variables are the quarters that were not felt to be affected by the events of 9/11; that is, all prior quarters.

Once the data have been collected and placed in a statistical computer program, the program will return the values for the regression. Most programs require the user to define which variable is the dependent variable and which are the independent variables. In this example, the number of passengers is the dependent variable and the remaining variables are all independent. The regression is then run and the output is displayed. While the output varies from program to program, they all contain the same basic characteristics. For our example, all regression output is from SPSS.[8]

The first major chart displayed in all regression output is a summary of the model. The model summary from SPSS for the regression is contained in Table 10.22. One of the most important statistics contained in any model summary is the R-square value. As mentioned previously, the R-square value, or coefficient of determination, determines the percentage of variation in the dependent variable that is explained by the independent variables. In this model, approximately 88 percent of the demand for travel between Orlando and Los Angeles can be explained by the independent variables.

Table 10.22   Model Summary of Demand Forecast for Orlando to Los Angeles Flights from SPSS

Model Summary[b]

Model	R	R Square	Adjusted R Square	Std. Error of the Estimate	Durbin-Watson
1	.940[a]	.883	.858	3702.89391	1.885

a. Predictors: (Constant), 9/11 Dummy Variable, Q2 Seasonality, Q3 Seasonality, Q1 Seasonality, Average Ticket Price, GDP

b. Dependent Variable: Number of Passengers

The adjusted R-square value is similar to the R-square value, but takes into consideration the degrees of freedom of the model. By way of definition, the degrees of freedom are the number of observations beyond the minimum needed to calculate a regression statistic (Hirschey, 2006). The degrees of freedom are determined by taking the total number of observations minus the number of independent variables. Higher degrees of freedom are created through more observations or less independent variables. In this example, the degree of freedom is 28.[9] Since a forecast is usually more accurate with an increased number of observations or with a smaller number of independent variables, the adjusted R-square value takes this into account. Therefore, the ordinary R-square value is adjusted downward to account for the degrees of freedom in the particular model. Models that contain low degrees of freedom have a greater difference between the ordinary R-squared and the adjusted R-square value. Since this model has relatively high degrees of freedom (28), the difference between the adjusted R-square and the ordinary R-square is not large.

8   SPSS (originally, Statistical Package for the Social Sciences) program is used for statistical analysis.
9   Degrees of freedom = Number of Observations – Number of Independent Variables = 34 – 6 = 28

Autocorrelation

Another major statistic to analyze in the model summary output is the "Durbin-Watson statistic."[10] The Durbin-Watson statistic measures autocorrelation, which can severely distort the accuracy and significance of the regression model. Autocorrelation occurs when the residuals are not independent and have an underlying trend, which violates one of the major underlying assumptions used in performing regression analysis (that is, in the derivation of the parameters of OLS, it is assumed that the error terms or residuals are independent of each other). While the Durbin-Watson statistic detects autocorrelation, the residuals can also be plotted against time to detect if any patterns exist in the residuals. Potential patterns that could exist include linear lines, fanning, or cyclical movements where the residual alternates from positive to negative. Figure 10.8 provides the residual plot for the Orlando to Los Angeles regression.

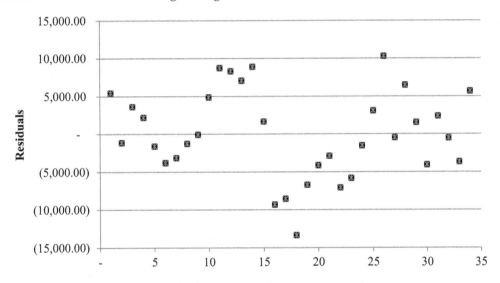

Figure 10.8   Residuals of the Regression Plotted against Time from Excel

Since Figure 10.6 does not display any trend in the residuals, it can be safe to say that autocorrelation does not exist in the regression. This is confirmed by a Durbin-Watson statistic of 1.885. Durbin-Watson statistics can range from 0 to 4, however, only values less than 1.5 or greater than 2.5 suggest that autocorrelation may exist in the regression. Therefore, the Durbin-Watson statistic of 1.885 falls within the acceptable range of 1.5 to 2.5.

The second major table contained in all regression outputs is an ANOVA table.[11] The ANOVA table for the Orlando to Los Angeles demand forecast is provided in Table 10.23. The ANOVA table provides the overall significance of the regression equation. As might be expected, there is a direct mathematical relationship between R-squared and the ANOVA F value for the overall significance of the regression.[12] The difference between them is the

---

10   It should be noted that the Durbin-Watson statistic is not displayed in the regression output obtained through Microsoft Excel.
11   ANOVA stands for analysis of variance.
12    $F = R^2 / K-1 / (1-R^2) / N-K$ where K stands for the number of independent variables and N stands for the number of observations. For a further discussion of this see Hanushek and Jackson (1977) pages 127 and 128.

Table 10.23   ANOVA for Demand Forecast from Orlando to Los Angeles from SPSS

ANOVA[b]

Model		Sum of Squares	df	Mean Square	F	Sig.
1	Regression	2.81E+09	6	467649874.1	34.107	.000[a]
	Residual	3.70E+08	27	13711423.28		
	Total	3.18E+09	33			

a. Predictors: (Constant), 9/11 Dummy Variable, Q2 Seasonality, Q3 Seasonality, Q1 Seasonality, Average Ticket Price, GDP

b. Dependent Variable: Number of Passengers

fact that the F-statistic allows us to pick a level of significance for the overall equation and compare this to a predetermined F distribution. Put more simply, it helps determine if the model is a sound representation of reality, or if the sample data is just an abnormality. This is accomplished by comparing the F-statistic of the regression to a predetermined level of significance. Conventional levels of significance are ordinarily set .90, .95 and .99 and these mean, respectively, that we can be 90 percent, 95 percent, and 99 percent sure that our regression results are due to a true relationship between the independent and dependent variables and not due to random chance. Most statistics text books contain complete tables of F distributions against which the regression F value can be compared; however, it is also true that most computer programs for regression contain (as part of the output) the level of significance of the independent variables for the given number of observations and degrees of freedom of the specific regression. So that, for example, the level of significance might be reported at .001 or .02; this means, respectively, that we can be 99.9 percent and 98 percent sure that our results are not due to chance.

The third major table contained in all regression output is a table of coefficients. This is displayed for the demand forecast from Orlando to Los Angeles in Table 10.24. The coefficients table allows the researcher to construct a linear equation that can be used for forecasting and it also determines if the individual variables are statistically significant. The first column of the coefficients table lists all the independent variables used in the analysis plus the constant. The constant term is usually interpreted as the value of the dependent variable when all the other independent variables are set to zero. Columns two and four both display values for the coefficients. The standardized values (column four) are generally used to compare the respective size of the impacts of the independent variables on the dependent variable. This is accomplished by calculating them in standardized units; that is, the standardized coefficient is the unstandardized value of the coefficient multiplied by the ratio of the standard deviation of the independent variable to the standard deviation of the dependent variable. Therefore, a standardized coefficient of 1.14, as the one for GDP, means that a 1 standard deviation change in the independent variable will lead to a 1.14 standard deviation change in the dependent variable. Similar interpretations apply to the other standardized coefficients. But, since the unstandardized values are the coefficients that are directly applicable to forecasting actual values, the unstandardized beta values are the coefficients that are used in the forecast equation. However, and as a final step prior to forming a demand equation, each independent variable needs to be tested to see if it is statistically significant.

The *t-statistic* is similar to the F-statistic discussed earlier except that it applies to a single individual variable rather than to the whole (or some subset) of the independent variables. The t-statistic is a measure of how accurate a statistical estimate is. More specifically, a t-value is calculated for each independent variable and this value is compared to a standardized t distribution. At this point a probability statement can be made about the significance (at some predetermined level of confidence) of the independent variable. So that, for example, if the predetermined level of significance is .90 or .95 and the t-value for the independent variable selected exceeds the t-value for the standardized table (at the degrees of freedom for the specific regression), then we can say that we are 90 percent or 95 percent sure that the relationship between the individual variable and the dependent variable is not due to chance. Therefore, a significant t-value indicates that the variable in question influences the dependent variable while controlling for other explanatory variables. Quantitatively, the t-statistic contained in column five is simply the unstandardized coefficient divided by the standard error of the coefficient. For example, the t-statistic for the independent variable GDP is found by dividing the beta value of 8.586 by the standard error of .869. This produces a t-statistic of 9.880.

$$t_i = \frac{b_i}{S_{b_i}} = \frac{8.586}{0.869} = 9.88$$

Generally speaking, if an independent variable passes the predetermined t-test, then it should be included in the model; however, if the variable fails the t-test, then it should be considered for exclusion from the model unless there are strong theoretical reasons to include the variable or there is a clear problem of multicollinearity (discussed in the next section). As a rule of thumb, if the value of a parameter is more than twice the size of its corresponding standard deviation (error), we can conclude, under a two tailed test, that the estimated coefficient is significantly different from 0 at 5 percent confidence level. Furthermore, if the estimated coefficient is greater than three times the estimated standard error, we can conclude the estimated value is significantly different from 0 at a 1 percent level of significance.

Table 10.24   Coefficients Significance for Demand Forecast from Orlando to Los Angeles from SPSS

Coefficients[a]

Model		Unstandardized Coefficients		Standardized Coefficients	t	Sig.	Collinearity Statistics	
		B	Std. Error	Beta			Tolerance	VIF
1	(Constant)	23409.230	16123.163		1.452	.158		
	GDP	8.586	.869	1.149	9.880	.000	.319	3.131
	Average Ticket Price	-211.815	52.474	-.456	-4.037	.000	.339	2.952
	Q1 Seasonality	778.912	2032.159	.036	.383	.705	.502	1.993
	Q2 Seasonality	5218.855	1857.769	.238	2.809	.009	.600	1.666
	Q3 Seasonality	3840.672	1869.170	.169	2.055	.050	.642	1.559
	9/11 Dummy Variable	-19971.5	2216.165	-1.017	-9.012	.000	.339	2.950

a. Dependent Variable: Number of Passengers

Multicollinearity

Another potential major problem that the coefficient table, Table 10.24, highlights is multicollinearity. Multicollinearity occurs when two or more independent variables are highly correlated with each other. If two independent variables are perfectly correlated, then the estimates of the coefficients cannot be computed. Intuitively, the problem arises since the regression cannot separate the effects of the perfectly correlated independent variables. Quantitatively, it arises because there is a term in the denominator for the variance of the individual independent variables that contains the correlation factor between the independent variables. As this term approaches one (perfect correlation), the variances of both of the independent variables approach infinity. And, as we have seen above, the t-statistic is calculated by dividing the numerical value of the coefficient by its standard deviation. Since the standard deviation is simply the square root of the variance, then the larger the variance, the larger the standard deviation and the smaller the t value. Thus, a high degree of multicollinearity between independent variables can cause a low level of significance for either one or both of the independent variables

Two different methods can be used to detect for multicollinearity. SPSS contains collinearity diagnostics in the coefficients table and these are contained in the last two columns in Table 10.24. The Variance Inflation Factors (VIF) statistic helps detect multicollinearity and, while the threshold of an acceptable VIF values varies (similar to confidence levels discussed earlier), a conventionally accepted level is that a VIF statistic above four indicates the presence of high multicollinearity.[13] The tolerance statistic is simply the inverse of the VIF score (1/VIF); therefore, smaller tolerance values indicate higher degrees of multicollinearity.

The other method for detecting a high degree of correlation between independent variables is to simply create a correlation matrix. Table 10.25 displays a correlation matrix for all the hypothesized independent variables in the example regression.

From the Table 10.25, the correlations between all the independent variables are presented. The key statistic is the Pearson correlation statistic, and any correlations greater than 0.90 are of concern. While the correlation between GDP and the 9/11 dummy variable is sizeable at 0.783, both variables are still highly significant from their independent t-tests (see Table 10.23) so it is clearly not enough to reject either variable from the regression analysis. Therefore, based on both the collinearity diagnostics and the correlation matrix, multicollinearity does not appear to be a problem in this particular demand forecast.

If multicollinearity is found to be a problem in a particular regression, then conventional methods for dealing with the problem are acquiring more data or eliminating one or more of the highly collinear independent variables. Since it is rarely possible to acquire more data for a given regression (time constraints and so on), attention shifts to elimination of variables. If all of the variables are still significant at conventional levels of significance, then it is generally advisable to retain the original model, since it was our best initial theoretical formulation of the relationship. If, on the other hand, one or more of the collinear variables are not significant at conventional levels, then consideration should be given to dropping the non-significant variable and rerunning the regression. In this case the researcher is implicitly assuming that the two highly collinear variables are providing the same information with respect to the dependent variable.

---

13   Variance Inflation Factors (VIF) is a statistic used to measuring the possible collinearity of the explanatory variables.

Table 10.25   Correlation Matrix for Independent Variables from SPSS

Correlations

		GDP	Average Ticket Price	Q1 Seasonality	Q2 Seasonality	Q3 Seasonality	9/11 Dummy Variable
GDP	Pearson Correlation	1	-.659**	-.023	.042	-.039	.783**
	Sig. (2-tailed)	.	.000	.895	.811	.828	.000
	N	34	34	34	34	34	34
Average Ticket Price	Pearson Correlation	-.659**	1	.370*	.062	-.272	-.668**
	Sig. (2-tailed)	.000	.	.031	.729	.119	.000
	N	34	34	34	34	34	34
Q1 Seasonality	Pearson Correlation	-.023	.370*	1	-.360*	-.333	-.040
	Sig. (2-tailed)	.895	.031	.	.036	.054	.823
	N	34	34	34	34	34	34
Q2 Seasonality	Pearson Correlation	.042	.062	-.360*	1	-.333	-.040
	Sig. (2-tailed)	.811	.729	.036	.	.054	.823
	N	34	34	34	34	34	34
Q3 Seasonality	Pearson Correlation	-.039	-.272	-.333	-.333	1	.041
	Sig. (2-tailed)	.828	.119	.054	.054	.	.816
	N	34	34	34	34	34	34
9/11 Dummy Variable	Pearson Correlation	.783**	-.668**	-.040	-.040	.041	1
	Sig. (2-tailed)	.000	.000	.823	.823	.816	.
	N	34	34	34	34	34	34

**. Correlation is significant at the 0.01 level (2-tailed).
*. Correlation is significant at the 0.05 level (2-tailed).

Once all the regression issues have been checked, the last step is to quantify the demand function so that forecasts can be created. While computer statistical packages provide a wide array of regression results, the default regression performed is the linear OLS regression discussed earlier in the chapter. Therefore, the forecasted demand function is a typical linear equation. Using the unstandardized coefficients for statistically significant variables, the forecasted demand function for air travel between Orlando and Los Angeles is:

$$D_{MCO-LAX} = 3409 + 8.59(GDP) - 211(P) + 778.91\ DUM_{Q1} + 5{,}218\ DUM_{Q2} +$$

$$3{,}840\ DUM_{Q3} - 19{,}971 DUM_{9-11}$$

Based on this equation, the demand for Orlando to Los Angeles air transportation can be estimated. Moreover, the forecast demand function also displays the impact that a change in one of the independent variables has on the demand. For example, a one dollar increase/decrease in the average ticket price will cause demand to decrease/increase by over 200 seats. As we might expect, this kind of information is extremely useful to aviation managers of all types. The seasonality dummy variables also have a large impact on demand. For example, if the flight is in the second quarter, then the demand for the flight will increase by over 5,200 passengers as compared to the fourth quarter. Again, this information is critically important to successful fleet mix planning. The other variables in this equation can be analyzed in a similar fashion.

Therefore, and by way of summary, air transportation industry demand forecasting is critical to strategic planning and the ultimate success of the airline. Regression analysis is a powerful tool that can be extremely useful in forecasting and other strategic decisions. While this chapter has merely provided an overview of various methods for forecasting, and a somewhat more detailed presentation of regression analysis, in-depth discussions of all the topics can be found in the reading list provided at the end of the chapter.

DATA SOURCES

In order to perform successful forecasting in the aviation industry, various data are required. This section outlines some of the data sources commonly used in aviation applications. It also indicates whether the data is freely accessible or can only be obtained through subscription fees and, where appropriate, the web addresses are provided.[14] While the majority of data sources described are from the US, data sources for international aviation are also provided. The major data sources discussed are:

- US Department of Transportation (DOT)/Bureau of Transportation Statistics (BTS)
- Federal Aviation Administration (FAA)
- International Air Transport Association (IATA)
- International Civil Aviation Organization (ICAO)
- Official Airline Guide (OAG)
- Airports Council International (ACI)
- Flightglobal (formerly Air Transport Intelligence (ATI))
- Airline Monitor
- UK Civil Aviation Authority (CAA)
- Transport Canada (TC)
- Eurocontrol
- The Aircraft Owners and Pilots Association (AOPA)
- Bureau of Economic Analysis (BEA)
- Bureau of Labor Statistics (BLS)
- Organization for Economic Cooperation and Development (OECD).

US Department of Transportation (DOT)/Bureau of Transportation Statistics (BTS)

One of the best sources for aviation-specific data for US aviation activity is the US Department of Transportation (DOT), through the Bureau of Transportation Statistics (BTS). There are multiple DOT databases that provide a wealth of information for the airline industry.

One database that is used throughout this book is the Form 41; this form provides a wealth of information concerning US airlines ranging from general airline financial data, specific airline cost data, general traffic data, and airport activity statistics. All US-registered airlines are required to provide the data to the DOT and it can be useful for evaluating airlines.

Another useful database is O&D, which stands for Origin and Destination. Using a 10 percent sample of actual tickets, various statistics are provided for individual US domestic city pairs. The O&D database shows on what airline the passengers traveled, the average ticket price, and a large amount of other data. As might be expected, the O&D data is very useful for demand estimation. The T100 database is similar to the O&D, but for international city-pairs. However, the data is presented in a slightly

---

14   Website addresses were current as of May 2012.

differently format and is not as extensive. These data enable demand estimation for international routes.

In addition to these three major databases, the DOT also provides additional databases such as schedules, fleet, and commuter.[15] While DOT statistics are technically public information and can be obtained for free, unless the user has advanced Excel and Access skills, the data is very difficult to access. Therefore, in order to use most of the DOT data, airline database packages such as Back Aviation are required. Unfortunately, these products require a paid subscription.

Federal Aviation Administration (FAA)

The Federal Aviation Administration (FAA) is another good source for US data. In particular, the FAA is a good source for information and data concerning aviation accidents and safety. The FAA also provides data concerning aviation forecasts and other issues such as terminal space usage, passenger facility charges, and airline service indexes. All this data can be obtained without charge through the Federal Aviation Administration's website, www.faa.gov.

International Air Transport Association (IATA)

The International Air Transport Association (IATA) highlights issues and provides information concerning issues affecting airlines globally. The free economic analysis section provides information concerning the industry outlook, cost comparisons, traffic analysis, and fuel prices. In addition to the free data, IATA provides a wealth of additional subscription information that compares international carriers and provides airline rankings in terms of a variety of statistics. Through IATA's website, www.iata.org, a wealth of information, (particularly concerning global aviation issues) can be collected.

International Civil Aviation Organization (ICAO)

The International Civil Aviation Organization (ICAO), an arm of the United Nations, is the source for pertinent legal issues, particularly international air service agreements. However, probably the most valuable source from ICAO is ICAOdata, a subscription database; this database provides international data, including origin and destination passenger statistics, airline financial data, and airport activity statistics. ICAOdata is a useful backup source to fill in any data not covered by DOT O&D and T100 databases. Information concerning ICAO and ICAOdata can be obtained through the website, www.icao.int.

---

15   It should be noted that additional aviation data is provided through the Department of Transportation (http://www.dot.gov) and the Bureau of Transportation Statistics (http://www.bts.gov) websites.

Official Airline Guide (OAG)

The Official Airline Guide (OAG) is a compilation of over 1,000 airline schedules, creating the definitive source on airline schedules. Users can access date-specific schedule information through www.oag.com without charge. However, for airlines, a complete historical OAG database is more useful. Through this database, ASMs can be easily determined for a large number of city pairs (www.oag.com).

Airports Council International (ACI)

Airports Council International (ACI) is a community of international airports that collectively lobbies on various issues concerning airports. Through ACI's website, www.airports.org, data and rankings can be obtained concerning the number of passengers handled by various airports, the cargo movements through the airports, and the number of international passengers, to name just a few, ACI helps collate information concerning airports worldwide (www.airports.org).

Flightglobal (formerly Air Transport Intelligence)

Flightglobal is a database encompassing a wealth of information on the aviation industry. Flightglobal provides a database of aviation-specific journal articles from such publications as *Airline Business* and *Flight International* and these can be quite helpful in any qualitative analysis. Flightglobal also provides searchable databases on information concerning airlines, airports, aircraft, suppliers, and schedules. While Flightglobal does not provide quantitative data, it is a valuable resource when initially researching specific areas. Flightglobal is only available to subscribers, and more information can be gathered at www.flightglobal.com.

Airline Monitor

Another subscription database is Airline Monitor, which reviews trends in the airline and commercial jet aircraft industries. Airline Monitor provides a variety of reports, in a variety of formats, over issues such as block hour operating costs, airline financial results, and commercial aircraft production. Airline Monitor also provides historical data and this is especially helpful in constructing time-series data with numerous observations. More information concerning the products offered by Airline Monitor can be found at www.airlinemonitor.com.

UK Civil Aviation Authority (CAA)

The UK Civil Aviation Authority (CAA) provides a function similar to the FAA, except in the UK. Using www.caa.co.uk, information covering the entire UK aviation industry can be obtained. Through the economic regulation and statistics portion of the CAA's website, a wealth of statistical data can also be accessed.

Transport Canada (TC)

Transport Canada (TC) is the governing body for all transportation-related activities in Canada. Statistics, data, and regulations concerning the commercial aviation industry can all be obtained through Transport Canada and StatsCan. More information concerning Transport Canada can be found at www.tc.gc.ca.

Eurocontrol

Eurocontrol, standing for the European Organization for the Safety of Air Navigation, is the primary provider of air traffic control services throughout Europe. While specific data can be difficult to obtain from Eurocontrol, their website does provide a variety of information concerning the aviation industry in Europe.[16] More specifically, Eurocontrol can provide detailed information pertaining to airport traffic, delays, and capacity management initiatives (www.eurocontrol.int).

The Aircraft Owners and Pilots Association (AOPA)

The Aircraft Owners and Pilots Association (AOPA) is a membership community that promotes and advocates for the general aviation industry. Recently, AOPA has been involved in the fight over fuel surcharges and restrictions concerning the use of general aviation aircraft in congested airspace. The AOPA website, www.aopa.org, is split into two sections: public and members. While the general public can receive basic information from AOPA, members can obtain a more thorough investigation of issues facing the general aviation community. Additionally, members receive information pertaining to weather and flight planning.

Bureau of Economic Analysis (BEA)

The US Bureau of Economic Analysis (BEA) is an essential source when forecasting demand for air transportation services. The BEA provides detailed statistics of the state of not only the US economy, but regional economies. Since GDP is a suitable proxy for consumer income, data from BEA can help in any regression analysis. The BEA provides additional macroeconomic indicators such as balance of payments, unemployment, and industry-specific economic accounts. Data can be freely obtained at www.bea.gov.

Bureau of Labor Statistics (BLS)

The US Department of Labor's Bureau of Labor Statistics (BLS) is the definitive source concerning the labor force in the US. BLS provides data on such factors as unemployment, consumer price indices, wages, and labor demographics. The level of data can be quite

---

16   Eurocontrol – www.eurocontrol.int.

detailed, with the various statistics broken down into industries and regions. For any analysis involving labor, www.bls.gov should be consulted.

Organization for Economic Cooperation and Development (OECD)

The Organization for Economic Cooperation and Development (OECD) is comprised of 30 member countries which have active relationships with over 70 countries and multiple non-governmental organizations (NGOs). OECD is primarily concerned with social and macroeconomic issues; therefore the selection of data from www.oecd.org encompasses these categories. Statistics are sorted into various industries and enable comparisons between countries. Unfortunately OECD does not publish any reports concerning the aviation industry; therefore, much of the useful data from the OECD will be general macroeconomic data, usually displayed on a monthly or quarterly basis (www.oecd.org).

SUMMARY

Since demand is not deterministic and varies with time, forecasting tools need to be developed to help researchers to estimate future demand. In order to run an airline efficiently, it is very important to generate and maintain an accurate demand forecast, based on a combination of historic data and statistical modeling. Naturally we want to have the most accurate estimate possible, which means continually updating the forecasting process and techniques. Forecasting has many applications in the aviation industry, and probably the chief among these is the forecasting of demand. Airlines could use both qualitative and quantitative tools in forecasting. Qualitative forecasts do not use econometrics tool, and do not provide measures of forecast accuracy since they are based on opinions, surveys, and beliefs. On the other hand, quantitative forecasts use statistical relationships to help forecast future events and, while they are more mathematical in nature, they may or may not be any more accurate than qualitative forecasts.

DISCUSSION QUESTIONS

1. Explain the differences between qualitative forecasting and quantitative forecasting, and provide an advantage and disadvantage of each.
2. Explain how a dummy variable can improve the accuracy of a forecast model.
3. What is multicollinearity in a regression model?
4. What is autocorrelation in a regression model?
5. Explain the difference between seasonal and cyclical variations. Give an example of how airlines respond to seasonal variations.
6. Describe the four different components of a time-series statistics.
7. If a regression analysis had a R^2 of .89, what does this mean?
8. In a regression analysis how would you incorporate seasonal variations and other important events (shock) like September 11 and such?
9. Formulate a multiple regression model showing how the quantity demanded of an airline depends on the ticket price, the income of passengers, and the ticket price of other airlines operating in the same market. What are the anticipated signs of the coefficients?

10. In a simple regression model is it possible that all the actual Y values would lie above or below the true regression line? Explain.

11. List about ten questions you would ask a group of passengers in order to estimate their demand function for a specific airline and a specific route.

12. Discuss the different methods of obtaining a trend projection from past observations to estimate the future demand.

13. Write the demand equation, in a general form, for an airline and identify the following terms:

a   Dependent variable
b   Independent variables
c   Y-intercept
d   Slope

Logically, one can often expect a multi-collinearity effect between the independent variables in a regression model. Which variables in this regression are expected to be highly correlated among each other?

14. EZjET, a small regional airline, wishes to predict sales for its business travelers between two cities for the year 2014. It has recorded data for its past 14 years' demand, and has obtained information on the number of businesses within its market area. This information is listed in the table:

Year	Number of Seats Sold	Number of Businesses
1999	74,970	4,664
2000	76,500	4,759
2001	78,795	5,480
2002	83,000	6,100
2003	80,000	6,940
2004	92,000	8,300
2005	94,760	9,850
2006	101,000	10,800
2007	103,000	12,100
2008	106,000	13,300
2009	104,880	13,034
2010	115,256	13,890
2011	118,713	14,307
2012	122,274	14,736
2013	119,829	14,442

a   Plot the annual sales data against the number of businesses in the market area and draw in the "line of best fit" that seems to be visually appropriate.

b   Measure the intercept and slope of the above line of best fit, state the approximate function relationship between the two variables.

    c   Suppose the number of businesses is projected to increasing to 14,500 in 2014. Use the above functional relationship to forecast the demand for the number of seats sold in business class in 2014.

    d   Comment on the probable accuracy of your forecast.

15. Quantico Australian Airlines faces the following annual demand function for its Los Angeles–Sydney route:

$$Q(p) = 38{,}658.235 - 8.667\,P$$

Where Q is the number of tickets sold, and P is the average ticket price. The regression analysis also produced the following statistics: coefficient of determination, 0.73; and standard error of the estimate, 4,200. Quantico's marginal cost per seat is $150 for all foreseeable levels of output.

    a   What is the profit-maximizing price for this route (assume single pricing)?

    b   What is the sales revenue-maximizing price?

    c   Calculate the price elasticity of demand at the profit-maximizing price and comment on the value obtained.

    d   At the profit-maximizing price, what is the 95 percent confidence interval for sales?

    e   What other qualifications and assumptions underlie your prediction?

16. DirectJet is a small regional airline based in the northern Europe. It operates a small number of jet aircraft and could initially produce only 400,000 ASM per day. However, the output rate picked up as soon as management learned to schedule their flights more efficiently, save turnaround time on the ground, decrease fuel consumption, and so on. Suppose that the average cost per seat mile during the first year were as shown below and that the variable factors will continue to become more productive as total output measured in ASM continues to increase.

ASM	CASM
400,000	15.82
350,000	16.85
500,000	12.15
450,000	10.25
400,000	10.95
500,000	10.01
490,000	9.75
530,000	9.00
580,000	9.65
380,000	12.89
480,000	9.25
380,000	15.00
340,000	16.00

a   Obtain the estimated regression function. Does a linear regression function appear to give a good fit?

b   Test the overall significance of the regression parameters.

c   Write the estimated equation for this relationship.

d   Based on the above statistics, forecast the average cost per seat mile when total output reaches 1 million ASM and 1.2 million ASM.

e   Interpret slope and intercept in your estimated regression function. Do they provide any relevant information? Explain.

f   Calculate MSE and MAD for the independent variable.

g   Perform an F test to determine whether or not there is a lack of fit of a linear regression function.

REFERENCES

Allen, W., Doherty, N., Weigelt, K. and Mansfield, E. (2005). *Managerial Economics* (6th ed.). New York, NY: W.W. Norton & Company.

Anderson, D. Sweeney, D. and Williams, T. (2006). *Quantitative Methods for Business* (10th ed.). Mason, OH: Thomson Higher Education.

Anderson, E. W. and Mittal, V. (2000). Strengthening the Satisfaction-profit Chain. *Journal of Service Research* Vol. 3, No. 2, 107–120.

Barreto, H. and Howland, F. (2006). *Introductory Econometrics: Using Monte Carlo Simulation with Microsoft Excel*. New York: Cambridge University Press.

Brandt, R. (1987). A procedure for identifying value-enhancing service components using customer satisfaction survey data, in Surprenant, C. (Ed.), *Add Value to your Service*, Chicago, IL: American Marketing Association, 61–65.

Doane, D. and Seward, L. (2007). *Applied Statistics in Business and Economics*. New York: McGraw Hill.

Durbin, J. and Watson, G. (1951). Testing for Serial Correlation in Least Squares Regression, II. *Biometrika* Vol. 38, No. 1–2, 159–179.

Hanushek, E. and Jackson, J. (1977) *Statistical Methods for Social Scientists*. New York: Academic Press.

Hirschey, M. (2006). *Managerial Economics* (11th ed.). Mason, OH: Thomson Higher Education.

Ovedovitz, C. (2001). *Business Statistics in Brief*. Cincinnati, OH: South Western College Publishing.

Spirtes, P., Glymour, C. and Scheines, R. (2001). *Causation, Prediction, and Search* (2nd ed.). Cambridge, MA: The MIT Press.

Truett, J. and Truett, B. (1992). *Managerial Economics* (4th ed.). Cincinnati, OH: South-Western.

11

Dynamic Pricing Policy and Revenue Management

I believe that revenue management is the single most important technical development in transportation management since we entered the era of airline deregulation in 1979.

We estimate that revenue management has generated $1.4 billion in incremental revenue in the last three years by creating a pricing structure that responds to demand on a flight-by-flight basis.

Robert L Crandall, Chairman and CEO, AMR, 3 1992

This chapter will introduce the reader to the concepts of airline pricing policy and revenue management. Revenue management is essentially the combination of methods, analysis, and techniques which an airline applies to the types of services it offers in order to maximize the aircraft revenue. Airlines employ revenue management not only to sell as many high-priced seats as efficiently as possible, but to also keep airplanes full. A short section on airlines' past pricing practices will be followed by the current pricing structure. Further, we show that segmenting the market or "price discrimination" based on the elasticity of demand for different types of passengers can increase revenues. We then discuss strategies an airline can use to segment its market directly and indirectly including the practices such as advanced purchase restrictions and Saturday night stay requirements. The general outline for the chapter is:

- Dynamic Pricing Policy
- Cost-based Pricing
  - Markup pricing
- Markup and Price Elasticity of Demand
  - Optimum markup
- Bundling
- Unbundling and Airline Ancillary Revenue
- Market Skimming and Penetration Pricing
- Peak-load Pricing
- Price Discrimination
- Consumer surplus
- Necessary Conditions for Price Discrimination
  - Market segmentation
  - Different elasticities in sub-markets

- – Market separation
- Degrees of Price Discrimination
    - – First-degree (perfect) price discrimination
    - – Second-degree (quantity discounts) price discrimination
    - – Third-degree (multi-market) price discrimination
- Uniform Pricing versus Price Discrimination
- Importance of Revenue Management
- Scenario: Uniform versus Multiple Pricing
- Revenue Management "Fences"
    - – Advance purchase restrictions
    - – Saturday night stay requirement
    - – Frequent flier mileage
    - – Refundability
    - – Change Fees
    - – Airline schedule
- Revenue Management Control Types
- Spoilage and Spillage
- Leg-based Expected Marginal Seat Revenue Model
- Overbooking
    - – Forecasting overbooking levels
- Other Issues Associated with Revenue Management
- Summary
- Discussion Questions
- Appendix: Derivation of Overbooking Probability Equation

DYNAMIC PRICING POLICY

Pricing policy and practices in the airline industry refer to how an airline sets its ticket prices based on demand, cost, and the competition in the market. In the airline industry, pricing policies are often dynamic; prices change daily and vary from airline to airline. Pricing is not always an exact science. Low-cost and start-up airlines offer services at lower prices to increase market share and stay in business in response to retaliation by incumbents. Today, airline ticket prices are highly volatile and airfare prices fluctuate widely day by day. There are more than ten different fares for any one flight, from the lowest discounted fare to first class. A passenger who buys his or her ticket at the right time can pay considerably less than other passengers on the same flight.

> "The most volatile airline ticket prices in America are between Atlanta and Las Vegas, a new survey by Yapta has found. Fares between those cities changed an astonishing 2,472,916 times since the beginning of the year. That's roughly once every six seconds."
>
> Christopher Elliot

The Civil Aeronautical Act of 1938 established a policy of economic regulation of the domestic US airline industry. This Act created the Civil Aeronautics Board (CAB) which had authority over the level and structure of airfares within the US. Prices were set by the CAB according to industry average costs, which disbarred low-cost airlines

Table 11.1   Top Ten Domestic City-pairs with Volatile Airfare Prices, 2010*

Rank	Origin City	Destination City	Number of Price Changes
1	Atlanta (ATL)	Las Vegas (LAS)	2,472,916
2	New York (JFK)	Las Vegas (LAS)	2,412,759
3	New Jersey (EWR)	Las Vegas (LAS)	2,377,668
4	Chicago (ORD)	Las Vegas (LAS)	2,215,994
5	New York (JFK)	San Francisco (SFO)	1,959,873
6	San Francisco (SFO)	New York (JFK)	1,862,270
7	Los Angeles (LAX)	Honolulu (HNL)	1,740,380
8	New Jersey (EWR)	Orlando (MCO)	1,725,727
9	Los Angeles (LAX)	New York (JFK)	1,641,397
10	Boston (BOS)	Chicago (ORD)	1,490,271

* Christopher Elliot, May 13, 2010

from offering lower prices since it was deemed unhealthy for the industry. Even though airfares were regulated, the policy of setting airfares above average costs is a dynamic pricing policy; in this case, a cost-based pricing. The only exceptions to this rule were in the states of California and Texas where airlines were able to set their own prices on intra-state routes where the CAB did not have authority. This led to the rise of low-cost airlines AirCal in California, and Southwest Airlines in Texas.[1] Within the regulated environment, airlines were provided with a protected route structure and guaranteed revenues that exceeded costs. This meant that airlines rarely failed in the domestic marketplace, and in the event that an airline incurred losses, federal subsidies were available to bail them out (Spiller, 1981). Hence, airline ticket pricing was simple and fixed. There was a small range of available fares for a route, and every so often there might be a sale fare.

The Airline Deregulation Act of 1978 abolished the CAB's authority over airlines and the market was permitted to decide airline fares and routes. After the deregulation process airlines were allowed to elect to fly any domestic route, without legal restrictions, and could offer any fare on any flight.[2] Deregulation also allowed low-cost carriers (LCCs) like Southwest to expand outside of its Texas market and encouraged start-ups of new discount carriers like People Express. At the beginning of deregulation, about 50 percent of total traffic traveled on a discount fare, yet by 1990 nearly 90 percent of traffic was on discounts. The increased competition and liberalized pricing structure have led industry analysts to claim that today's air fares are 20 to 30 percent below what they would have been had regulation remained in place.

---

1   AirCal was a California based airline and was eventually bought out by American Airlines.
2   *Business Week*, Airline Deregulation, Revisited, January 20, 2000.

COST-BASED PRICING

Cost-based pricing is one of the simpler approaches to pricing. It refers to the process in which prices are determined on the basis of costs plus an additional profit expressed as a percentage of the cost. This was the approach utilized during the period of airline regulation where the CAB set prices based on the average costs of the route with an added percentage. This pricing method does not consider the elasticity of demand or the extent of competition.

Markup pricing

Markup is the amount that a business firm charges over and above the total cost of producing its product or service. Firms use this strategy in order to make a certain profit and it is generally expressed as a percentage. In industries where there are thousands of products varying in values, including the wholesale and retail industries, firms tend to utilize this pricing model as a guideline for determining sale prices. This allows the company to cover all costs associated with the production and sale of the products and still make a profit. For example, if the total cost of producing a product is $200, but it's sold for $250, then the extra $50 is the markup and can be expressed as 25 percent. The following equation shows price as a function of cost and markup.

$$P = AC + Markup$$
$$P = AC + MU \times AC$$
$$P = AC \times (1 + MU)$$

Where:

- P = price
- AC = this may include the direct, indirect and fixed costs of the production of a product or service
- MU = markup percentage.

For example, a product is costing you $650 per unit. Therefore, if you want a markup of 25 percent (a profit equal to 25 percent of total cost) the selling price must be set at $812.50.

$$P = \$650 \times (1 + 0.25)$$
$$P = \$812.50$$

Markup pricing has several advantages as well as disadvantages. The advantages stem from its simplicity and relative ease of use. It is also useful to the buyer if there is enough information available about the supplier's costs. Working backwards through the equation, the buyer can determine what the markup is on a particular product as a percentage of the costs. The simplicity of markup pricing also leads to its disadvantages; if the elasticity of demand is ignored then a firm may increase the price in an elastic market which decreases total revenue and vice versa in an inelastic market. Markup pricing also

Table 11.2   Advantages and Disadvantages of Markup Pricing

Advantages	Disadvantages
Easy to calculate	Ignores elasticity of demand
Relatively quick and simple to apply	Based on historical costs
Useful if buyer has information about supplier's costs	May ignore opportunity costs
	A lack of incentive for efficiency

does not reflect the cost of replacing a product which often is less than the initial cost of acquiring/manufacturing the product initially.

MARKUP AND PRICE ELASTICITY OF DEMAND

Depending on the industry and product, the amount of markup that is preferred varies significantly. The amount of markup depends on several factors, but the key is identifying and utilizing the specific relationship that exists between price elasticity and markups. If demand is relatively inelastic, then a company can set a high price. At the other extreme, if demand is perfectly elastic then a firm has no choice but to accept the market price. This is the reason why markup pricing is not optimum if market structure and demand sensitivity are not considered. If the elasticity of demand is taken into account, then markups can vary significantly from industry to industry. With elastic demand, businesses may not recognize that a higher markup may not lead to higher revenue or greater profits. However, in some industries, such as health care, the markup may be a large percentage of the total cost of the product or service because the demand for these services is inelastic.

> Higher elasticity leads to a price closer to the competitive price.

Optimum markup

Every business faces two questions: how much to produce and what price can be charged to maximize profit. The profit-maximizing rule is to increase quantity as long as marginal revenue (MR) is greater than marginal cost (MC). Conversely, profit will decline if MC exceeds MR; hence, the profit-maximizing level of output is found by equating its MC with its MR.

Therefore, since the objective of markup policy is to maximize profit by charging the optimum price (not too high, and not too low), a product's price elasticity is a key element in setting its markup and price. The formula below describes the relationship between optimal markup, price, and maximum profit. It should be noted that the formula below only works when E is greater than 1 since it is already assumed that the firm is pricing beyond the inelastic portion of the demand curve.

The profit–maximizing condition is:

$MC = MR$

Where

$$MR = P \times \left(1 + \frac{1}{E_P}\right)$$

Rearranging the above, we obtain:

$$MC = P \times \left(1 + \frac{1}{E_P}\right)$$

$$P = MC \times \frac{1}{\left(1 + \dfrac{1}{E_P}\right)}$$

$$P = MC \times \left\{\frac{E_P}{1 + E_P}\right\}$$

$$Optimal\ Markup = \left\{\frac{E_P}{1 + E_P}\right\}$$

According to the above formula, as long as the price elasticity demand is very low, it is advantageous to increase the price; the seller receives more money for fewer goods. Conversely, if the price elasticity of demand is very high (elastic), it is beneficial to cut the price; the seller gets more money from selling even more goods. Hence, assuming that E= -5, we can see from the above formula that the price is equal to 1.25 times the MC.

In Table 11.2, a variety of elasticities are presented along with the resulting optimum markup and effect on selling prices of a particular good where the MC is $100. From this table we see that the higher the elasticity of demand, the smaller the optimum markup and consequently, the lower the selling price.

As mentioned and, as Table 11.3 opposite shows, the more elastic the demand for a product, the smaller the optimal markup. Products with relatively inelastic demand will be optimally priced with higher markups.

BUNDLING

On many occasions the pricing arrangement includes purchasing groups of complimentary products. Bundling is the combination of services offered within the cost of the ticket. The products are bundled or sold as a block which has become common practice for airlines, fast food, banking, cable companies, and theatrical or sporting events. Bundling often results in greater revenue for the producers across industries. Virgin Atlantic bundles door-to-door limousine service, in-flight massage, and drive-through check-in with its transatlantic flights (Ovans, 1997). Bundling can be good for consumers and may provide the following advantages:

- reducing the transaction costs;
- simplifying billing;
- lowering prices;
- integration of products and services.

Table 11.3   Effect of Elasticity on Optimum Markup and Price

Case	Elasticity	Optimal Markup	Marginal Cost (MC)	Price (P)
1	-10	1.11	100	111
2	-9	1.13	100	113
3	-8	1.14	100	114
4	-7	1.17	100	117
5	-6	1.20	100	120
6	-5	1.25	100	125
7	-4	1.33	100	133
8	-3	1.50	100	150
9	-2	2.00	100	200
10	-1.5	3.00	100	300
11	-1.4	3.50	100	350
12	-1.3	4.33	100	433
13	-1.2	6.00	100	600
14	-1.1	11	100	1,100
15	-1.01	101	100	10,100

Airlines routinely bundle vacation packages, combining air travel, life insurance, with car rentals and lodging. For example, United Airlines had bundled hotel reservation, car rental, and airline services with the coordination of Allegis Corporation but eventually had to abandon the practice. The US Department of Justice on May 18, 1998 sued Microsoft alleging that it attempted to monopolize the market by bundling its Internet Explorer web browser software with its Microsoft Windows operating system (Washington Post, 2000). Some common examples of bundling in other industries include computer hardware and software bundles (Microsoft bundles its Access, Excel, and Word software) and utility companies.

UNBUNDLING AND AIRLINE ANCILLARY REVENUE

> Ancillary performance is about identifying what added services and products your customers place value on and how much that value is worth to them.
>
> Doug Hesley, Norwegian Cruise Lines

Many airlines have been busy unbundling services that traditionally came with a seat, such as bag check fees, seat selection charges, fees for carry-on luggage, fees for pillows, fuel surcharges, overhead bins, food, seat reservations, and even the ability to pay by credit card. The unbundled revenue is considered airline ancillary revenue; any revenues generated beyond the direct sale of airline tickets. The list of commission-based selling is growing with the inclusion of any duty-free items and products purchased on board. Airline ancillary revenue can be separated into three primary categories:

- *A la carte features*: pillows, overhead bins, baggage fees, seat selection, and so on.
- *Commission-based*: hotels, rental cars, and travel insurance.
- *Frequent flier activities*: sale of miles to program partners including hotel chains and co-branded credit cards.

United Airlines was the first US airline to start charging separately for bags checked. Low-cost Spirit Airlines was the first US Airline to begin charging passengers for carry-on bags that didn't fit under a seat, and currently charges $45 for a carry-on bag. As of November 6, 2012 the passengers who pay the fee at the boarding gate will fork over $100 (Bomkamp, 2012). Some airlines even have started to charge additional fees for the window, aisle, or emergency-exit row seats. Ryanair and AirAsia sell the external surface of their aircraft as advertising space. Other airlines engage in commission-based selling for third parties with products and services that are relevant to fliers including hotel rooms, rental cars, and travel insurance (Airline Weekly, 2010). Unbundling may also be a forced event in the wake of anti-trust violations. For example, The European Court of First Instance ruled in 2004 that Microsoft must provide the European market with a version of Windows operating system without their media player to prevent a monopoly (The Economist, 2004).

MARKET SKIMMING AND PENETRATION PRICING

Upon entering a market, an initial pricing strategy is selected. This strategy depends on the nature of demand for the product/service and takes the form of either market skimming or penetration pricing. One of the best examples of penetration pricing comes from the Irish-based carrier Ryanair. In 1991, Ryanair transformed itself into a LCC with an aggressive market penetration strategy and rapid expansion. In new markets, Ryanair frequently employs penetration pricing; a pricing technique where a new entrant into a market sets its initial prices relatively low in order to attract new customers and gain market share. This is evident through fares as low as £1 and €1 in new markets, followed by an increase in the prices following the entry period. Penetration pricing as a strategy is based on the principle that passengers for leisure travel are sensitive to price (demand is elastic) and will switch over to the new brand solely based on the price.

Market skimming takes the opposite approach; a relatively high price is set for a new product or service and the price is lowered over time. Market skimming, as the name implies "skims" the maximum that the consumer is willing to pay for the product before competition and other market forces lower the price. This is successful when demand is inelastic. This approach has been largely utilized in technological and high-end markets where a relative high price is charged when the product is introduced to the market. As the product approaches maturity, the price is lowered to appeal to a greater number of consumers who are willing to pay the lower price.

Today, many businesses have focused on ancillary revenue. For instance, the cruise industry has been enjoying ancillary revenue for years by selling tours, photos, on-board activities, restaurants, nightlife, casino, special events, and car parking services.

PEAK-LOAD PRICING

Peak-load pricing has been utilized by cell-phone and electrical companies for years; it is the policy of charging higher prices at times when demand for a service/product is highest.[3] If most of the air traffic is concentrated around the peak times, this will exceed the airport capacity resulting in significant delays. The basis for peak-load pricing is that it reflects the cost to the supplier to meet demand at peak times. Higher landings fees during peak hours or lower fees during off-peak can provide incentives to more efficiently use existing runway capacity. This pricing policy has also been used by several airports around the world through peak-load pricing for landing fees when, during peak hours, the landing charges increase significantly. Uniform pricing does not generally work for landing charges because it does not reflect the willingness to pay higher prices for peak-load landing slots. Uniform pricing also neglects the cost of increasing capacity to meet peak demand periods which may be under-utilized during off-peak times. The British Airports Authority (BAA) uses different landing charges for peak, shoulder and off-peak times at London Heathrow (LHR) and Gatwick Airport (LGW). For instance, in Heathrow peak fees exceed off-peak charges by 230 percent, while in Gatwick this difference goes up to 300 per cent (Ewers, 2001). Peak-load pricing of landing slots presents the following advantages to airports:

- shifts demand to off-peak periods thereby making better use of facilities and resources—this reduces over and under-utilization;
- the cost of expanding capacity to meet demand is distributed to the airlines that utilize the airport during that period and require the additional capacity;
- peak-load pricing better reflects cost-based pricing since it does not utilize average costs across all periods but focuses solely on peak times.

PRICE DISCRIMINATION

Price discrimination is the practice of charging different prices to different customers for the same product. The different prices charged are based on the different price elasticities of demand between consumers. While the practice of price discrimination may seem unfair on the surface, price discrimination is legal and very common in modern business. For example, grocery stores practice price discrimination by offering coupon discounts to consumers who are not time-sensitive but price-sensitive, and who are willing to search out and bring the coupons to the store. Universities, especially state universities, practice price discrimination by offering different tuition levels for international, out of state, and in-state students. The practice of discounted calling rates on evenings and weekends by telephone companies is also price discrimination. In many flea markets there are no set prices for the goods offered, but customers bargain with the merchants. In fact, bargaining is the oldest form of price discrimination and has existed since commerce began.

> The reservation price is the highest price a buyer is willing to pay for goods or a service.
> Likewise, it is the lowest price at which a seller is willing to accept for a good or service.

---

3   Consumption of these services is high during peak periods and lower in off-peak periods.

Price discrimination is essentially the flip side of yield management and is a requirement for its practice in the airline industry. And, as is now commonly appreciated, every flight has numerous fare classes for essentially the same seats and service. The airline industry is one of several industries that expend great effort on practicing price discrimination.

CONSUMER SURPLUS

Consumer surplus is the difference between the highest amount the passenger is willing to pay and the amount he or she actually pays. In essence, consumer surplus is the perceived "deal" that consumers receive when they purchase a good or service. The goal of price discrimination and therefore yield management is to reduce the amount of consumer surplus.

Consider a flight with six passengers who are all willing to pay various prices for the same flight. Their maximum willingness to pay is contained in Table 11.4. Based on this data, a demand curve can be created for this flight, which is done in Figure 11.1 opposite. If an airline charges a single fare of $250, consumer surplus, the difference between the maximum willingness to pay and the actual ticket price, would exist for five of the six passengers. Only passenger F would receive no consumer surplus. The ultimate goal of price discrimination, and revenue management, is to minimize consumer surplus; therefore, six individual fare categories would have to be created to maximize airline revenue and minimize consumer surplus. The catch for the airlines is that knowing every passenger's willingness to pay; that is, the flight's exact demand curve can be difficult (if not impossible) in practice.

Table 11.4   Consumer Surplus under Uniform Pricing

Passenger	Demand	Ticket Price	Maximum Ticket Price Willing to Pay	Consumer Surplus
A	1	250	500	250
B	2	250	450	200
C	3	250	400	150
D	4	250	350	100
E	5	250	300	50
F	6	250	250	0

NECESSARY CONDITIONS FOR PRICE DISCRIMINATION

In practice, there are a number of ways that a business can institute at least some form of price discrimination. However, in order for it to be successful, three necessary conditions

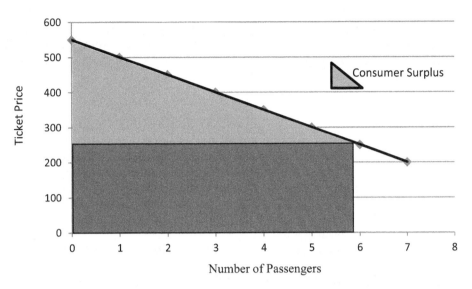

Figure 11.1   Consumer Surpluses under Uniform Pricing

need to applied to the market: *market segmentation, different elasticities in different sub-markets* and *market separation*.

Market segmentation

The first requirement for price discrimination to exist is that the markets must be segmented. By this we mean that different groups of consumers who do not have the same interests should exist. In the aviation industry, a common method of market segmentation is between leisure and business travelers. Since there may be extensive overlap in these categories, a more accurate segmentation would be between time-sensitive and price-sensitive travelers. Time-sensitive travelers are typically business travelers who demand to travel on certain days and certain times. These passengers will typically ignore the ticket price in order to satisfy their demand for traveling at a certain time and date. Additionally, certain types of leisure travelers may also be contained in this category, especially vacationers who may be leaving and returning on a set schedule. Price-sensitive travelers are the opposite in that their selection of flights is based on the ticket price. These are travelers who are willing to travel at inconvenient times and by longer routings if this results in lower fares.

Different elasticities in sub-markets

The second requirement for price discrimination is that different elasticities must exist for different sub-markets. This requirement is closely related to the first one in that the market can be segmented by price elasticity, but the first requirement deals with how the passengers can be grouped, while this requirement deals with the passengers' willingness to pay. If all passengers had the same price elasticity, then the airline would not be able to charge different prices. In the air travel industry, both these requirements are easily met, since every market contains a variety of different people who are willing to fly at different times and at different prices.

Market separation

The third requirement of price discrimination, market separation, is that the airline must be able to effectively isolate the market and be successful at charging different prices to different passengers. Airlines achieve market separation through pricing "fences" (this practice will be covered in detail in a subsequent section). Briefly, some examples of fences are the non-transferability of tickets and Saturday night stay requirements; these allow the airline to keep the customer from reselling or using the ticket in some other way than for the flight—hence the term "fences."

DEGREES OF PRICE DISCRIMINATION

Price discrimination is separated into degrees: *first degree*, *second degree* and *third degree*.

First-degree (perfect) price discrimination

First-degree price discrimination, also called perfect price discrimination, involves charging different prices for every unit up to the point where consumer surplus does not exist. Bartering is the classic case of perfect price discrimination, as are car dealerships to a certain extent. In both these cases, every consumer pays a different and unique price for the same product. Another example is an auction where consumers will keep bidding up until they reach their maximum willingness to pay. With regard to the airline industry, perfect price discrimination is practically non-existent for reasons mentioned earlier.

Second-degree (quantity discounts) price discrimination

Second-degree price discrimination is simply the existence of quantity discounts. The practice of second-degree price discrimination is common in industrial sales where large quantities of a product are purchased at once. The commercial airline industry has had limited experience with second-degree price discrimination although charter fights and corporate travel deals are examples of second-degree price discrimination. In these situations, the companies can receive discounted prices by agreeing to buy a large proportion of seats on a flight. Airlines typically favor this as it provides them with a certain amount of guaranteed revenue for a flight, albeit at a reduced rate.

Third-degree (multi-market) price discrimination

Third-degree price discrimination is the kind most typically practiced by the airline industry. It involves dividing consumers into different groups based on a set of characteristics, and estimating their respective demand curves. At this point each group is charged a different price: the group with the most inelastic demand (typically the most time-sensitive group) is charged the highest price. This price discrimination is identifiable by the different fare classes that can be observed in the market. With third-degree price discrimination, a certain amount of consumer surplus will exist since the prices are not set for every individual but for the group as a whole. However, creating additional fare classes reduces the amount of consumer surplus and ultimately increases revenues for the airline.

> Price discrimination can increase the profit, since the airlines can charge a higher price to those passengers with less elastic demand, and a lower price to those passengers with more elastic demand.

UNIFORM PRICING VERSUS PRICE DISCRIMINATION

Despite the negative connotation, price discrimination is a common and generally efficient procedure. To see this, consider the impact of a uniform pricing policy as shown in Figure 11.2.

With the typical cost structure of imperfect competition, we see long run equilibrium where $P^* = ATC$ at output Q^*. Note that this leaves a huge segment of demand unsatisfied and results in a great degree of wasted capacity (that is, an awful lot of empty airline seats). Many consumers would be willing to pay a price higher than the MC of serving them, though lower than P^*. However, with uniform pricing, the price cannot be set below average total cost (ATC) in the long run since the airline is barely breaking even at a price equal to ATC. The only option available to airlines to increase their revenues is selective

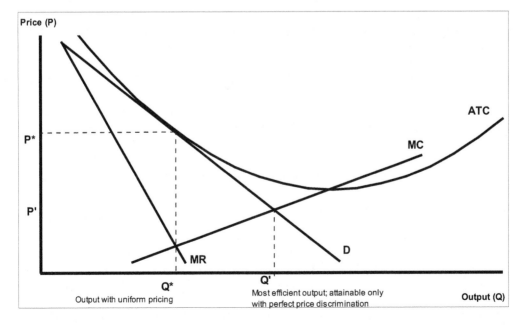

Figure 11.2   Pricing policy and Price Discrimination

price cuts. If it were possible to read consumers' minds, we could reduce price just enough below P* to induce them to buy a ticket. In this case, each consumer pays exactly the highest price he or she is willing to pay; MR is now equal to this personalized price so that every customer willing to pay a price greater than MC can be served. Under this "perfect price discrimination" regime, output can be increased to Q'.[4]

For airlines, perfect price discrimination is not possible, but sophisticated third-degree price discrimination can allow airlines to move closer to the efficient output level of Q' as they fill otherwise empty seats with selective price cuts. For instance, if most students are on the demand segment below P*, then a discount for students will bring in new revenues. Similarly, demographically-based price cuts for senior citizens or families can also achieve the desired effect. The airlines can also keep the price cuts limited through the use of the revenue management "fences" mentioned earlier and discussed in a later section.

Of course, in moving from uniform pricing to the more tailored approach of price discrimination, it may be that some segments of demand will face a price higher than P*. An interesting question is whether price discrimination leads to an average fare lower than P*, the theoretical uniform price. Most economists agree that price discrimination typically does lead to a lower average price, producing fewer empty airline seats and greater economic efficiency. The key reason for this being competition. It is always easier to cut prices than it is to raise them because of the threat of being undercut by a competitor.

Even if price discrimination reduces average price, it may still raise price for particular consumers in the upper portion of the demand curve. A commonly asserted complaint is that certain business travelers end up paying high fares to "subsidize" consumers receiving discounts. However, there are problems with this theory. As already explained, a uniform price would drive many discount customers completely out of the scheduled airline market. This would reduce airline revenues, and ultimately lead to both price increases and reductions in available seat miles (ASMs), both of which would be unpleasant to business travelers. Thus, in a sense, one could just as easily argue that discount fliers "subsidize" business travelers. In order to continue the present service standards, airlines need every penny of revenue they can get, just as a restaurant may be financially viable only with revenues from regular, full-paying customers combined with revenues from patrons that dine there only occasionally when they have a discount coupon. Each set of consumers benefits from the other since only their combined revenues are enough to sustain the product they both enjoy.

Price discrimination based on other characteristics, either when firms offer policies at a fixed price or when they charge according to some consumption variable that is correlated to costs, has been studied by Buzzacchi and Valletti (2005). For instance, consider an airline serving only leisure vacationers (therefore no business or time-sensitive travelers). This airline would have a very different product design from other carriers. Their consumers care little about the exact time of departure, are willing to commit to a schedule way in advance, and are very price-sensitive. This type of airline would look pretty much like today's charter airlines: it would operate with infrequent service, would employ large aircraft in a high-density seating configuration with very high load factors (probably over 90 percent), and would routinely cancel any flight well in advance if it was substantially undersubscribed. Under such conditions, costs and average prices could be kept very low. However, if the airline wanted to accommodate business travelers, the airline would have

---

4   Beyond Q', MC is greater than the price that could be charged so that segment of demand will not be serviced.

to offer multiple flights with varying departure times and adhere to a schedule published a few months in advance. The airline would also keep some seats open for late, even last-minute, travelers. Since business travelers require the design of a more expensive product, it seems reasonable to argue, as Frank (1983) does, that it is philosophically appropriate to charge them more. In essence, the appearance that business and leisure travelers are sometimes paying very different prices for the same service is an illusion. In reality, the typical time-sensitive traveler demands a very different and much more expensive sort of service than price sensitive travelers.

IMPORTANCE OF REVENUE MANAGEMENT

The financial performance of airlines, like most other businesses, depends mainly on their sales strategy within a competitive industry. The knowledge of varying demand conditions, different classes of passengers, degrees of price sensitivity (elasticity of demand) among various groups of passengers, and the significance of the stochastic nature of demand by the traveling public (for example, number of reservations and actual trips may differ) will influence the airlines' ultimate performance. Therefore, it is not surprising that airlines, recognizing all these dynamic factors, charge different fares to effectively respond to varying elasticities, different passengers' income, competitors' pricing policy and market conditions. This practice is termed revenue (or yield) management.

Revenue management is a quantitative technique which allows an airline manager to handle the supply of aircraft seats and passenger demand to maximize revenues. The basic theory behind revenue management is that it may be beneficial to not sell something today at a low price if it can be sold tomorrow at a higher price, or allowing something to be sold today at a low price if it is otherwise likely to remain unsold. In essence, revenue management is a game of probabilities with the goal of extracting the largest amount of revenue a passenger is willing to pay. An effective revenue management system requires:

- the establishment of a differential fare structure;
- a system of constraints (or fences) on the use of lower-fare seats to limit their availability to passengers who might otherwise be willing to a pay a higher fare;
- a system of seat allocation which maximizes expected revenue in the face of stochastic demand;
- a reliable forecast of demand, no-shows, cancellations, overbooking, and inventory limit.

The importance of revenue management cannot be overstated. The example of People Express in the 1980s is probably one of the starkest examples of the importance of revenue management to the industry. Donald Burr, former CEO of People Express explains:

We were a vibrant, profitable company from 1981 to 1985, and then we tipped right over into losing $50 million a month. We were still the same company. What changed was American's ability to do widespread Revenue Management in every one of our markets. We had been profitable from the day we started until American came at us with Ultimate Super Savers. That was the end of our run because they were able to underprice us at will and surreptitiously. There was nothing left to defend us. What you don't know about revenue management could kill you.

People Express was a fledgling discount airline that was born out of deregulation in 1978. Initially the discount airline flew niche markets that competed mostly with buses and cars, which the major carriers were happy to leave to it. With a cost structure that was $1 billion below other airlines such as American Airlines, People Express began to compete on the routes of the major US domestic carriers. With People Express operating at a 75 percent load factor and a 72 percent breakeven load factor (BLF) on some of America's busiest routes in 1983, the discount airline with its simple pricing structure and extensive cost advantage seemed unstoppable, yet it had one significant disadvantage. In order for People Express to save money during start-up, its reservations system was simple and unable to practice revenue management. People Express's information technology system could offer peak and off-peak fares, but each flight had to be either one or the other (peak or off-peak). This meant that on each flight, People Express was able to offer only one fare. Its reservations system did not allow People Express to offer multiple fares on a single flight (Cross, 1995).

On January 17, 1985 American Airlines became the first airline to take advantage of this problem in People Express's reservation system by launching "Ultimate Super Saver" fares that were priced at People Express's lowest prices (Cross, 1995). American placed 21-day advance purchase restrictions on the "Ultimate Super Saver" fares in order to allow only the most price-sensitive travelers to be eligible for the discounted fares (Cross, 1995). Additionally, American controlled the number of "Ultimate Super Saver" fares available on each flight in order to save space for high-revenue passengers. In essence, American Airlines was able to generate revenue from both low-revenue and high-revenue passengers, while People Express could only accommodate low-revenue passengers with its single fare class reservations system. As a result of American Airlines revenue management practices, People Express's load factor dropped from 70 percent in 1984 to 57 percent in 1986. Additionally, as Table 11.5 displays, People Express's BLF jumped 10 percent from 1985 to 1986. Ultimately, with People Express's BLF significantly above its actual load factor, the airline lost money.

The example of People Express not only displays the importance of revenue management to an airline, but also introduces the concept of revenue management by any business. While modern revenue management has its roots in the airline industry, revenue management is also widely used in the car rental, hotel, and cruise ship industries to name a few. The same theory and practice of revenue management applies to all these industries and therefore, its use and understanding is an important management tool.

Table 11.5   People Express's Breakeven Load Factor, 1984–1986

	1981	1982	1983	1984	1985	1986
Load Factor	58%	61%	75%	70%	61%	57%
Break-even Load Factor	71%	60%	72%	70%	62%	72%

Source: OAG Form 41 data

SCENARIO: UNIFORM VERSUS MULTIPLE PRICING

As the People Express example highlighted, revenue management in the airline industry was used first in 1985 by American Airlines with its "Ultimate Super Saver" fares. Since then, almost every airline in the world has adopted a revenue management scheme to some degree. The benefits of doing so are immense, as Delta Air Lines attributed $300 million in profits to revenue management when it first started implementing it. To highlight the benefits and provide a better understanding of revenue management, a quantitative example follows. Assume that DirectJet Airlines operates a short-haul route where the maximum daily demand for the flight is 100 passengers and the maximum any passenger is willing to pay for the flight is $250. This information helps us construct a demand curve for the flight, which in this case is assumed to be linear. A depiction of the demand curve is shown in Figure 11.3.

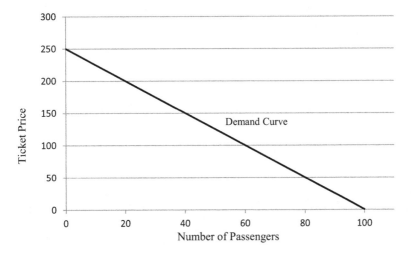

Figure 11.3   Demand Curve for DirectJet Airlines Flight

Under a uniform pricing strategy, DirectJet Airlines sets one single price for all passengers on a single flight. Recall that People Express was an airline that operated with a uniform pricing strategy. In our example, let's assume that DirectJet Airlines charges a uniform price of $100 for this particular short-haul flight. Based on the estimated demand function and at $100 airfare, 60 passengers are willing to buy tickets from the airline. This would generate $6,000 in total daily revenue for DirectJet Airlines. This is graphically represented in Figure 11.4 with the shaded area under the curve representing total daily revenue for the flight.

The second pricing scenario available to DirectJet Airlines is a multiple-pricing strategy where the airline uses segmented (differential) pricing to maximize revenue. In our example, DirectJet Airlines decides to adopt a new four-tier pricing structure where it offers fares ranging from $200 to $50. Based on the estimated demand function for this particular flight, 20 passengers are willing to pay the $200 fare, 40 passengers are willing to pay the $150 fare, 60 passengers the $100 fare, and 80 passengers the $50 fare. Figure 11.5 graphically displays the revenue potential for the DirectJet Airlines flight using a multiple-pricing strategy. It is immediately apparent that the shaded area under the

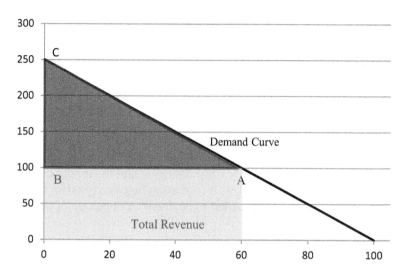

Figure 11.4   Uniform Pricing for DirectJet Airlines Flight

multiple-pricing policy is greater than the shaded area in Figure 11.4 under the uniform-pricing policy. This is confirmed numerically where the four-tier pricing structure generates $10,000 in total daily revenue:

New Aircraft Revenue = 20 Seats($200) + 20 Seats($150) + 20 Seats($100) + 20 Seats($50) = $10,000

which is greater than the uniform pricing policy of $6,000. Therefore, the major benefit of multiple-pricing policy is to increase total flight revenues, but it also enables DirectJet Airlines to offer cheap, discounted air fares that could undercut the competition. If we assume that DirectJet Airlines is operating a 100-seat aircraft, then price-sensitive passengers would simply be occupying an otherwise empty seat. This is exactly what American Airlines was able to achieve with its "Ultimate Super Saver" fares. The end result of a multiple-pricing strategy is that total revenue and the number of passengers would most likely increase versus a uniform pricing strategy. In this example, revenues increased from $6,000 to $10,000.

Another way of looking at uniform versus multiple pricing is through the eyes of the passengers. In our example, under uniform pricing, many passengers who purchased the $100 airfare were willing to pay more than that. There were many passengers who were willing to pay over $200, but ended up only having to pay $100. As mentioned, this difference between what a passenger is willing to pay and what the passenger actually paid is called consumer surplus. Consumer surplus can also be easily calculated by finding the area of the unshaded triangle region that lies beneath the demand curve. Under uniform pricing consumer surplus amounted to $4,500 or $[\frac{60\,seats \times \$150}{2}]$.[5] Conversely, under a multiple-pricing strategy, the amount of consumer surplus is the area of the multiple unshaded triangles that lie beneath the demand curve. From Figure 11.3 we are able to determine that there exists only $2,000 in consumer surplus for DirectJet's four-tier pricing structure.[6] Since the goal of revenue management is to maximize revenue from

---

5   The dotted area in Figure 11.4.
6   Consumer surplus = .5(20*50) + .5(20*50) + .5(20*50) + .5(20*50) = $2,000.

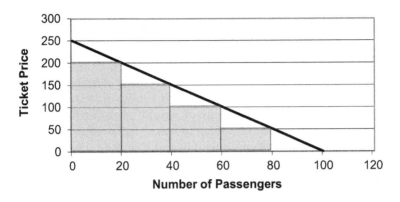

Figure 11.5   Multiple Pricing for DirectJet Airlines Flights

every passenger, it can also be said that revenue management's objective is to minimize consumer surplus. At the limits of this strategy, if DirectJet could sell each customer a ticket at the maximum fare that the customer would pay, then DirectJet would get the maximum possible revenue and there would be no consumer surplus. In the real world this is never possible since the information requirements are too large and there is too much uncertainty in consumer behavior itself. However, it is possible to present a menu of prices to consumers based on estimates of their likely price sensitivity and that leads us directly to our next topic—price discrimination.

REVENUE MANAGEMENT "FENCES"

One of the most important factors for the implementation of a revenue management system is the effective use of "fences," or barriers that limit the use of discounted seats to only passengers who are price-sensitive, rather than passengers who might otherwise be willing to pay a much higher fare. The airlines do not want a business traveler who is willing to pay full fare actually obtaining a deeply discounted fare. The way this is accomplished is through the use of "fences." In practical airline pricing policy there are six major "fences" and each will be discussed in greater detail. The six major "fences" are:

- advance purchase requirements (restrictions);
- Saturday night stay;
- frequent flier mileage;
- ticket refundability;
- change fees;
- airline schedule.

Advance purchase restrictions

Advance purchase restrictions are some of the oldest "fences" implemented in the airline industry. Advance purchase restrictions simply limit the amount of time before the day of departure that a ticket can be purchased. American Airlines's "Ultimate Super Saver" fares for example, had a 21-day advance purchase restriction on them. Delta Airlines's "deeply discounted fares" may require advance purchases of three, seven, 14, or 21 days. Advance purchase restrictions were implemented believing that passengers who were more price-sensitive (and less time-sensitive) would book further out. Conversely, if passengers show up at the airport and wish to travel on the next flight, they are clearly extremely time-sensitive and price-insensitive, and therefore, the fare that they would be willing to pay should be quite high. A typical fare class structure relating to advance purchase restrictions is illustrated in Table 11.6. It should be noted that no industry standard currently exists for fare codes but Y is generally considered to be unrestricted economy travel. Some carriers may use as many as 15 different fare codes, and LCCs tend to use only a few codes.

Table 11.6   Fare Class Restrictions

Fare Type	Fare Code	Restrictions
Economy (lowest fares)	L, U, T, H, Q, K	Most restrictive
Economy (discounted)	B, M	Some restrictions
Economy (flexible)	Y	Minimal restrictions
First/Business	P, F, A, C, D, J, I	Least restrictive

Source: Compiled by the authors from Delta Airlines's Fare and Ticket Rules, January 2011

Saturday night stay requirement

One of the most infamous revenue management "fences" is the Saturday night stay requirement. The Saturday night stay requirement was implemented to try and keep business travelers from receiving cheaper air fares. Since most business travelers are time-sensitive and want to depart Monday morning and return Friday evening, the Saturday night stay requirement was used to help segment the business travelers from the leisure travelers. Many LLCs have eliminated the Saturday night stay requirement from their pricing policy because business traveler trends have changed slightly and the rule seems archaic. LLCs have used the abolishment of this "fence" in many marketing campaigns as well.

Frequent flier mileage

While frequent flier programs have been around for quite some time, only recently have they begun to be used as a revenue management "fence." Since frequent flier programs have been successful at attracting and retaining loyal customers, the number of miles offered for a fare class can be an important factor to passengers. For instance, a passenger may be willing to pass-up a fare class if the next fare class offers more frequent flier miles. However, in order for this fence to be effective, full transparency of the fare classes/options is required. That is, if only one fare option appears when a passenger wishes to purchase, then this "fence" will not be effective since the passenger will not know about other options. The most effective marketing use of frequent flier mileage as a yield management "fence" is a matrix approach with various fare types available to the passenger. Alaska Airlines and Air Canada's websites are good examples of airlines that offer matrices of fare types to their customers. Another related frequent flier benefit is complimentary first class upgrades. Depending on the fare class booked, a passenger may be entitled to a complimentary first-class upgrade or redeem miles for the upgrade. For most airlines the passenger must be booked above a certain fare class level to be eligible for these perks. Table 11.7 updates the fare structure to include a percentage of actual miles flown that a passenger would receive as frequent flier mileage.

Table 11.7   Fare Class Frequent Flier Mileage Credit

Delta Air Lines		
Fare Type	Fare Code	Miles Earned
Economy (Lowest Fares)	L, U, T, H, Q, K	100%
Economy (Discounted)	B, M	150%
Economy (Flexible)	Y	150%
First / Business	P, F, A, C, D, J, I	150%
United Airlines		
Fare Type	Fare Code	Miles Earned
Economy (Lowest Fares)	V, W, T, S, K	100%
Economy (Discounted)	G, Q, M, L, E, H	100%
Economy (Full Fare)	Y, B	125%
Business (Discounted)	Z, P	150%
Business (Full Fare)	J, C, D	175%
First	F, A	250%

Source: Compiled by the authors from Delta Airlines' Fare and Ticket Rules & United Mileage Earnings

Refundability

Ticket refundability is another major "fence" implemented by airlines worldwide to help segment the market. The general rule is that higher-fare classes will have full ticket refundability; this enables a passenger to cancel a reservation and receive a full refund. Thus, the refundable ticket provides the passenger with greater flexibility; this of course is usually more desired by time-sensitive travelers. Lower-fare classes usually do not provide a refund unless there are extenuating circumstances.

Change fees

Similar to ticket refundability, change fees are another important revenue management "fence" used to differentiate travelers based on their time-sensitivity. Travelers who require great time-flexibility like the option to be able to change flights at will or for a nominal fee. Usually the highest-fare class allows full flexibility and is desired by business passengers whose schedule can change with short notice. On the other side of the spectrum, some of the lowest-fare classes may not even allow schedule changes. However, most fare classes require a change fee to be paid in addition to the difference in fare. While the change fee may be minimal, the difference in fare could be extensive, especially from lower-fare classes. In essence, the difference in fare charge is the difference between the fare class paid by the passenger and the lowest available fare class on the flight the passenger wants to change to. Since higher-fare passengers have less fare classes above them, the difference in fare charge is usually not as large. International carriers like Delta have adopted a multi-pronged approach to change fees; that is, provided all fare and ticket rules are met and seats are available, there is no change fee for a refundable ticket. However, based on the particular fare class rules, a service fee, and/or a difference in fare may be applicable.

Airline schedule

The final "fence" to discuss relates to the timing of an airline's schedule. Since different types of passengers have different traveling patterns, airlines can more profitably allocate high- and low-fare seating if they are aware of the likely composition of the passengers for a flight. For instance, time-sensitive travelers might desire an early morning departure and an evening return so that they can conduct a full day's business. Therefore, the airline will choose to limit the number of low-fare flights for a same day round trip. Leisure passengers on the other hand exhibit different travel patterns; these might include the ability to be flexible with regard to departure and return dates. Hence we observe mid-week sale specials and last-minute discounts to various locations.

REVENUE MANAGEMENT CONTROL TYPES

Before the various fare class allocation methods (or control types) are presented, two key terms need to be explained. Booking limit is the maximum number of seats that can be purchased for each fare class and protection level is the number of seats that are left unsold so they may be purchased by a higher-fare class. Depending on the control type implemented, there may only be a few seats distinctively retained, or protected, for Y fares or there may be a substantial number. For yield management analysts, booking limits and protection levels are extremely important concepts as they attempt to maximize revenue for every flight.

There are two main types of control limits used in yield management: distinct and nested. In distinct control, a fixed number of seats are allocated to each fare class and the fare can only be purchased if there remains inventory in the fare bucket. Under distinct control, protection level and booking level are equal since there is no provision for shifting fare classes, as shown in Figure 11.6. From the airline's point of view this is obviously an inefficient scheme since it amounts to a rather inflexible form of price discrimination. For

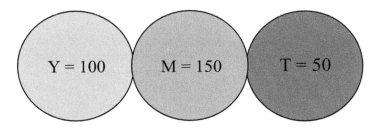

Figure 11.6   Distinct Fare Control

example, there may be many passengers who are price-sensitive and would purchase a lower-priced fare if it were available. If the airline has guessed wrong on the number of seats allocated at the lower fare, and if the passengers are unwilling to pay the next higher fare, then there are likely to be unsold seats on the flight. This of course results in lower overall revenue. Because of these inefficiencies, distinct control is very rarely implemented in airline yield management.

The predominant scheme utilized in yield management is some derivation of nested control. Nested control schemes can be customized to suit the individual characteristics of the flight, but the basic principle is that lower-fare classes are embedded in higher-fare classes booking limit. Therefore, under a pure (or serial) nested control scheme, a higher-fare bucket will never be closed out prior to a lower-fare bucket. Figure 11.7 highlights a serial nesting scheme for the same 300-seat aircraft, with the number representing the booking limit for each class. Under this scenario, the total aircraft capacity could be booked in Y class, but only 100 seats are protected for Y class. In this case, protection level is calculated by simply finding the difference between each of the fare classes. Another example of a nested control structure is a parallel nesting scheme as presented in Figure 11.8. While similar to serial nesting, a parallel structure allows for the M class fare to be closed prior to the T class, yet still allows the entire aircraft to be booked in full Y class. Such a structure, or derivation thereof, may be used to provide a set inventory reserved for frequent flier mileage redemption or corporate travel arrangements (Vinod, 1995).

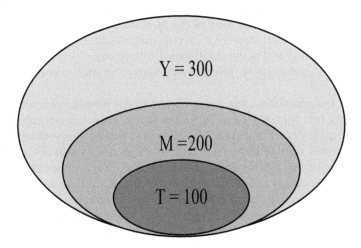

Figure 11.7   Serial Nesting

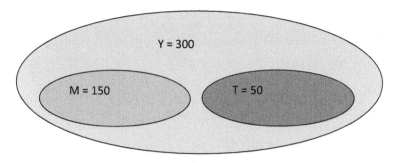

Figure 11.8   Parallel Nesting

A major type of nested control is virtual nesting. Virtual nesting deals from a total revenue and total network perspective as it helps determine if selling a seat in a high-fare class on a single sector might be sub-optimal relative to selling that same seat to a connecting passenger in a lower-fare class. In essence, since many airline itineraries involve a change of planes through a hub, virtual nesting looks at the total revenue the booking would generate. For instance, if a full unrestricted economy fare on a short-haul sector generates less revenue than a discounted fare on a long-haul international flight, the longer itinerary would have priority. This is accomplished by "clustering" the various itinerary fare classes that flow over a flight leg into a manageable number of buckets, based on customer value (Vinod, 1995). Thousands of potential itineraries can be grouped into a few virtual buckets, but the variance in each of these buckets can be considerable (Vinod, 1995).

SPOILAGE AND SPILLAGE

Using the various control types, revenue management analysts are able to open and close fare buckets to adjust to the demand for the flight. Prices for flights are adjusted with respect to the normal booking curve for the flight, which is based on historical demand. A normal booking curve is generated by assuming that ticket purchases follow a normal historical distribution and the last seat of the aircraft is purchased just before the time of departure.[7] If such a situation were to occur, this would represent complete revenue management effectiveness and maximize the airlines' revenues (Littlewood, 2005). Figures 11.9, 11.10, and 11.11 each display different situations that may occur with respect to the normal booking curve.

In Figure 11.9, the actual booking curve results in the number of bookings at the date of departure being less than the capacity of the flight. This difference between capacity and actual bookings is called "spoilage." Spoilage is visually represented as empty seats on an aircraft. Airlines want to reduce spoilage since an empty seat does not provide any additional revenue for the airline, and in all likelihood, that seat could have been sold if the price was right. Spoilage is a result of prices that are too high for the market for the given fare class. If this occurs, it will be evident from the fact that the booking rate is less than the normal booking curve for the flight. At this point the revenue analyst can lower the average ticket price, or open lower-fare classes for sale, to reduce potential spoilage.

---

7   If there is no overbooking.

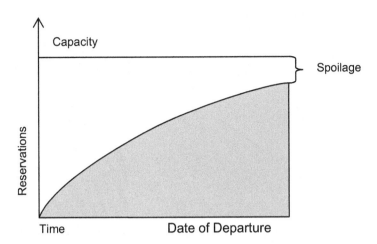

Figure 11.9   Spoilage

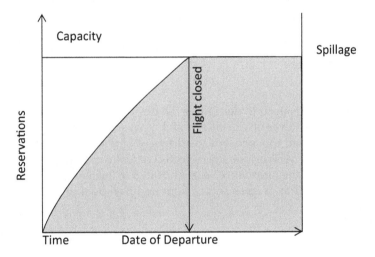

Figure 11.10   Spillage

Figure 11.10 depicts the reverse situation of spoilage, which occurs when all the seats are purchased prior to the departure of the flight. This situation, called "spillage," is also a problem because the airline generally wants to hold a few seats available for last-minute travelers who might be willing to pay full fare for the flight. By already having the flight fully booked, the airline is incurring a potential loss of revenue for the flight. Spillage is the result of average fares that are too low for the flight, which leads to a booking curve that lies above the normal booking curve. Both spillage and spoilage are of concern for revenue analysts, and they must balance the fine line between both problems to reach an ideal normal booking curve.

The third figure of the series, Figure 11.11, depicts a situation similar to spillage when the airline books more passengers for the flight than capacity. This situation, called "overbooking," is a normal occurrence, since a probabilistic percentage of passengers usually do not show up for their flight. Airlines routinely set booking limits that exceed the capacity of the aircraft in order to maximize revenue. The overbooking issue is discussed later in the chapter.

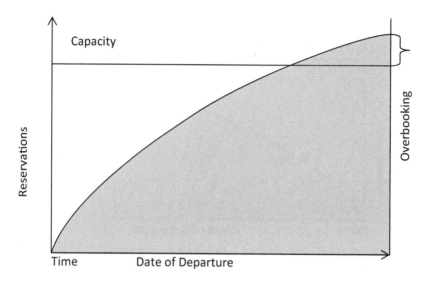

Figure 11.11   Overbooking

LEG-BASED EXPECTED MARGINAL SEAT REVENUE MODEL

One of the major methods used to determine the desired booking limits for a flight is the Expected Marginal Seat Revenue (EMSR) model. It was developed by Littlewood (1972); later, Belobaba (1987) used this concept to address a single-leg flight with multiple fare classes. Simply put, the EMSR means the expected revenue contribution of one additional seat. In an EMSR model the number of seats allocated to each fare class is determined by using historical information of fares and current and past booking figures. The EMSR of the ith seat sold is:

$$EMSR_i = f_i \times P(S_i)$$

EMSR is the product of the fare level, f_i, and the probability that there will be at least n passengers willing to buy "i" class tickets for the flight under consideration.[8]

Figure 11.12 provides the cumulative probability distribution for two unique fare classes. Since the underlying assumption is that the probability of booking is based on a normal distribution, the average probability of demand for each fare class is 50 percent. In both fare classes there is close to a 100 percent probability that at least some seats (about 15 and 12 respectively) can be sold at the given fare classes, while it is unlikely that more than a certain number will be sold (about 25 and 50 seats respectively).

Using the formula, the EMSR for every seat can be calculated by simply multiplying the ticket fare by the cumulative probability of demand for that seat. For instance, assuming that a ticket costs $500 and the cumulative probability of demand for that seat is 50 percent (or 0.5), then the EMSR for that seat is $250. This formula is applied to every seat for every

---

8   The probability for a passenger's willingness to buy is assumed to be a normal distribution. Values for a cumulative normal distribution can be calculated by using the "normdist()" function in Microsoft Excel, or similar functions in other spreadsheet packages.

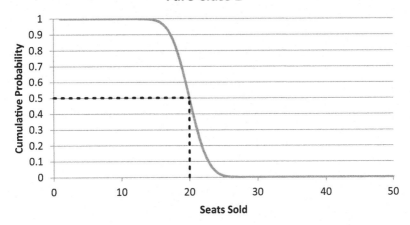

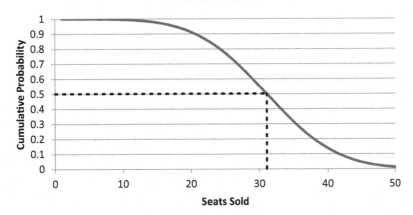

Figure 11.12   Cumulative Probability Distribution for Fare Class 1 & 2

fare class, so that it is now possible to graph the EMSR curve. This has been done for the
two cumulative probability distributions that were presented above, assuming the fare in
the first fare class is $500 and the fare in the second fare class is $400.

   As shown in Figure 11.13 overleaf, the EMSR curves for both fare classes are similar
to the cumulative probability distributions, except that the vertical axis is no longer
probabilities, but actual dollars. The average EMSR for fare class 1 occurs at $250 and
roughly 20 seats, while in fare class 2 the average EMSR is $200 at 31 seats.

   The final step in the analysis is to combine the two EMSR curves to determine the
protection level and booking limits for the fare classes. In Figure 11.14, the two EMSR
curves intersect at a point close to 19 seats. This point represents the protection level for
fare class 1 over fare class 2. If we assume that these are the only two fare classes for the
flight, the booking limit for the higher-fare class would be the capacity of the aircraft, since
if all the seats can be sold at the highest fare, the airline would of course be happy with

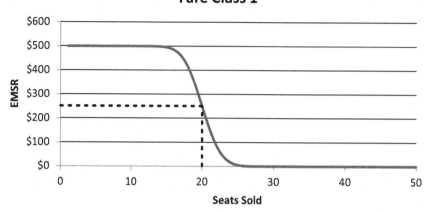

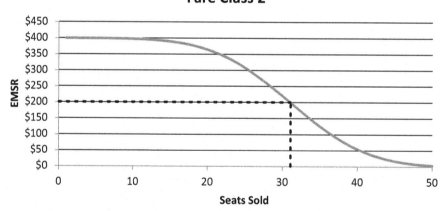

Figure 11.13   Expected Marginal Seat Revenue for Fare Class 1 & 2

this outcome. However, for this example, the actual protection level would be 19 seats for the first fare class because it is at this point that the expected revenue of the second class exceeds that of the first class. At this point the airline should begin selling tickets in the second class. As Figure 11.14 clearly shows, this should continue until all seats are sold, since the expected revenue of the second class exceeds that of the first class for the remaining seats.

The application of the EMSR to revenue management can be presented through the use of a decision tree, as shown in Figure 11.15. In essence, every seat has some probability of being booked, and an airline revenue management analyst must choose the option that provides the airline with the greatest expected seat revenue. Since demand is not deterministic, revenue management analysts must use probability to foresee the future and be able to protect some number of seats for higher-paying customers. This protecting of seats is in addition to the "fences" that were described earlier

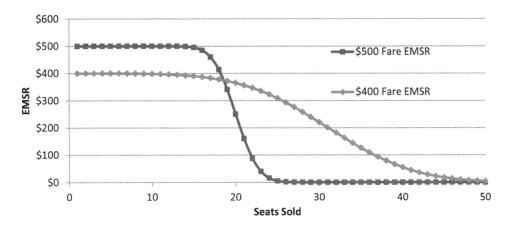

Figure 11.14 **Optimal Booking Limit and Protection Level in Two Nested Fare Classes**

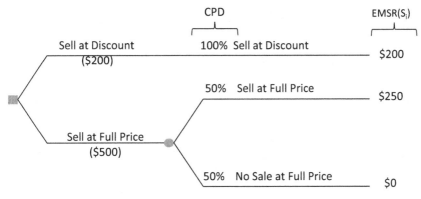

Figure 11.15 **Decision Tree**

The decision tree scenario presented in Figure 11.15 assumes the airline is presented with the situation that it can either sell a ticket at the discounted price or it has some probability of selling the same ticket at a higher fare. The key in determining which situation to choose is based on the actual probability of selling the ticket at the higher price. In our scenario, the probability of selling the discounted ticket at $200 is 100 percent, and the probability of selling the full price ticket at $500 is 50 percent. This of course means that there is also a 50 percent chance that the ticket will not be sold at the full price. In order to make a decision, the expected revenue needs to be computed. This is done by simply multiplying the probability of selling the ticket by the ticket price. This provides expected revenue of $200 for the discounted ticket and expected revenue of $250 for the full price ticket; therefore, the discounted ticket should not be sold in hope of selling the full price ticket. Thus, the formula for computing EMSR can be stated as:

$$EMSR(S_i) = f_i * P(S_i) + 0*[1-P(S_i)]$$

Where $EMSR(S_i)$ is the product of the average fare level, f_i, and the probability of selling the ith seat $P(S_i)$ plus the fare associated with the alternative event of not selling the seat (0) times the probability of that event $[1-P(S_i)]$.

In a situation where the capacity of the aircraft is increased by one seat, the revenue management analyst must choose the highest marginal EMSR that the seat would generate.

A concrete example to determine the appropriate booking limits and protection levels for some assumed fare classes should help make this clear. Assume that DirectJet is a new airline that operates an 80-seat regional jet aircraft. The airline utilizes a nested three-tier fare structure ($300, $400, $500) and demand for all three fare classes is assumed to be normally distributed. The airline has historical demand data for the past 30 days for one of its routes which is presented in Table 11.8.

Table 11.8   Historical Demand for DirectJet's Three Fare Classes

Flight History Day	$500 Fare	$400 Fare	$300 Fare
1	20	30	36
2	17	40	34
3	18	35	33
4	22	25	32
5	24	18	31
6	20	45	40
7	20	32	42
8	21	22	29
9	22	29	32
10	19	34	38
11	18	38	40
12	20	31	38
13	19	22	36
14	22	24	30
15	18	29	26
16	23	30	26
17	24	36	30
18	23	35	31
19	17	26	33
20	20	42	34
21	19	22	37
22	20	18	35
23	17	34	45
24	18	33	27
25	23	48	28
26	21	16	34
27	20	26	36
28	20	42	33
29	18	38	37
30	17	32	29

The first step required in determining the optimum booking limit and protection level for DirectJet's flight is to determine the mean and standard deviation of the demand for each fare class. These data will be required when determining the probabilities of purchasing the ticket in each fare class. Both the mean and standard deviation for the three fare classes are contained in Table 11.9.

Table 11.9   Mean and Standard Deviation

	$500 Fare	$400 Fare	$300 Fare
Mean	20.00	31.07	33.73
Standard Deviation	2.13	8.17	4.70

The next step in the EMSR process is to create a normal distribution for each fare class in order to assign a probability that a given number of seats will be purchased for each fare class. Knowing the mean and standard deviation, a cumulative probability distribution for each fare class can be created using the NORMDIST() function in Microsoft Excel. The cumulative probability of selling each seat is then multiplied by the fare to produce the EMSR for each fare class. The data are presented in Table 11.10.

The final step is to determine the appropriate booking limit and protection level for each fare class based on the EMSR values calculated. The goal when choosing the appropriate level is to select the highest EMSR value, regardless of which column the value lies in. This process continues until all 80-seats (assuming no overbooking) are allocated. The shaded region of Table 11.10 represents the greatest 80 EMSR values for the flight.

In this case, we see that the $500 fare class results in the highest EMSR for the first 18 seats. Continuing down the table, we see the initial intersection point between the $500 fare EMSR and the $400 fare EMSR which occurs roughly between seat 18 and 19. Seats in the $400 fare class then have the highest EMSR up to 22 seats whereupon the next highest EMSR once again belongs to the $500 fare class. As such, the next seat is once again allocated back to the $500 fare class. Continuing on, the $400 fare now has the highest EMSR for the next three seats until the $300 fare takes over for 24 seats. We then switch back and forth in sequence until the entire 80 seats have been allocated between the fare classes.

For DirectJet, the appropriate protection levels for the flight would be 20 $500 fares, 30 $400 fares, and 30 $300 fares. The appropriate booking limits for each class would be 80 for the $500 fare, 60 for $400 fares, and 30 for the $300 fares. These values are expressed in Table 11.11. As we might expect, the booking limit always includes the protection level for the class under consideration plus all classes below that class—another example of the fact that the airline is always prepared to sell all seats at a higher fare.

The EMSR curves can also be graphed, and are displayed in Figure 11.16. Graphing of the EMSR curves provides a visual method of determining the protection level and booking limit for each fare class. Note that the points of intersection between the EMSR curves correspond to the values in Table 11.10 where the EMSR for one fare exceeds that of another. This is an example of serial nesting which is also displayed in Figure 11.16.

The example of DirectJet helps show how booking limits and protection levels are determined in revenue management. Of course, airlines utilize many more fare classes than the three DirectJet uses, making the process all the more complicated, but the

Table 11.10   Optimal Booking Limit for DirectJet's Three Fare Classes

Seat	$500 Fare Probability	$500 Fare EMSR	$400 Fare Probability	$400 Fare EMSR	$300 Fare Probability	$300 Fare EMSR
1	1.0000	$500.00	0.9999	$399.95	1.0000	$300.00
2	1.0000	$500.00	0.9998	$399.92	1.0000	$300.00
3	1.0000	$500.00	0.9997	$399.88	1.0000	$300.00
4	1.0000	$500.00	0.9995	$399.81	1.0000	$300.00
5	1.0000	$500.00	0.9993	$399.71	1.0000	$300.00
6	1.0000	$500.00	0.9989	$399.57	1.0000	$300.00
7	1.0000	$500.00	0.9984	$399.35	1.0000	$300.00
8	1.0000	$500.00	0.9976	$399.04	1.0000	$300.00
9	1.0000	$500.00	0.9965	$398.61	1.0000	$300.00
10	1.0000	$500.00	0.9950	$398.01	1.0000	$300.00
11	1.0000	$499.99	0.9930	$397.18	1.0000	$300.00
12	0.9999	$499.96	0.9902	$396.06	1.0000	$300.00
13	0.9995	$499.75	0.9865	$394.58	1.0000	$300.00
14	0.9976	$498.79	0.9816	$392.64	1.0000	$300.00
15	0.9905	$495.27	0.9753	$390.13	1.0000	$299.99
16	0.9698	$484.90	0.9673	$386.94	0.9999	$299.98
17	0.9205	$460.25	0.9574	$382.94	0.9998	$299.94
18	0.8261	$413.06	0.9450	$378.01	0.9996	$299.88
19	0.6806	$340.32	0.9300	$372.02	0.9991	$299.74
20	0.5000	$250.00	0.9121	$364.84	0.9982	$299.47
21	0.3194	$159.68	0.8909	$356.37	0.9966	$298.98
22	0.1739	$86.94	0.8663	$346.53	0.9937	$298.11
23	0.0795	$39.75	0.8381	$335.25	0.9887	$296.62
24	0.0302	$15.10	0.8063	$322.53	0.9807	$294.22
25	0.0095	$4.73	0.7710	$308.40	0.9683	$290.49
26	0.0024	$1.21	0.7323	$292.92	0.9499	$284.97
27	0.0005	$0.25	0.6906	$276.23	0.9238	$277.14
28	0.0001	$0.04	0.6462	$258.49	0.8885	$266.55
29	0.0000	$0.01	0.5998	$239.92	0.8428	$252.85
30	0.0000	$0.00	0.5519	$220.76	0.7863	$235.88
31	0.0000	$0.00	0.5033	$201.30	0.7194	$215.81
32	0.0000	$0.00	0.4545	$181.82	0.6437	$193.12
33	0.0000	$0.00	0.4065	$162.61	0.5619	$168.58

Table 11.11   Optimal Booking Limit and Protection Level in Three Nested Fare Classes

Fare	500	400	300
Protection level	20	30	0
Booking Limit	80	60	30

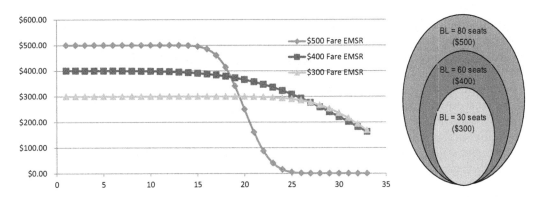

Figure 11.16   Optimal Booking Limit and Protection Level in Three Nested Fare Classes

principles behind it are exactly the same. The one glaring omission from the DirectJet example is the presence of overbooking. Overbooking is a real issue revenue management analysts face and is usually taken into account when setting the appropriate protection levels for a flight. The issue of overbooking is explored in the next section where the DirectJet example is expanded.

OVERBOOKING

Overbooking is practiced by airlines to combat spoilage since invariably some passengers do not show up for a flight or miss connections, especially for flights departing from hubs. While overbooking may cause headaches for a few passengers, the benefits of airline overbooking include increased seat availability, more access to the flight of first choice, and the reduced overall cost of travel through the more efficient use of airline seats (Dunleavy, 1995). Additionally, overbooking is practiced in other industries such as car rentals and hotels (Netessine and Shumsky, 2002).

Airlines are able to predict to some degree the "no-show" level, or percentage of passengers who will fail to show for the flight, based on the probabilistic nature of the situation. Additionally, passengers themselves have differing patterns of not showing up for a flight. Typically, business passengers have a higher probability of missing their flight than leisure passengers. Therefore, it is not surprising that flights to leisure markets have lower authorization levels for overbooking than flights to business-heavy markets. Additionally, revenue managers must also take into consideration the probability of passengers cancelling itineraries close to departure, misconnecting, or having ticket issues where they show up for the flight but do not have confirmed reservations. These ticket issues are coined a "go-show" in the airline industry (Dunleavy, 1995).

While the benefits of overbooking for the airline is reducing spoilage and increasing potential revenue, the tradeoff is that overbooking can also be costly for the airline. Costs associated with overbooking include meal and hotel vouchers, flight coupons for future flights, departure delays, passengers being rolled-over to other flights, staffing issues, and loss of goodwill. The level of these costs vary from flight to flight, since the costs associated with a daily (or less frequent) international flight are invariably higher than a short-haul flight where the airline offers ten daily flights.

In a situation where more passengers show up for a flight than there are seats, the airline will ask for volunteers to give up a seat. In order to entice passengers to do so, the airline will usually offer some form of compensation package that may include future travel discounts, meal vouchers, hotel accommodation, or first-class upgrades. Ideally, airlines want to solve their overbooking problem by asking for volunteers, as the next step, involuntary denied boarding, can be much more costly. If there are not enough passengers willing to give up their seat, some passengers will be denied boarding. This situation creates extremely negative goodwill against the airline and the airline is required to still transport the passengers to their destination. Since airlines want to avoid these overbooking situations as much as possible, accurate forecasting for any overbooking is desired.

Forecasting overbooking levels

There are several methods utilized to determine the overbooking level for a particular flight which are based on the probability that a passenger will not show up. Utilizing historical averages and probabilities of no shows for a flight, an analyst can weigh the accommodation costs of bumping passengers against the expected lost revenues from not overbooking. This situation has been well described as a classic news vendor problem wherein a vendor of newspapers must balance the chances of inventory going unsold (spoilage) with the chances of not having enough to sell (spillage). For airlines, these expected costs and probabilities can be manipulated to calculate the optimal number of seats to overbook. The following ratio of these expected costs gives us the probability that the opportunity cost of underselling is equal to the cost of overbooking. The optimal overbooking level is the smallest value of Q such that:

$$\Pr(Q \geq X) = \frac{B}{B+C}$$

Where:[9]

- B = opportunity cost of flying an empty seat
- C = negative goodwill and penalties of denying a passenger boarding
- Q = actual number of seats overbooked
- X = optimum number of seats to overbook (Winston, 2007).

The probability determined from this ratio can then be expressed as a z-score which takes the probability percentage and translates it into terms of standard deviations under the area of a normal distribution. Along with the historical mean and standard deviation from the sample, the resulting z-score can then be substituted into the following equation to determine the optimal overbooking limit.

$$Q_{overbooking} = \mu + Z \times \sigma$$

---

9   The derivation of this formula can be found in the Appendix.

Excel functions can also be used as demonstrated in this next example. Suppose that DirectJet has been able to determine the following data for one of its flights: the average number of no-shows for the flight is normally distributed with a mean of five and a standard deviation of three. Moreover, DirectJet estimates that it costs $200 to "bump" a passenger (the airline receives no revenue from the passenger and $200 is the cost of accommodating the passenger for his/her next flight). If the seat is not sold then the airline loses revenue equal to the price of a ticket at the discounted rate. In this example, the lowest discounted air fare is $80.

- Standard Deviation, σ 3
- Mean, μ 5
- Overbooking Cost, C $200
- Underbooking Cost, B $80

Applying the formula previously identified to DirectJet's analysis, the corresponding optimum number of seats to overbook can be determined.

$$Pr(Q \geq X) = \frac{80}{80 + 200} = 0.2857$$

From the normal distribution table (or using the NORMSINV() function in Excel) the z-score associated with a probability of 0.2857 is approximately -0.566. Plugging the z-score into the second equation yields an optimal overbooking limit of 3.30. Rounding down, we determine the number of seats to be overbooked to be three.

$$Q_{overbooking} = 5 + -0.566 \times 3 = 3.30$$

The other, and simpler option, is to use the NORMINV() function in Microsoft Excel (Netessine and Shumsky, 2002) which returns the value of seats to be overbooked after the probability, mean and standard deviation are input. For the above scenario, the NORMINV() function would be as follows:

$= NORMINV(probability, mean, standard\ deviation)$

$= NORMINV(0.2857, 5, 3)$

$= 3.302$

Regardless of the method utilized, the optimum number of seats for DirectJet to overbook is equal to three. This solution computes the tradeoff between overbooking passengers and the additional revenue that would be generated against the costs associated with overbooking. We can also determine the additional revenue that overbooking would generate. Overbooking by three additional seats would generate additional revenue of $240 (fare x number of seats). The following figures show the overbooking costs and spoilage costs as the number of aircraft seats sold increase (aircraft capacity is 150 seats). Figure 11.18 shows the overbooking costs as the number of seats sold begins to exceed the capacity of the aircraft. The spoilage costs in Figure 11.17 reflect the opportunity cost associated with not selling an aircraft seat; the fewer seats sold, the higher the costs.

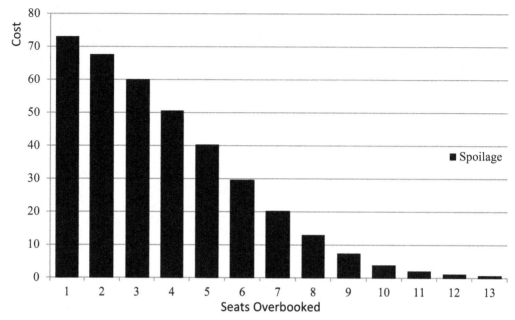

Figure 11.17   Spoilage Costs

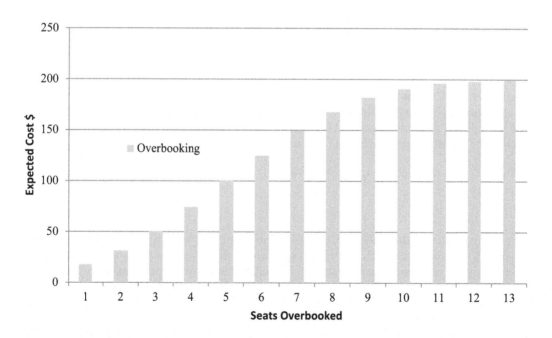

Figure 11.18   Overbooking Costs

OTHER ISSUES ASSOCIATED WITH REVENUE MANAGEMENT

Since revenue management was first introduced by American Airlines in 1985, it has grown considerably to take into consideration many issues and has become more complex to better minimize consumer surplus. While current revenue management practices have been effective in helping airlines' profits, revenue management is still largely based on historical, probabilistic demand. While this is not an ideal situation, it probably is the best way to forecast demand for revenue management purposes. Additionally, the creation of fare classes is done by grouping customers together based on their price elasticity, but even with multiple fare classes, customers are still going to be grouped into classes that they may not be part of. This can increase consumer surplus which reduces revenue for the airline.

A potential solution to both these problems is dynamic pricing where seats are priced based on existing (ongoing) passenger demand and other factors such as competitors' revenue management strategy (Burger and Fuchs, 2005). In essence, dynamic pricing is timelier since it allows a carrier to more closely match its normal booking curve. Basically, it allows the airline to change fares based on whether ongoing bookings are above or below the normal booking curve before the actual takeoff time. A few airlines, mostly LLCs, have implemented dynamic pricing into their revenue management models.

A recent trend in airline revenue management that was started by LLCs like Southwest is to reduce the complexity of fare structures. Instead of having a seemingly infinite number of booking classes and fares for a particular flight, airlines are reducing the number of booking classes for marketing reasons. Since the average passenger has been confused about all the different booking classes and restrictions, airlines like Delta and Air Canada have introduced simplified fare structures. These new simplified fare structures, termed restriction-free pricing, are forcing revenue management to use "weak" market segmentation through active management of fare availability instead of "strong" market segmentation, such as "fences" (Ratliff and Vinod, 2005). While generally more appealing to passengers, it remains to be seen if the increase in consumer surplus from simpler fare structures can be offset by any incremental increase in bookings. Additionally, the move by legacy carriers to allow one-way tickets has put further complexities in revenue management systems.

Revenue management is one of the most important business units in an airline organization. Revenue management is responsible for ensuring that customers purchase the airline's product and a revenue management analyst can, on a daily basis, impact the bottom line of the company more than almost any other employee. Ever since the theory of revenue management was introduced by American Airlines, it has been an effective tool used by airlines to maximize revenues and aid their profits.

SUMMARY

This chapter covers the topics of pricing policy and revenue management in detail. Various pricing models are discussed, and the crucial underlying assumptions of price discrimination are explained together with the conditions necessary to implement a policy of price discrimination. A quantitative example is presented to show how price discrimination can be used to increase revenues and sales. The various methods that are used in the airline industry to use price discrimination for revenue management are covered. The topics of spoilage and spillage are covered and explained with a graphical presentation. The more sophisticated EMSR model is explained using concrete quantitative examples and graphs to explain the ideas of nesting fare structures and the creation of a probabilistic normal booking curve. An actual method for allocating seats to various fare classes is explained in detail with the generation of the appropriate protection and booking levels for the example classes. The idea of dynamic pricing using the normal booking curve is discussed and explained. Finally, the important topic of overbooking is covered with a numerical example showing how an overbooking amount can be determined in practice.

DISCUSSION QUESTIONS

1. Define peak-load pricing and provide at least two examples.
2. Explain markup pricing and why it may not be a good tool to apply in the airline industry.
3. What is meant by the terms "reservation price" and "consumer's surplus" and how are these concepts related to price discrimination?
4. A major hotel offers 20 percent discounts for renting a room for more than three days. Therefore, in this case, the price paid by a traveler will be much lower than that of another traveler who stays less than three days. Is this an example of price discrimination? Why or why not?
5. How do airlines use price discrimination to maximize revenue based on elasticity of demand?

Passenger Type	Cost	Elasticity
First Class	$400	-1.25
Business Travelers	350	-1.1
Leisure	150	-3.5
Holiday	150	-2.5

6. List the conditions that are needed for successful price discrimination.
7. Calculate the optimal markup on cost for each service, based on the following estimates of point price elasticity of demand;
8. Why is price discrimination more widely used in service industries than in product industries?
9. What is the main problem with average cost pricing?
10. What is the basic idea behind EMSR models?
11. How are seats allocated using the EMSR model?
12. What are the merits of using peak-load pricing at commercial airports?
13. Can every firm with monopoly power discriminate?
14. List three revenue management fences that airlines typically use to practice price discrimination and briefly explain how each one works.
15. Discuss the advantages and disadvantages of price discrimination for airlines and passengers.

APPENDIX: DERIVATION OF OVERBOOKING PROBABILITY EQUATION

What is the expected payoff of overbooking? Assume the following:

- For each ticket sold as an overbooking, there is an incremental profit of $80
- For each bumped passenger, there is an incremental loss of $200
- A historical no-show mean of 5 with standard deviation of 3.

The incremental profits associated with selling X overbooked seats if the quantity demanded (Q) is *greater than or equal* to X is:

$$\pi_{Incr.} = IP_sX = 80X$$

The incremental profits associated with selling X overbooked seats if the quantity demanded (Q) is *less* than X is:

$$\pi_{Incr.} = IP_sQ - IC\pi(X - Q)$$

$$\pi_{Incr.} = 800Q - 200(X - Q)$$

Expected incremental profit of selling X overbooked seats is:

$$E_{\Pi Incr.} = \Pr(Q \geq X) \cdot 80X + (1 - \Pr(Q \geq X)) \cdot (80Q - 200(X - Q))$$

To determine the optimal number of high fare tickets to hold, take the derivative with respect to X and set it equal to zero:

$$E_{\Pi Incr.} = 80X \cdot \Pr(Q \geq X) - 200 \cdot (1 - \Pr(Q \geq X)) = 0$$

$$\Rightarrow 80X \cdot \Pr(Q \geq X) = 200 - 200 \cdot (\Pr(Q - X))$$

$$\Rightarrow \Pr(Q \geq X) = \frac{80}{80 + 200} = 0.2857$$

This probability can then be translated into a z-score, multiplied by the standard deviation, and added to the mean in order to calculate optimal overbooking as shown in the text.

REFERENCES

Airline Weekly (2010, February) *Ancillary Revenue: It's Not Non-Core Anymore.* Retrieved on March 23, 2011 from http://www.airlineweekly.com/AWSR1.pdf

Belobaba, P. (1987). *Air Travel Demand and Airline Seat Inventory Management.* Cambridge, MA: Flight Transportation Laboratory, Massachusetts Institute of Technology.

Bomkamp, S. (2012, May 2) Spirit Airlines raising carry-on bag fee to $100, *The Associated Press*, May 2, 2012.

Burger, B. and Fuchs, M. (2005). Dynamic Pricing—A Future Airline Business Model. *Journal of Revenue and Pricing Management*, Vol. 4, No. 1, 39–53.

Buzzacchi, L. and Valletti T. (2005). Strategic Price Discrimination in Compulsory Insurance Markets. *Geneva Risk and Insurance Review*, Vol. 30, No. 1, 71–96.

Cross, G. (1995). An introduction to revenue management, in Jenkins, D. (Ed.), *Handbook of Airline Economics*. New York: McGraw-Hill, pp. 443–458.

Dunleavy, N. (1995). Airline passenger overbooking. Jenkins, D. (Ed.), *Handbook of Airline Economics*. New York: McGraw-Hill, pp. 469–476.

The Economist (2004, December 29). Unbundled. Retrieved on April 2006 from http://www.economist.com/node/3523035.

Ewers, H. J. (2001). *Possibilites for the Better Use of Airport Slots in Germany and in the EU – A Practical Approach*. Berlin: Wirtschafts and Infrastruktur Politik.

Frank, R. (1983). When Are Price Differentials Discriminatory? *Journal of Policy Analysis and Management*, Vol. 2, No. 2, 238–255.

Littlewood, K. (1972). Forecasting and Control of Passenger Bookings. *12th AGIFORS Symposium Proceedings*, Nathanya, Israel, pp. 103–105.

Littlewood, K. (2005). Forecasting and Control of Passenger Bookings. *Journal of Revenue and Pricing Management*, Vol. 4, No. 2, 111–123. (Note: Originally written in 1972.)

Netessine, S. and Shumsky, R. (2002). Introduction to the Theory and Practice of Revenue Management. *INFORMS Transactions on Education*, Vol. 3, No. 1, 34–44.

Ovans, A. (1997) Make a bundle bundling, *Harvard Business Review*, Vol. 75, No. 6, 18–20.

Ratliff, R. and Vinod, B. (2005). Airline Pricing and Revenue Management: A Future Outlook. *Journal of Revenue and Pricing Management*, Vol. 4, No. 3, 302–307.

Spiller, T. (1981). The Differential Impact of Airline Regulation on Industry, Firms, and Market. *Journal of Law and Economics*, Vol. 26, No. 3, 655–689.

Vinod, B. (1995). Origin-and-destination revenue management in Jenkins, D. (Ed.), *Handbook of Airline Economics*. New York: McGraw-Hill, pp. 459–468.

Washington Post (2000) US v. Microsoft Timeline. Retrieved on November 12, 2011 from http://www.washingtonpost.com/wp-srv/business/longterm/microsoft/timeline.htm.

Winston, W. (2007). *Operations Research: Applications and Algorithms* (4th ed.). Belmont, CA: Duxbury Press.

12

Low-cost and Start-up Airlines: A New Paradigm

Ten years ago, network airlines almost universally declared that low-cost carrier operations simply wouldn't work in the Asia Pacific region, but today nearly all full-service airlines in the region have their own low-cost offshoots.

Tom Ballantyne, Chief correspondent for *Orient Aviation*, April 2012

One of the more recent developments in the aviation industry has been the emergence of a new breed of airlines—low-cost carriers (LCCs). LCCs have enjoyed significant growth and can now be found in virtually every market. Globally, LCC fleets are expected to grow at an annual rate of 5.7 percent over the next 20 years with the greatest concentrations occurring in the Asia-Pacific region where LCCs are expected to have as much as a 60 percent market share by 2030. The LCC model has overwhelmingly been the favored mode of airline start-up. India for example, is the ninth largest aviation market, and has experienced remarkable LCC growth from 0.1 percent in 2003 to 67.4 percent of domestic market share,[1] with airlines such as Air Deccan (which later became Kingfisher Red) in 2003 which have dramatically changed the aviation landscape in India. Other full-service carriers either established low-cost subsidiaries (for example, Air India established Air India Express in 2006) or merged/acquired low-cost airlines (for example, Kingfisher, which acquired the original Air Deccan). The Indian LCC market is now almost equally distributed between six LCCs: IndiGo, JetKonnect,[2] SpiceJet, GoAir, Air India Express, and Kingfisher Red, most of which began operations in post-2003 and have experienced continuous double-digit growth.

The overall objective of this chapter will be to introduce the student to the history of LCCs, the various strategies used by low-cost airlines to gain competitive advantages over legacy airlines and how these low-cost models have evolved over time. The chapter starts by providing information on "legacy" airlines and how the emergence of low-cost airlines affected the market share of legacy airlines. The general outline for the chapter is:

- The Evolution of the Industry
- Characteristics of Low-cost Carriers
  - Lower labor costs and higher productivity

---

1   CAPA—Centre for Aviation Low Cost Carrier Data, 2011.
2   Formerly JetLite as of March 2012.

- – Lower ticket distribution costs
- – No-frills service
- – Common fleet type
- – Point-to-point service
- – Use of secondary airports
- – Higher aircraft utilization
- Cost Structure Comparison
- Incumbent Carriers' Response to Low-cost Carriers
  - – Low-cost carrier creation
  - – Cost cutting
- The Future of Low-cost Carriers
- Summary
- Discussion Questions
- Appendix: Select Airline Two Letter Codes

THE EVOLUTION OF THE INDUSTRY

The evolution of LCCs (or budget airlines) is an interesting example of competitive market forces in action. There is now no part of the world which is not serviced by at least one low-cost airline. Across Australia, New Zealand, Latin America, and Asia there are many low-fare, low-cost airlines. While only the last decade has seen tremendous growth in this sector of the industry, the roots of LCCs can be traced back to 1971 when founder Herb Kelleher mapped out the route and cost structure for Southwest Airlines. Since the US aviation industry was still regulated, Southwest Airlines was able to only fly intra-Texas routes where the Civil Aeronautics Board (CAB) did not have authority. From these humble beginnings, and based on its simple low-cost strategy, Southwest Airlines was able to grow into one of the most successful airlines in the US.

> "...the twentieth century largely belonged to the traditional, high-cost airlines (with a few snipers, like me, upsetting their cozy cartel). The twenty-first century will be the preserve of the no-frills airlines..."
>
> Sir Freddie Laker, 2002

Following airline deregulation in the US, many new airlines entered the market; this caused airfares to plummet 40 percent in real terms between 1978 and 1997, while the number of passengers more than doubled. Airlines like Southwest were able to successfully expand, while new LCCs like America West, Reno Air, and People Express emerged with varying degrees of success. Today there are many successful LCCs in the US, such as AirTran, Frontier, the Las Vegas-based Allegiant Airline, and JetBlue that have taken away market share from the traditional legacy carriers. On the other hand, while there is a huge demand for low-cost travel, many LCCs have failed. Predictably, the remaining LCCs have been those with the lowest cost base. For example, an airline like Southwest has been tremendously successful at retaining a low cost structure while still expanding aggressively.

The emergence of LCCs is not just a North American phenomenon, but is a global trend in the airline industry. Today almost all markets contain at least some low-fare carriers.

While Southwest is the founder of the LCC model, the idea spread to the UK in the early 1970s when Sir Freddie Laker was able to secure the necessary route authorizations to launch cheap trans-Atlantic flights between Gatwick and New York. With North America experiencing a wave of LCC start-ups following US airline deregulation, Europe experienced a "second-wave" of LCCs following the liberalization of European airspace. This rapid entry of LCCs quickly became a key growth area for air traffic in Europe.

In Latin America, the growth of LCCs has been phenomenal. After stepping down as JetBlue's chief executive officer and founder, David Neeleman started a new low-cost airline (Azul) in Brazil in 2008. Azul is now Brazil's third largest domestic carrier after Gol and TAM.[3] Using the low-cost model, Easysky[4] started operations in Honduras in September, 2011. In Mexico, many LCCs such as Viva Aerobus, Volaris Aviacsa, Aero California, and Alma De-Mexico provide services to domestic markets as well as international destinations.

As shown in Table 12.1, not only have these airlines been successful at acquiring market share (it is expected that LCCs will increase their global market from 23 percent today to 34 percent by 2030),[5] but they have also been profitable. Table 12.2 below provides a brief synopsis of the financial situation for some of the world's major LCCs. As the table shows, the founding airline, Southwest, is still the most largest airline in the low-cost category in terms of operating revenues and passengers enplaned. Southwest is followed by European discounter Ryanair, Air Berlin, easyJet, and fledging Brazilian LCC Gol.

Table 12.1   Low-cost Carrier Capacity Share of Total Seats: 2001–2011

	World	North America	Europe	South Asia	South East Asia	Central & South America
2001	8.0%	17.6%	4.9%	-	3.3%	1.1%
2002	9.5%	19.8%	8.2%	-	4.6%	1.9%
2003	11.4%	21.9%	13.6%	0.1%	4.0%	2.4%
2004	13.5%	24.0%	17.7%	0.9%	9.8%	3.1%
2005	14.9%	24.9%	20.5%	5.8%	13.6%	4.3%
2006	16.7%	26.0%	23.7%	22.7%	18.1%	5.3%
2007	19.3%	27.1%	28.4%	38.6%	23.2%	6.8%
2008	21.1%	28.5%	31.4%	42.8%	26.8%	7.3%
2009	21.9%	28.0%	32.1%	47.3%	30.9%	9.3%
2010	23.4%	28.7%	35.0%	50.0%	30.7%	11.6%
2011	24.3%	29.7%	35.9%	50.0%	32.4%	13.5%

Source: Compiled by the author from CAPA-Centre for Aviation data

---

3   Center For Aviation (CAPA), November 28, 2011.
4   Easy Sky Aerolinea de Bajo Costo.
5   Airbus Long Term Passenger Aircraft Forecast 2011–2030.

Table 12.2   **Low-cost Carrier Operating Revenue and Passenger Numbers, 2011**

Rank	Airline	USD millions	Pax (m)
1	Southwest Airlines	$15,658	135.30
2	Ryanair	$5,952	76.00
3	Air Berlin	$5,909	35.30
4	easyJet	$5,552	54.50
5	Gol	$4,515	36.20
6	JetBlue Airways	$4,504	26.40
7	WestJet	$3,116	16.00
8	Jetstar	$2,604	13.80
9	Norwegian	$1,892	15.70
10	Frontier	$1,760	14.90
11	AirAsia Berhad	$1,464	18.00

Source: Airliner Business, May 2012

The aviation industry in South and South-East Asia has witnessed explosive growth in LCCs since 2000. Malaysia's AirAsia has rapidly expanded to become the major carrier in the region and largest by fleet size.[6] AirAsia's rivals, notably Singapore's Tiger Airways and Australia's Qantas-owned Jetstar compete aggressively with each other on Asia to Australia routes as well the domestic Australian market. Their growth is also evident in their aircraft orders which are rapidly shaping the aircraft manufacturing landscape; for example, in 2011, AirAsia, placed a record order for 200 A320neo alongside 77 orders for previous A320 models, making it Airbus's largest client for its single-aisle product line. All three of the big Asian LCC groups—AirAsia, Jetstar, and Tiger—are currently expanding at rates exceeding 20 percent per annum as LCCs in Europe and North America are beginning to experience a more mature market. As the region's economy is expected to significantly outpace the world's average growth (at 4.7 percent per year for the next 20 years), more than half of the world's air traffic growth will be driven by travel to, from, or within the Asia-Pacific region.[7] Over the next 20 years, 78 percent of the demand for new airplanes will come from outside North America, with about 34 percent of deliveries going to the Asia-Pacific region. Intra-Asia, LCCs are expected to have as much as 60 percent market share by 2030.[8]

While many LCCs have been successful, the list of failed LCCs is long. In North America, four major legacy carriers launched LCC brands between 1998 and 2004 and all four failed within five years: MetroJet by US Airways (1998–2001); Tango by Air Canada (2001–2003);

---

6   AirAsia's subsidiaries include Thai AirAsia, AirAsia X, and Indonesia AirAsia.
7   Boeing Current Market Outlook 2011–2030.
8   Airbus Global Market Forecast 2011–2030.

Song by Delta Air Lines (2003–2006); and Ted by United Airlines (2004–2009) (Shannon, 2011). Moreover, not only have the majors been unable to replicate the success of LCCs, but also independent LCCs like Reno Air, People Express, and Independence Air all failed. In Europe, Snowflake, a low-cost airline subsidiary of Scandinavian Airlines System (SAS), ceased operations partially in 2004. KLM divested itself of Buzz while British Airways did the same with its LCC Go. Additionally, the list of failed LCCs in Europe includes Airlib Express, BerlinJet, Fresh Aer, and Goodjet, just to name a few. This brings up the obvious question of why some LCCs were successful while others were failures.

Part of the problem stems from the fact that many airlines (particularly major carriers) that have tried to imitate the low-cost model never truly adopted it. That is, they tried to establish the LCC image without changing their existing cost structure; this inevitably led to their failure. Additionally, regardless of the type of carrier, any new airline entrant will face tremendous competition from the incumbent airlines, making any new start-up airline's probability of success slim. While every carrier is unique, there are certain common characteristics that have enabled some LCCs to succeed where others have failed. In general, these characteristics allowed the successful carriers to maintain a low-cost structure, and the specific characteristics will be discussed in greater detail in the following sections.

CHARACTERISTICS OF LOW-COST CARRIERS

The concept of LCCs isn't a new one. About 41 years ago, Herb Kelleher and Rollin King founded a new airline, Southwest, based on a business model that combined low cost and high productivity. In 1977 Laker Airways was one of the first long-haul, no frills, airlines to adopt the low-cost business model, with flights between London and New York (Flight International, 1961). In North America, People Express was a no-frills airline that operated from 1981 to 1987 (Frank, 2006). In spite of the fact that LCCs operate all over the globe in different environments, they all exhibit a few basic general characteristics. In Europe, Ryanair started in 1985 and has become one of the oldest and most successful low-cost airlines. By charging very low ticket prices, Ryanair has opened the continent to air travel just as Southwest did for many years in the US.

> "Most people think of us as this flamboyant airline ... but we're really very conservative from the fiscal standpoint ... We never got dangerously in debt and never let costs get out of hand."
>
> Herb Kelleher the co-founder, Chairman Emeritus

Lower labor costs and higher labor productivity

Since labor costs are one of the largest costs for any airline, it is imperative for LCCs to keep their labor costs under control and/or increase labor productivity. While many LCC airlines simply pay lower than industry average wages, Southwest has proved that LCCs can pay competitive rates, yet still have low labor costs per hour of productivity. By having high employee productivity, Southwest has been able to pay high salaries yet remain very competitive on a per block hour basis. For example, Southwest pilots are some of the highest paid in the industry, yet they fly many more hours per month than their counterparts at other airlines and they also pitch in to help do things such as clean the aircraft or carry bags.

While employee productivity tells some of the story, the importance of low labor costs cannot be overstated. Any airline with high labor rates and moderate to low productivity will ultimately be unsuccessful. Both Air Canada and Delta launched LCCs that utilized employees from the mainline carrier. In essence, the new carrier was supposed to be low cost, but had the same employee group as the "high-cost" airline. Since these new LCCs were uncompetitive on the labor front and had no new productivity increases, it is not surprising that they disappeared shortly after their start-up.

Figure 12.1 provides a comparison of labor costs per block hour for various US airlines, which include both legacy carriers and LCCs. While LCCs AirTran and Frontier are the two leading carriers in terms of crew cost per block hour, Southwest Airlines is in the middle of the pack.[9] The relatively low labor costs for the first two airlines represent a significant competitive advantage, while Southwest Airlines, a more mature airline, must focus on employee efficiency to help offset the relatively higher crew costs per block hour. Not surprisingly, American Airlines had the highest crew costs in 2011, which coincides with the year that the airline entered bankruptcy protection.

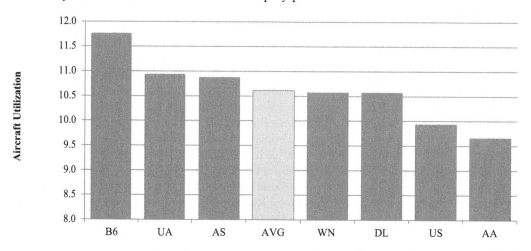

Figure 12.1   Crew Costs per Block Hour, 2011

Source: Compiled by authors from OAG, Form 41 data

Lower ticket distribution costs

Ticket distribution costs are another major area where the entire airline industry is attempting to reduce costs. The initial step airlines took to reduce ticket distribution costs was to cut travel agent commissions. Then, through the Internet, airlines moved to electronic ticketing and pushed ticket sales through their online websites. However, most

---

9   The appendix contains the two letter airline identification codes for various airlines.

of the major carriers still rely on GDS (Global Distribution Systems), such as Sabre and Worldspan, to distribute their tickets worldwide. While GDS provide an airline with a global reach, it costs close to $13 to distribute a ticket through a GDS as opposed to a mere few dollars through Internet e-ticketing (Ionides and O'Toole, 2005).

> Low-cost airlines generally have many characteristics that differentiate them from the legacy airlines. These characteristics encompass: a simple fare structure, one aircraft type, direct ticket sales, flying to cheaper and less congested airports, no frequent flier programs, no free food and beverages, and fast aircraft turnaround times.

LCCs have been far more successful at selling tickets through their online websites than the major carriers. For example, Southwest books about 60 percent of its revenues through its website, while JetBlue brings in 75 percent of their revenues through jetblue.com (Field and Pilling, 2005). Compare those percentages to Continental (25 percent) and Delta (28 percent) and there are clear cost savings achieved by the LCCs in ticket distribution costs (Field and Pilling, 2005; Ionides and O'Toole, 2005).

One successful strategy used by LCCs is to initially align themselves with multiple GDS providers and then, as their brand becomes stronger, to slowly end their agreements with the GDS providers. This enables the carrier to have a wide distribution network initially, and then narrow its distribution (and lower its costs) as it pushes ticket sales toward its website. Ryanair in Europe was very successful with this strategy (Field and Pilling, 2005). On the other hand, LCC Independence Air decided to begin operations without any GDS distribution, but its website had no brand awareness, and ultimately the airline went out of business (Field and Pilling, 2005).

It is worth noting that a universal push toward online ticket distribution by LCCs is not typical for all LCCs. Since the Internet is not as readily available to consumers in South America and Asia, airlines in these areas still need to rely on travel agencies for ticket distribution.[10] In general though, LCCs have been the catalyst in the e-ticketing/website distribution world, and this has provided them with a significant ticket distribution cost advantage. Finally, the International Air Transport Association (IATA) had announced earlier that it would not accept paper tickets as of 2008. Therefore, issuing e-tickets is one of the most important tasks of not only LCCs but also legacy airlines.

No frills service

Historically, one of the clearest examples to consumers of the difference between LCCs and legacy airlines was a "no-frills" service. In the US on a legacy carrier's flight, passengers received a complimentary hot meal with an extensive beverage service whereas on a Southwest flight a passenger would receive peanuts and a soda. However, with the cost-cutting measures implemented by legacy carriers, all economy-class service in North America has turned into "no-frills." In Europe, LCCs have gone one step further where everything, including beverages, is on a buy-on-board basis. Therefore, the in-flight food

---

10   According to the International Air Transport Association (IATA) only 13 percent of airlines in the Middle East issued e-tickets in 2006.

service that used to easily distinguish low-cost airlines from "full-service" carriers is no longer applicable.

However, no-frills service does not just pertain to in-flight service. Many LCCs also do not have frequent flier programs or expensive business lounges; these amenities are not offered in order to cut costs. Another cost-cutting measure that has recently been implemented by LCCs is the restriction of luggage allowances. Particularly in Europe, LCCs have strict rules concerning luggage allowance weights per passenger; this conserves fuel and generates extra marginal revenue.

The underlying premise behind the LCCs' no-frills service strategy is ultimately a "pay as you go" approach, where the ticket price entitles you to just a seat on the aircraft. As a result of this strategy, LCCs can offer attractive airfares. While these service cuts may seem minimal, when they are compounded over the number of flights, it can actually make the difference between profitability or loss.

Common fleet type

Another major characteristic of successful LCCs is the use of a common fleet type. Southwest Airlines was the pioneer of this strategy building its entire fleet around the Boeing 737. A single fleet type provides many advantages for an airline; these include a reduction in spare parts inventories, reduced flight crew training expenses, and increased operational flexibility. Additionally, bulk purchase discounts from suppliers (including aircraft manufacturers) can be negotiated when using a single fleet type. However, it is economies of scale that are the most important cost reduction elements underlying the common fleet type strategy. That is, the airline is required to spend fixed fleet costs only once. For example, all the specialized equipment that might be needed for a 737 only needs to be purchased once.

In addition to savings from economies of scale, a single fleet provides increased operational flexibility. In the event of irregular operations, a single fleet type makes it easier to find a replacement aircraft or usually, and more importantly, replacement flight crew. Since airlines usually have a reserve pilot pool for each fleet type, limiting the number of fleet types limits the number of reserve pilots the airline requires.

Using one fleet can also have advantages and disadvantages concerning markets served. Depending on aircraft choice, the aircraft used by the airline may not be the optimal aircraft for some markets. Thus, if the aircraft has a relatively short range, many intercontinental markets will not be feasible. AirTran for example had this problem with the 717s and therefore had to purchase another fleet of 737s[11]. The flip side is that a single fleet contains aircraft that have the same pilot requirements and maintenance standards. For LCCs, the two most widely used generic aircraft types are the 737NG and the A32X. Both these aircraft types enable a carrier to have planes with as few as 120 seats and as many as 200 seats. This enables them to be able to change aircraft seating capacity to better meet demand on any given day.

While a single common fleet has been the LCC standard, both JetBlue and easyJet bucked the trend by creating fleets with two aircraft types. In doing so, both airlines felt that the economies of scale benefits on their initial fleet had reached a threshold. This

---

11   After being acquired by Southwest, the 88 Boeing 717s formerly operated by AirTran were agreed to be sub-leased to Delta in May 2012, as part of Delta's plan to remove older DC-9-50s and smaller regional jets from its fleet.

occurs when the benefits of the first large fleet type are outweighed by the benefits of a second large fleet type. Based on these examples, it appears that the minimum number of aircraft needed to achieve full economies of scale benefits is probably slightly under 100.

Table 12.3 clearly shows that LCCs in North America have less diverse fleets than the legacy carriers. Part of the reason for legacy carrier's more complex fleets is that international flying requires larger aircraft, and legacy carriers have undergone more mergers, thereby combining fleets. LCCs are also generally younger companies, and have emphasized a single fleet strategy. And, while Southwest Airlines is shown as having two fleets (B737CL and B737NG), the airline still operates only one aircraft type since these are two different generations of the same aircraft.

Table 12.3   Aircraft Fleets for Major North American Operators

North American Fleets as of February 2012		
Carrier	Fleet Types	# of Fleet Types
Low Cost		
AirTran †	B717-200, B737NG	2
Frontier	A32X	1
JetBlue	A32X, ERJ 190 LR/AR	2
Southwest	B737-CL/NG	1
Spirit	A32X	1
Westjet	B737NG	1
Legacy		
AeroMexico	737 NG, 777, 767	3
Air Canada	EMB 175/190, A32X, B767, A330-300, B777	5
Alaska	B737CL/NG	2
American	MD80, B737NG, B757, B767, B777, A300-600	6
Delta	A32X, A330, DC-9, MD80, MD90, B737NG, B747, B757, B767, B777	10
United	A32X, B737NG, B747, B757, B767, B777	6
US Airways	A32X, B737CL, B757, B767, A330, ERJ 190 LR/AR	7

Source: Air Transport Intelligence
† AirTran Airways was acquired by Southwest Airlines

Regardless of aircraft type, low-cost operators configure their aircraft in a high-density all-economy configuration. In some cases closets and washrooms are removed in order to squeeze more passengers onto the flight. The seats on easyJet and Ryanair also are all non-reclining in order to accommodate more passengers. Obviously, since every flight is largely a fixed cost (once it has been launched), the more people on board, the more revenue the airline can obtain. This in turn enables the airline to offer a few seats at highly discounted prices. In the North American market, the LCCs have pursued different marketing practices, as both JetBlue and Westjet have removed seats from their aircraft to provide additional legroom in an effort to attract more business clientele. Also, very few discount carriers operate any sort of a premium cabin, believing that additional economy revenue would exceed any premium cabin revenue. Tables 12.4 and 12.5 provide seating capacities of airlines based in North America. All the LCCs put more seats in their aircraft than the industry average, thereby spreading costs per seat over a greater number of seats. For example, a Southwest 737-300 has 7 percent more seats than the industry average, while a jetBlue A320 has roughly 6 percent more seats than the North American industry average.

Table 12.4   Seating Capacity of North American Boeing Operators

Aircraft Type	WN	FL	CO	UA	US	AS	AA	DL	Industry Average
717-200	-	117	-	-	-	-	-	-	117
737-300	137	-	-		126	-	-	124	129
737-400	-	-	-	-	144	144	-	-	144
737-500	122	-	114	104	-	-	-	-	113
737-600			-	-	-	-	-	-	119
737-700	137	137	124	-	-	124	-	-	130
737-800	-	-	157	-	-	160	148	160	157
737-900	-	-	173	-	-	172	-	-	173
747-400				374	-	-		393	384
757-200	-	-	175	182	176	-	188	182	181
757-300	-	-	216	-		-	-	224	220
767-200	-	-	174	-	204	-	168	-	182
767-300				244	-	-	225	224	223
767-400	-	-	235	-	-	-	-	246	241
777-200	-	-	276	269	-	-	243	273	268

Source: Compiled by the Authors using seatguru.com

Point-to-point service

Since deregulation, the legacy carriers have adopted a hub-and-spoke route structure; this means that all spoke flights come into one hub airport and this airport provides the connecting feed for the spoke flights that depart shortly thereafter. While the hub-and-spoke system has been effective for legacy carriers in providing a large number of city pair connections, a hub is also an extremely expensive operation. Hubs usually have peaks to

Table 12.5   Seating Capacity of North American Airbus and Embraer Operators

Aircraft Type	B6	F9	UA	US	AC	DL	NK	CO	Industry Average
A318	-	120	-	-	-	-	-	-	120
A319	-	138	120	124	120	126	145	-	129
A320	150	164	138	150	146	148	174	-	153
A321	-	-	-	183	174	-	214	-	190
A330-200	-	-	-	258	-	243	-	-	251
A330-300	-	-	-	254	265	298	-	-	272
EMB 120	-	-	30	-	-	31	-	-	31
ERJ 135	-	37	-	-	-	-	-	-	37
ERJ 145	-	-	50	50	-	50	-	50	50
ERJ 170	-	-	70	76	-	70	-	-	72
ERJ 175	-	-	-	86	73	76	-	-	78
ERJ 190	100	99	-	99	93	-	-	-	98

Source: Compiled by the Authors using seatguru.com

minimize passenger transfer time, but this also means that the hub will have downtimes where it is not fully utilizing many of its facilities. This is extremely costly as employees may be idle for extended periods of time and assets (such as gates and ground equipment) may be left unused. Since employees and gates must be paid over the full working day (and not only when flights are arriving and/or departing) this represents some level of inefficiency. Moreover, the flip side of this is equally true, and that is that the hub carrier must also have adequate staffing for its peak number of flights, thereby further increasing the cost of unproductive time.

> Hub-and-spoke is taking one airport as a hub, and connecting all other airports (spoke) through the hub. For example, with a six-airport network, one as the hub, we can have fully connectivity with only five flights.

Beside the undoubted revenue advantage of many city pair choices and the ability to increase load factors by consolidating passengers at the hub, one of the main cost benefits underlying a hub is the ability to realize certain economies of scale. Consolidating operations in one place reduces fixed overhead costs such as required reserve labor pools, maintenance operations, and terminal-related expenses. The problem is that the peak flight scheduling necessary for passenger convenience also means that these economies of scale are not always achieved. Moreover, once the numbers of flights reaches a critical level, diseconomies of scale will set in. That is, any additional flight added will actually increase average costs instead of reducing them. The reason for this is the fact that the added congestion increases costs. For example, as the airport becomes busier and busier, aircraft have to wait longer to land and takeoff, and this is reflected in higher fuel and labor costs. Arguably, many of the hubs in the US have surpassed the critical inflection point where economies of scale have turned into diseconomies of scale. Unfortunately, since there are many different and interactive decision makers, it is nearly impossible to determine an optimized flight level for a hub airport.

The prevalence of diseconomies of scale mentioned above is one of the major reasons why LCCs typically operate a point-to-point or origin and destination (O&D) route structure. Under a point-to-point route structure, the airline will operate a more spread out route network and typically will offer non-stop flights between city-pairs. Under this route structure, airlines will still operate bases where economies of scale are realized, but will not have any peak level of flights. This allows the airline to continually use airport facilities and more evenly use employee services. This increased utilization of airport assets allows a point-to-point airline to operate more flights with less facilities and personnel, and this ultimately reduces costs. Southwest Airlines has sizeable operations at many airports across the US, but these bases have not grown to the size of the legacy carriers' hubs. Also, Southwest Airlines generally operates at least eight to ten flights out of any city to experience some level of economies of scale, spread fixed costs over a greater number of flights, and increase the frequency of flight choice for the passengers.

In North America, while airlines like Southwest generally operate a point-to-point route structure, a good number of passengers still connect on Southwest flights through some of Southwest's larger bases. In Europe by contrast, LCCs typically do not allow any connecting flights, thereby relying solely on O&D demand for all its flights. By not connecting passengers, the airline does not have to worry about transferring luggage between aircraft and compensating passengers for misconnections and this further reduces operating costs. The European LCC model generally bases a few aircraft at one airport and then flies to various destinations from there. This enables the carrier to receive some of the benefits of economies of scale at these bases. Thus, European LCCs operate a base-and-spoke network with no connections or synergies with the airline's other bases. Both Ryanair and easyJet have been successful using this strategy.

Use of secondary airports

Similar in nature to using a point-to-point route structure, the use of secondary airports is another characteristic of LCCs. A secondary airport is usually under-utilized and further from a city center like Malmo Airport compared to Copenhagen. Congested primary airports usually mean more time on the ground and higher airport fees so LCCs avoid them where possible. For example, Southwest does sometimes fly into busy airports, such as Los Angeles, but it has avoided flying into Chicago O'Hare (instead serving Midway airport). While Southwest was the pioneer of utilizing secondary airports (saving costs by being able to turn aircraft around quicker), Ryanair has been the most aggressive in serving secondary airports (Doganis, 2001). Examples of Ryanair using secondary airports include Hahn airport for Frankfurt, Charleroi airport for Brussels, Beauvais airport for Paris, and Weeze airport for Dusseldorf. Flying to secondary airports allows lower landing fees, less congestion, and quicker aircraft turnaround. Hence, at these secondary airports low-cost airlines are able to operate more efficiently and more cost-effectively.

The benefits an airport receives from an LCC starting service are immense; therefore, airports have become very aggressive in attracting new LCC service. The classic example of this is an agreement Charleroi airport made with Ryanair[12] whereby Ryanair received a reduction in airport charges of around €2 per passenger, a reduction in ground handling charges to €1 per passenger, one-time incentive bonuses for starting new routes, and marketing promotion (Smith, 2005). Additionally, airports have begun to design airports that cater specifically to the needs of a LCC (that is, low operating costs). Marseille Airport has designed a dedicated low-cost terminal with cheaper passenger service charges, while Geneva airport has opened a terminal for "simplified aviation" (Buyck, 2005). These actions obviously lead to lower overall costs and these can in turn be passed to passengers in the form of lower fares

Higher aircraft utilization

Another central focus of success for low-cost airlines is a high level of aircraft utilization. Since an aircraft is not earning money while sitting on the ground, the more an aircraft is flying, the more passengers the airline can carry. There are two central ways an airline can increase its daily average aircraft utilization: turn the aircraft around quicker or fly longer routes. Figure 12.2 provides a comparison of average daily block hours per aircraft between LCCs and network carriers in the US for domestic operations only.[13] JetBlue, United, and Alaska Airlines all enjoyed higher block hours than the average (AVG).

Since high aircraft utilization rates are one of the major strategies for any LCC, it is not surprising that the top airline in terms of aircraft utilization is JetBlue. Airlines like JetBlue are able to achieve high aircraft utilization rates since they focus on having quick turnarounds. By having free seating, passengers tend to enplane and deplane faster, and

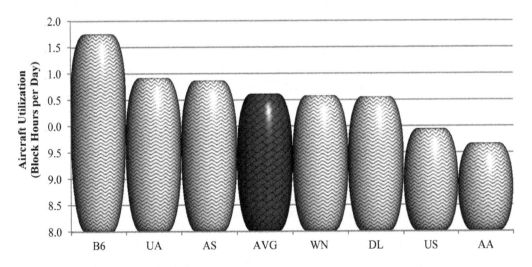

Figure 12.2   Aircraft Utilization (Block Hours per Day) in 2011

Source: Compiled by authors using Back Aviation Form 41

---

12   Ryanair launched a new base of operation in Charleroi Airport in 2001.
13   The legacy carriers' aircraft utilization statistics are slightly distorted since their international flights would increase aircraft utilization.

by having considerably fewer hub-and-spoke operations, the ground baggage handling situation is less complex. Additionally, the use of secondary, less-congested airports allow the airlines to schedule more flights since there is less delay in the schedule. These efficiencies enable LCCs to operate more flights, thereby providing more revenue. The downside to increased utilization is that maintenance costs will increase since the aircraft are being flown more often, yet this is a tradeoff most airlines are willing to make.

The other method of increasing aircraft utilization is by flying longer routes. Figure 12.3 provides a comparison of the average domestic stage length for major US carriers. JetBlue accomplishes high aircraft utilization rates by flying transcontinental and Florida flights from its New York JFK base, while Southwest obtains high aircraft utilization by operating short flights with quick turnarounds (like intra-Texas flying). Since a Southwest aircraft will be landing and departing more frequently in a day than a JetBlue aircraft, this highlights Southwest's tremendous efficiency in its ground handling operations. Only recently has Southwest ventured into the transcontinental flying market in an effort to increase its average stage length and increase aircraft utilization.

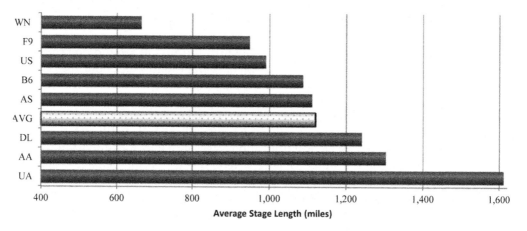

Figure 12.3   Average Aircraft Stage Length in Miles, 2011

Source: Compiled by authors using Back Aviation Form41

Figure 12.4 displays the results of a correlation analysis between operating Cost per Available Seat Mile (CASM) and aircraft utilization. CASM is expressed in cents to operate each seat mile offered, and is calculated by dividing operating costs by Available Seat Miles (ASM). The regression line is plotted in the figure and it is evident that a strong negative relationship exists between increased aircraft utilization and reduced operating costs. This trend is confirmed statistically by having a significant R-squared value of 0.454. Airlines that lie below the trend line have a lower operating CASM for their level of aircraft utilization than the industry trend. Not surprisingly, LCCs such as JetBlue and Southwest lie at the bottom-end of the trend line, while legacy carriers are generally at the top. More specifically, Figure 12.4 shows clearly that legacy airlines need to reduce their operating CASM for their level of aircraft utilization as they lie significantly above the trend line.

The number of aircraft departures per day is also directly related to aircraft utilization; that is, the more departures per day per aircraft, the higher the utilization and the lower the CASM. Figure 12.5 shows this negative relationship between operating CASM and

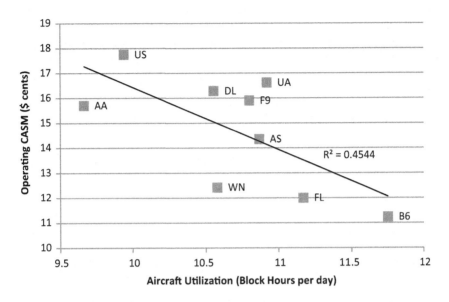

Figure 12.4   Correlations between Operating Costs per Available Seat Mile and Aircraft Utilization, 2011

Source: Compiled by authors using Form 41 data

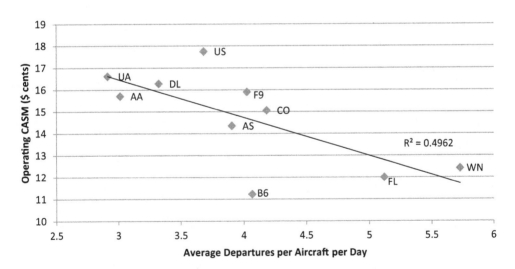

Figure 12.5   Correlation between Operating Cost per Available Seat Mile and Departures per Aircraft, 2011

Source: Compiled by authors using Form 41 data

departures per aircraft per day. This relationship has a significant R-squared value of 0.496. The majority of the legacy carriers are all grouped in the upper-left quadrant above the trend line. The major outlier in this analysis is JetBlue. However, this can be explained by the fact that JetBlue has pursued a strategy of lowering operating CASM by flying fewer departures per aircraft but with much longer average stage lengths.

A third correlation to operating CASM is fuel efficiency. Fuel efficiency is the number of seat miles flown with one gallon of fuel. As expected, the general relationship between operating CASM and fuel efficiency is downward sloping, since the more fuel efficient an airline is, the lower its operating costs. Figure 12.6 displays the relationship, which contains an R-square value of roughly 0.35. Based on the trend line, all the LCCs outperform the market in terms of fuel efficiency since their observations all fall below the trend line.

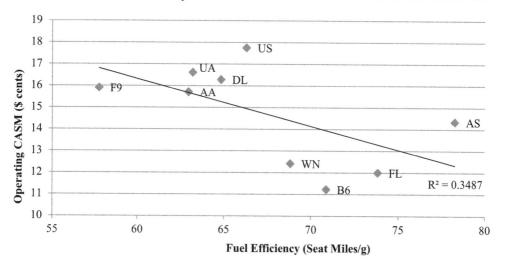

Figure 12.6   Correlation between Operating Cost per Available Seat Mile and Fuel Efficiency, 2011

Source: Compiled by authors from Form41 data

COST STRUCTURE COMPARISON

In the modern competitive aviation world, the most successful airlines are those with the lowest cost structure. Southwest, for example, has a long record of continually maintaining low operating costs, so that it can offer attractive air fares. Figures 12.7 through 12.10 provide a comparison of direct operating CASM for four different aircraft that are commonly used throughout the US by legacy and LCCs. The direct operating costs of the aircraft include fuel, labor costs of flying, all maintenance costs, and all ownership or leasing costs.[14] Figure 12.7 provides a comparison of airlines using the older 737-300 aircraft. Southwest is the largest operator of 737-300 aircraft with 163 in active service; the only other major carrier with 737-300 still in active service is US Airways with seven aircraft.[15] Delta retired their 737-300 fleet, with the last aircraft leaving the fleet in the second quarter of 2006. United has also used their 737-300 aircraft in its Shuttle subsidiary operation.

Figure 12.8 provides a similar comparison of CASM for airlines using the newer Boeing 737-700 aircraft and again Southwest Airlines is the cost leader. It is interesting to note that the CASM for the -700 series is at least two cents less than the CASM for the -300 series, which is a similarly-sized aircraft. This could be the result of two factors: longer flights

---

14   All associated costs and ASM are for domestic US operations only.
15   Vision Airlines, a low-cost charter airline operates one 737-300 and Sky King, which charters services to professional sports teams and long-term public charters, operates two 737-200s.

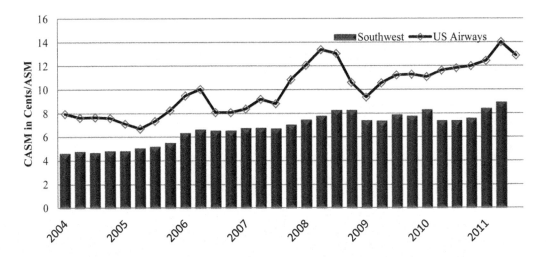

Figure 12.7   Cost per Available Seat Mile for USA 737-300 Operators (2004–2011)

Source: Compiled by authors using Form41 data

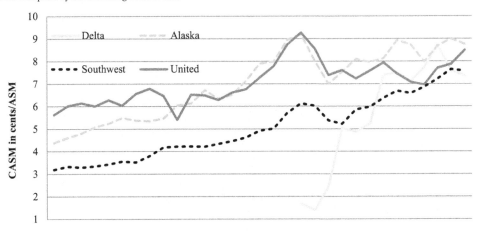

Figure 12.8   Cost per Available Seat Mile for USA 737-700 Operators (2004–2011)

Source: Compiled by the authors from Form 41*

* Note that some data for Southwest, Continental, and Alaska was unavailable

and/or increased efficiency. Since the -700 aircraft have greater range, they typically fly longer flights, thereby increasing the ASMs while keeping flight costs relatively fixed. Also, the new generation of aircraft is much more fuel efficient so that fuel costs are decreased, resulting in an overall decrease in CASM.

Figure 12.9 shows similar CASM cost comparisons for A319 operators. Frontier is the primary low-cost operator of the A319 in North America but it has a cost structure that is similar to United, America West, and Northwest. This is probably due to the fact that Frontier started as a short-haul legacy carrier and is yet to make the complete transition to an LCC. Only US Airways has a cost structure that is well above the rest of the industry.

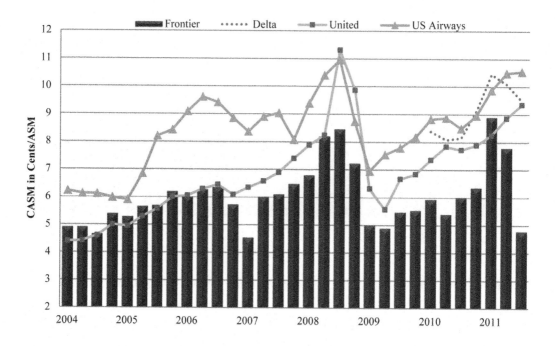

Figure 12.9   Cost per Available Seat Mile for USA Airbus 319 Operators (2004–2011)

Source: Compiled by the authors from Back Aviation Form 41

The Airbus 319 also enables comparison to the Boeing 737-700 since they are both modern aircraft with similar seating capacity. Comparisons are complicated since the aircraft are operated by different airlines, but it appears that the general trend is that the Airbus 319 initially has a lower CASM, but the costs for the Boeing 737-700 have not increased as quickly as the Airbus 319. This could be a result of the introduction of blended winglets on the majority of Boeing 737-700 aircraft.

The final CASM cost comparison (Figure 12.10) is for the larger A320 aircraft; this shows LCC JetBlue as the cost leader. The airline has on average been a whole cent lower than the competition, and this could easily be the difference between profitability and non-profitability in any given quarter. This highlights the fact that the scheduling and operating procedures for individual carriers can have a significant impact on the CASM for any given aircraft. United's statistics for this figure include numbers for both mainline A320 flights and Ted[16] flights so it appears (from the figure) that the introduction of Ted has not impacted United's CASM figures dramatically.

With the exceptions noted above, the general trend of comparisons by individual aircraft type show that LCCs have been able to achieve lower operating costs. This trend is further reinforced in Figure 12.11. Here the average domestic CASM is compared between LCCs and legacy carriers since 2000. For this figure, the LCCs include US airlines such as Southwest, JetBlue, America West, and AirTran, while the legacy carriers include American, United, Delta, Continental, US Airways, and Northwest.

---

16   Ted was a United LCC subsidiary from 2004 to 2009.

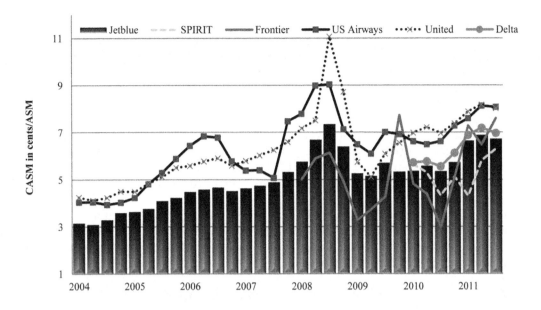

**Figure 12.10   Cost per Available Seat Mile for USA Airbus A320-200
Operators (2004–2011)**

Source: Compiled by the authors using Back Aviation Form 41

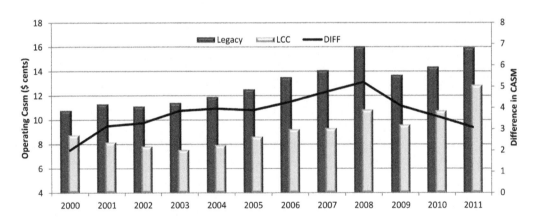

**Figure 12.11   Comparison of Low-cost Carriers' and Legacy Carriers'
Operating Cost per Available Seat Mile**

Source: Compiled by the authors using Back Aviation Form 41

As the figure clearly shows, the cost-structure difference between LCCs and legacy carriers was sizeable. Fortunately for them, the legacy carriers were still able to obtain high revenues during a boom time in the economy. However, following the tragic events of September 11, 2001 and the subsequent dramatic drop in air traffic, the LCCs responded more quickly to the new economic reality; that is, they adjusted their cost structures while legacy carriers struggled to adjust. Since then the difference in CASMs has remained somewhat constant, with LCCs on average having a 20 percent cost advantage over legacy carriers. While the airline industry's profitability as a whole is highly correlated to the economy's strength, this comparison shows that legacy carriers definitely experience greater volatility in their profits and/or losses than LCCs. The figure also shows that rather small changes in average costs can produce dramatic returns in revenues.

Table 12.6 provides a common-size income statement[17] that compares the LCCs to the legacy carriers from 2007–2010. The table shows that LCCs have financially outperformed legacy carriers since 2007. LCCs have been much more successful in terms of operating profit and profit margin than the legacy carriers and one of the main reasons for this has been their lower cost structure. While the LCCs' total aircraft operating expenses consume a larger percentage of total operating expenses, they have had comparably lower direct operating costs.

Table 12.6 also shows that LCC carriers acquire a greater proportion of their revenues from passengers than other sources. This is largely because many LCCs do not carry any additional cargo on their flights (which could cause aircraft turnarounds to be slower) while legacy carriers do carry cargo.

Table 12.6   Common-size Income Statement Comparison between Low-cost and Legacy Carriers

	2007		2008		2009		2010	
	LCC	Legacy	LCC	Legacy	LCC	Legacy	LCC	Legacy
Revenues								
Passenger revenues	92.4%	72.5%	92.0%	70.5%	90.6%	56.1%	90.6%	78.1%
Other revenue	7.6%	27.5%	8.0%	29.5%	9.4%	43.9%	9.4%	21.9%
Total operating revenues	100.0%	100.0%	100.0%	100.0%	100.0%	100.0%	100.0%	100.0%
Expenses								
Labor	31.9%	23.8%	27.8%	20.1%	31.6%	25.4%	30.2%	24.8%
Fuel	30.2%	24.9%	35.4%	30.0%	28.7%	21.9%	30.9%	24.2%
Flight equipment and maintenance	8.9%	8.3%	8.4%	7.1%	9.5%	8.8%	9.1%	8.6%
Transport-related expense	1.1%	18.3%	0.9%	18.0%	0.9%	18.4%	1.5%	18.1%
Other expenses	27.9%	24.6%	27.5%	24.8%	29.3%	25.5%	28.3%	24.3%
Total operating expense	100.0%	100.0%	100.0%	100.0%	100.0%	100.0%	100.0%	100.0%

Source: Compiled by the authors from Form 41

Comparing personnel costs, LCCs spend a reduced portion of total operating expense on personnel than legacy carriers. This could be caused by either of two factors: a quantity or a price effect. Under the quantity effect, legacy carriers would have the same wage levels as LCCs, but simply have more staff (proportionately) for their flights. The price effect would be the opposite; that is, where the proportionate staffing levels are the same,

---

17   A common-size income statement essentially displays the income statement merely in percentage terms. All revenue categories are stated as a percentage of total revenue. Conversely, all expense categories are stated as a percentage of total operating expenses. The benefit of a common-size income statement is that it enables a better comparison of cost structures and helps recognize where money is being spent.

but LCCs simply pay their employees less. In all likelihood, the difference in personnel expense proportions is probably caused by a mixture of the two effects.

Finally, LCCs spend a greater proportion of their operating expenses on fuel and maintenance as compared to legacy carriers. A possible explanation for these higher cost proportions is that Southwest and AirTran have shorter average stage lengths. Since takeoff consumes the largest amount of fuel for any stage of flight, and most maintenance programs revolve around the number of takeoffs, the more departures per aircraft there are, the higher the expected proportion of fuel and maintenance costs. This theory is supported in Figure 12.12. The figure shows the average number of departures per aircraft in 2006 for major US airlines. Southwest and AirTran lead this list, indicating that this could be the driving factors behind the slight rise in the LCC's percentage of fuel and maintenance costs compared to the legacy carriers.

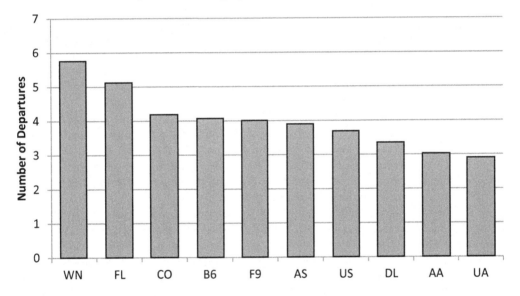

Figure 12.12   Average Departures per Aircraft per Day, 2011

Source: Compiled by authors using Back Aviation Form 41

INCUMBENT CARRIERS' RESPONSE TO LOW-COST CARRIERS

Legacy airlines are nibbling at amenities that passengers have long taken for granted, and are charging passengers for services or those who want to cancel or change their tickets. They have also implemented two major strategies to combat LCCs: creation of their own LCCs and unilateral cost cutting.

Low-cost carrier creation

Most major carriers in both North America and Europe have experimented in creating their own LCCs, but with little success. In the early 1990s, Continental launched its own

LCC, Continental Lite, which was configured with all economy seats, served no food, and operated flights under 2.5 hours (Bethune and Huler, 1998). Continental committed over 100 aircraft to the operation, but ceased operations shortly thereafter (Bethune and Huler, 1998). "There was only one problem: People said, 'I don't want to buy that. That is not what I want'" (Bethune and Huler, 1998). Since Continental Lite did not offer a competitive product, passengers did not choose the airline. Additionally, since Continental's name was closely associated with the new carrier, brand confusion occurred and Continental mainline also lost passengers (Bethune and Huler, 1998). Continental's experiment with Lite was one of the first of a long list of failed attempts by legacy carriers to develop their own LCCs.

Delta Air Lines has had two attempts at creating a LCC. Delta Express began operations in 1996 utilizing 737-200's in a high-density layout (O'Toole, 1999). Delta Express was based out of Orlando International and operated flights principally along the north-east corridor. It was created to compete with Southwest, Air Tran, and eventually JetBlue (O'Toole, 1999). A 1999 study put Delta Express's CASM at 10.86 cents, which was considerably lower than Delta mainline, but still well above Southwest's CASM of 7.75 cents (O'Toole, 1999). Following the tragic events of September 11, 2001, Delta Express's operations were significantly reduced as leisure travel declined sharply (Johnston, 2001). Express ceased operations in November of 2003, shortly before Delta started its second LCC, Song. Song launched services in April 2003 amid much fanfare. The carrier attempted to obtain a hip, style-conscious brand by operating larger 757-200's in an all-economy configuration with leather seats and an excellent in-flight entertainment system. The airline largely served leisure routes from Florida and transcontinental flights from the Northeast. Delta shutdown the Song operation in April 2006, only three years after its start, as the carrier attempted to restructure under bankruptcy protection.

US Airways launched MetroJet in 1998 to respond to low-cost competition from Southwest and Delta Express (Henry, 1998). The airline's base was Baltimore/Washington International (BWI) where Southwest also had a large operation. The airline operated 737-200's in an all-economy configuration and the operation received labor concessions from the unions (Henry, 1998). MetroJet's main focus was on northeast and Florida flights, but MetroJet mostly reduced its own mainline passengers and it faced fierce competition from Southwest. Much of the cost savings achieved by MetroJet were the result of lower pay rates for employees and economies of density achieved through all-economy seating. However, while MetroJet did have lower costs than mainline US Airways, its cost structure was still high. This was as a direct result of the fact that it was imbedded within the mainline carrier. Following the September 11, 2001 terrorist attacks, MetroJet's operations were shut down and much of US Airways presence at BWI was never restored (Johnston, 2001).

United Airlines is another carrier that has operated two LCCs. United's first LCC, Shuttle by United, was an all-economy service utilizing 737-300/500's out of San Francisco International Airport (SFO) (Flint, 1996). During the late 1990s the Shuttle operation appeared quite successful for United, but actual statistics and data for Shuttle were never publicly released (Flint, 1996). Shuttle lasted for several years and provided United with a focus on operational efficiencies, but like other first-generation legacy LCCs, United ended Shuttle in 2001 with the aircraft being folded back into mainline service.

In 2004 United Airlines relaunched its low-cost model in the form of Ted (standing for the last three letters in United), which operates all-economy A320's from Denver. The airline operates to leisure markets such as Orlando and Phoenix replacing mainline service to such cities. All Ted flights were operated by United Airlines crew, since it did not have its own operating certificate. As a result of spiking fuel prices, TED's operations were folded back into the mainline brand on January 6, 2009. Of all the LCCs spun off by legacy carriers, none are still operating. In Canada, Air Canada holds the dubious distinction of operating two LCCs at the same time, neither of which has survived. In 2001, Air Canada launched Tango utilizing A320s that were configured in an all-economy layout. While Tango was operated by Air Canada crews, the airline was totally autonomous from Air Canada mainline flights. This created a problem in that Tango's flights relied solely on O&D demand. Tango was created to respond to LCCs such as Canada 3000. However, shortly after Tango's launch, Canada 3000 fell into bankruptcy.

Air Canada's second discount carrier called Zip was launched in 2002 and operated as a totally separate airline with its own operating certificate, labor force, and management, and code-shared on all its flights with Air Canada. The carrier was based in Calgary to compete heavily against Calgary-based LCC Westjet. The airline operated 737-200's in an all-economy layout, but following Air Canada's entry into bankruptcy in 2004, both Zip and Tango disappeared. In Europe a similar phenomenon occurred with both British Airways and KLM setting up their own LCCs. In 1997, British Airways launched Go Fly using Boeing 737s that were based at London Stansted Airport (Goldsmith, 1998). The airline highlighted its ties with British Airways and posted a profit in 2000. When new management took over British Airways, Go became a liability as it was reducing the airline's core business (Goldsmith, 1998). In a move that is not in the business model for a LCCs in the US, easyJet bought Go in May 2002 (Clark, 2002). Go's network was subsequently integrated into easy Jet's.

KLM launched Buzz in 2000 to compete with other low-cost operators such as easyJet, Ryanair, and Go in the UK market (Dunn, 1999). Unlike most LCCs, Buzz operated two small fleets of Bae 146s and 737-300s. Since both fleets were small, no economies of scale could be realized and operating costs were not "low cost." Additionally, from Buzz's base in London Stansted, the airline flew into busy airports such as Amsterdam and Paris Charles de Gaulle (Dunn, 1999). Similar to the easyJet deal, Ryanair bought Buzz in 2003 for £15.1million, but with Buzz having close to £11million of cash on hand, the true cost of the purchase was much less (BBC News, 2003). Ryanair operated Buzz as a separate unit for a year, but eventually dissolved the operation and had Ryanair take over all operations. Therefore, KLM's experience with a low-cost carrier was short-lived and, like most other legacy carriers, unsuccessful. In 1997, Deutsche Lufthansa set up a low-cost department, which became a separate company under the name Germanwings on 27 October, 2002.

In both Europe and North America, legacy carriers' experiments in creating their own LCCs have largely failed. Part of the problem lies in the operation never truly being low cost, especially with regard to labor costs. The legacy carriers have also been very concerned about the new operator reducing its own core business. As the discussion above clearly shows, the legacy carrier strategy of creating LCCs of their own has been a complete failure (at least to date).

Cost cutting

While under Chapter 11 bankruptcy, United Airlines reached an agreement with pilots saving the airline $1.1 billion. In 2007, Finnair and the Finnish Flight Attendants' Association concluded negotiations by cutting the cost and improving the work productivity of cabin staff. The other major response by legacy carriers has been unilateral cost cutting of mainline service. In the US, on-board food service has been reduced to the point where almost no food is served in economy on any domestic flight. US Airways, by removing in-flight entertainment systems on domestic flights, has saved about $10 million annually in fuel and other costs (Cassels, 2008).

Legacy carriers have also begun to charge for such amenities as pillows, blankets, and in-flight entertainment. While these measures enable legacy airlines to reduce costs, it also introduces the problem of lack of product differentiation. When legacy carriers reduce their service product to equal LCCs, they are largely competing just on cost. And, as we have seen above, competing solely on cost is risky, since LCCs have much lower cost structures than legacy carriers.

> "Passenger traffic among the leading 75 percent low cost carriers in 2011 jumped by nearly 12 percent to 727 million passengers."
>
> *Airline Business*, May 2012

Legacy carriers have also attempted to reduce their cost structure by retiring older aircraft, receiving labor concessions, and reconfiguring aircraft seating layouts. In general, however, the legacy carriers' response to LCCs has largely been ineffective and many carriers have attempted to avoid LCCs by focusing on international flying (where they have definite competitive advantages due to legal restrictions). The future remains uncertain and is discussed in the following section.

THE FUTURE OF LOW-COST CARRIERS

While there have been many failures in the low-cost sector, there has also been tremendous success. And, it must be pointed out that many of the failures did not retain the major characteristics of LCCs and therefore they were not truly cost leaders. The message seems clear: the LCCs that were successful focused solely on reducing costs and being efficient.

The future of LCCs looks promising. On the one hand, the line between a LCC and a legacy carrier is blurring. LCCs such as JetBlue are providing free live television entertainment, while a legacy carrier may have no audio/visual entertainment. Some LCCs have adopted leather seats and increased legroom to provide additional amenities for their passengers. Also, airlines such Southwest have loyalty programs just like the legacy carriers. From a passenger standpoint, the differences are becoming very hard to distinguish. On the other hand, LCCs will face tremendous competition from legacy carriers and other LCCs alike. Since much of LCCs' strategy is based on growth, the LCCs need to develop new markets to continue growing. Additionally, with carriers such as Ryanair and easyJet having over 100 aircraft on order, they may be hard pressed to find routes to fill their planes. It has been calculated that each of these aircraft will have to carry 250,000 passengers a year to breakeven (Turbulent Skies, 2004).

The last domain for legacy carriers has been international flying. The legacy carriers have dumped capacity into international markets, since this is where profits are being made. It is also the only area where legacy carriers do not face fierce competition from LCCs. This is principally a result of international air treaty regulations and the fact that international flying diverges from the low-cost model (since it generally requires different and larger aircraft types). While it is unlikely Southwest will begin flying transatlantic, the carrier has recently decided to build a new $100 million international terminal at Houston Hobby airport which is expected to start service in 2015. The success of such a model is unknown (although the example of People Express shows that it is possible), but there is no doubt that the legacy carriers will protect their turf fiercely. If a carrier can find success using the low-cost model on international flights, then the world is truly the limit for low-cost air transportation. As quoted earlier from Sir Freddie Laker, the twenty-first century could be that of the LCC.

SUMMARY

This chapter introduces the history of LCCs, the various strategies used by low-cost airlines to gain competitive advantages over legacy airlines and how these low-cost models have evolved over time. A low-cost airline generally has many characteristics that differentiate it from the legacy airlines and these characteristics are discussed and analyzed in some detail in the chapter. The growth and evolution of LCCs from a North American phenomenon to a popularized airline business model, particularly in the Asia-Pacific region, is discussed and compared to the more mature markets of Europe and North America. The response of legacy carriers, from creating many failed low-cost competitive brands to a more calculated response of matching some of the offerings of the LCC carriers is also covered in depth. Finally, the future of low-cost airlines and, in particular, their ability (or inability) to penetrate international markets is discussed.

DISCUSSION QUESTIONS

1. You are a manager at a low-cost airline planning to serve Charleroi Airport. Your airline has the option of serving Brussels Airport International Airport, the main airport serving the Belgium, or Charleroi Airport, a secondary airport. Brussels Airport is served by all of the major legacy, international, and LCCs, and is the closest to governmental offices and the city center. Charleroi Airport is located about 29 miles south of central Brussels and is served by a smaller low-fare airline and several European charter airlines. What are some of advantages and disadvantages of each airport for your airline?
2. What are some of the strategic policies of Southwest Airlines and Ryanair that have allowed them to be so successful?
3. What's the biggest difference between low-cost airlines and legacy airlines?
4. From a passenger points of view:
   a   What are the advantages of low-cost airlines?
   b   What are the disadvantages of low-cost airlines?
5. Explain how having a single type of aircraft in a fleet affects an airline's ability to recover from irregular operations such as diversion recovery or an extended airport shutdown due to bad weather.
6. Describe some of the differences of basic no-frills service between LCC carriers in the US and those in Europe.
7. What are the challenges and problems faced by LCCs to expand into international markets?
8. How can legacy airlines grapple with low-cost competition?

9. In the last 30 years there has been a surge in the low-cost market. Do you see any possibility for creation of airline alliances among LCCs? Explain you answer.

APPENDIX: SELECT AIRLINE TWO LETTER CODES

Airline	Code
Air Canada	AC
Airtran Airways	FL
Air Transat A.T.	TS
Alaska Airlines	AS
American Airlines	AA
America West Airlines	HP
Continental Airlines	CO
Delta Airlines	DL
Easyjet	U2
Frontier Airlines	F9
JetBlue Airways	B6
Northwest Airlines	NW
Ryanair	FR
Southwest Airlines	WN
United Airlines	UA
US Airways	US
Westjet	WS

REFERENCES

BBC News. (2003). *Q&A: Ryanair Wwoops on Buzz*. Retrieved on September 19, 2006 from http://news.bbc.co.uk

Bethune, G. and Huler, S. (1998). *From Worst to First: Behind the Scenes of Continental's Remarkable Comeback*. New York: John Wiley & Sons.

Buyck, C. (2005). Wooing Europe's New Breed. *Air Transport World*, Vol. 42, No. 9, 32–35.

Calder, S. (2002). *No Frills: The Truth Behind The Low-Cost Revolution in the Skies*. London: Virgin Books.

Cassels, K. (2008, July 10) *US Airways to Shut Down In-Flight Movies*. The Associated Press, Retrieved on November 3, 2010 from http://www.travelagentcentral.com/airline-policies/us-airways-shut-down-flight-movies.

Clark, A. (2002). easyJet Lines up Merger with Go: Shake-up of Budget Airlines Could Mean Higher Fares. *The Guardian*, May 4, 2.

Doganis, R. (2001). *The Airline Business in the 21ˢᵗ Century*. London: Routledge.

Dunn, G. (1999). KLM launches low cost airline—named "Buzz". *Air Transport Intelligence News*, September 22.

Field, D. and Pilling, M. (2005). The Last Legacy. *Airline Business*, Vol. 21, No. 3, 48–51.

Flight International (1961, April 13) Britain's New Board, Plain Man's Guide to the Air Transport Licensing Board, World Airlines Survey, p. 471.

Flint, P. (1996). The Leopard Changes its Spots. *Air Transport World*, Vol. 33, No. 11, 51, 54.

Frank, W. (2006, October 30) Leaders Should Beware of Uncontrolled Growth, *Denver Business Journal*.

Goldsmith, C. (1998). British Airways Launches No-Frills Unit—Move May Risk Diluting Brand Name, Some Say. *Wall Street Journal*, May 22, 5.

Henry, K. (1998). Aiming High with Lower Fares; US Airways' MetroJet Set to Debut, Battle Southwest, Boost BWI. *The Sun*, May 31, 1.

Ionides, N. and O'Toole, K. (2005). Points of Sale. *Airline Business*, Vol. 21, No. 3, 42–45.

Johnston, D. C. (2001). Airlines Are Cutting Their Discount Services. *New York Times*, November 18, 5.

O'Toole, K. (1999). Express Yourself. *Airline Business*, January 28. Retrieved on November 28, 2006 from Air Transport Intelligence.

Shannon, D. (2011, November 23). North American Legacies Take Another look at LCC Concept, *Aviation Week*.

Smith, S. (2005). The Strategies and Effects of Low-Cost Carriers. *Steer Davies Gleave*. Retrieved on September 20, 2006 from http://www.icea.co.uk/archive.htm.

Turbulent Skies. (2004). *The Economist*, July 10, 59–63.

13

The Economics of Aviation Safety and Security

> If you were born on an airliner in the US in this decade and never got off you would encounter your first fatal accident when you were 2300 years of age and you would still have a 29% chance of being one of the survivors.
>
> Les Lautman, Safety Manager, Boeing Commercial Airplane Company, 1989

Today's commercial airlines are the safest of all modes of transportation and according to a report issued by the International Air Transport Association (IATA), airlines enjoyed their lowest rate for major accidents ever in 2011. The report suggests that airline safety rates, measured by crashes and passenger fatalities, have decreased by 50 percent in 2011. For every 7.1 million fliers around the globe, there was about one passenger death, which broke the previous record of one per 6.4 million, set in 2004. However, this enviable record still needs to be put into perspective when we consider the topic of safety. In common usage, the term safety is often used incorrectly as an absolute value; that is, one is either safe or unsafe. Moreover, safety is never absolute since there is always some probability of an accident. Thus, safety depends on the given situation and the risks that are part of that situation. However, many people claim that safety should be maximized regardless of the cost. The reasoning goes something like this: If human life is deemed sacred then it may seem reasonable to consider human safety to be sacred—shouldn't our goal be to achieve as much safety as possible? But if this is true, then we should outlaw any activity where fatalities are even remotely possible. We need to ban swimming, skiing, fishing, flying, pregnancy, social gatherings (where disease may spread), driving more than 15 mile per hour, and virtually everything else that people enjoy doing!

The benefits of safety are undeniable—not only from a moral standpoint, but also from an economic standpoint. Some of the potential economic benefits of aviation safety include: strengthened consumer demand, strengthened labor supply, reduced insurance costs, lower cost of capital, lower liability risk, and reduced costs associated with government fines or penalties. While this chapter will generally analyze aviation safety from an economic standpoint, it will also provide some more specific facts and figures on aviation safety. The general outline for this chapter is:

- The Basics of Aviation Safety
- The History of Aviation Safety
- Incentives for Aviation Safety

- – Passengers' reaction
- – Labor reaction
- – Financial concerns
- – Insurance costs and liability risks
- – Government enforcement
- • Causes of Aviation Accidents
  - – Flight crew error
  - – Aircraft malfunction
  - – Weather related
  - – Airport/air traffic control
  - – Maintenance
  - – Miscellaneous/other
  - – Classification of accidents by phase of flight
  - – Classification of accidents by regions
- • Basic Economics of Safety
- • Politics and Safety Regulation
- • Accident Prevention
- • Summary
- • Discussion Questions

THE BASICS OF AVIATION SAFETY

Increases in safety are optimal only when the benefits of safety justify the costs; thus, minor increases in safety that impose major costs are never cost-efficient. This is the reason why people do not wear helmets all the time or the government does not establish a national speed limit of 15 mph, although clearly both of these actions would reduce the risk of accidents. In spite of the fact that expressions like "safety must be preserved at any price" are commonly used, safety still needs to be judged within the economic context of a simple cost-benefit analysis.

Prior to discussing the topic of aviation safety, it is useful to understand some basic aviation safety terminology, since a few terms are sometimes used synonymously, but may contain different meanings. The National Transportation Safety Board (NTSB) defines an aviation *accident* as:

> An occurrence associated with the operation of an aircraft which takes place between the time any person boards the aircraft with the intention of flight until all such persons have disembarked, and in which any person suffers death or serious injury as a result of being in or upon the aircraft or by direct contact with the aircraft or anything attached thereto, or in which the aircraft receives substantial damage.

On the other hand, the NTSB defines an aviation *incident* as "an occurrence other than an accident, associated with the operation of an aircraft, which affects or could affect the safety of operations" (Vasigh and Helmkay, 2002). In practicality, the difference between an accident and an incident is largely based on the severity of the situation. If damage occurs to an aircraft, then the situation would likely be deemed an accident. However, an aircraft landing on a parallel taxiway or a runway incursion may not provide aircraft damage, (and therefore be classified as an incident) but it may actually represent a more serious threat to safety.

A fatal aviation accident is one that results in fatalities, or deaths. The fatalities could involve passengers, crewmembers, or people on the ground. A hull loss occurs when an aircraft is a complete write-off from an accident and is no longer flown. Typically, aviation safety is measured in terms of accidents, fatal accidents, fatalities, and hull losses.

Other terms that may appear in aviation safety literature include a near midair collision (NMAC), a pilot deviation, and a runway incursion. While a midair collision involves two aircraft making contact while in flight, a NMAC is an incident associated with an aircraft flying within 500 feet of another airborne aircraft (Vasigh and Helmkay, 2002).

Pilot deviation refers to the actions of a pilot that result in the violation of a Federal Regulation or a North American Aerospace Defense Command (NORAD) Air Defense Identification Zone (ADIZ) Directive (Vasigh and Helmkay, 2002). Pilot deviation simply means that the aircraft goes into airspace that is either totally restricted or the aircraft has entered the airspace without taking the appropriate procedural steps; these deviations may be a result of equipment malfunctions, weather conditions, operational factors, and/ or pilot experience (Vasigh and Helmkay, 2002).

Finally, a runway incursion refers to any occurrence on an airport runway involving an aircraft and any object or person on the ground that creates a collision hazard or results in a loss of separation with an aircraft taking off, intending to takeoff, landing, or intending to land (Federal Aviation Administration (FAA), 2006). Although aviation accidents have been diminishing over the past few years, runway incursions continue to occur (Table 13.1). Runway incursions are further classified into four categories (A, B, C, D where A and B are considered "serious") based on available reaction time, evasive or corrective action, environmental conditions, speed of the aircraft and/or vehicle and proximity of aircraft/vehicle.[1]

Table 13.1   Number of "Serious" Runway Incursions, 2000–2010

Year	Total A and B Incursions	# Involving Commercial Aircraft
2000	67	34
2001	53	26
2002	37	11
2003	32	10
2004	28	9
2005	29	9
2006	31	10
2007	24	8
2008	25	9
2009	12	2
2010	6	3

Source: FAA Runway Safety Fact Sheet, 2000–2010

---

1   FAA Adopts ICAO Definition for Runway Incursions, FAA news release, October 1, 2007.

THE HISTORY OF AVIATION SAFETY

In 1903, the Wright Brothers made the first ever controlled and sustained, heavier than air flight. Today, more than a century later, tens of thousands of airplanes are in the air at any one time, with those aircraft spanning all shapes and sizes. While aviation technology has developed at a tremendous rate, so too has aviation safety. Once a highly risky method of transportation, aviation has developed into the safest mode of transportation available to the public. In fact, in terms of fatalities per passenger miles in the US from 1998 to 2008, air transportation was 70 times safer than passenger autos. Table 13.2 displays the fatality rate for various modes of transportation.

As shown in Tables 13.2 and 13.3, air transportation is an incredibly safe mode of transportation, with 2002, 2007, 2008, and 2010 actually producing zero fatalities. These statistics support the fact that one is more likely to be involved in a fatal accident while driving than while flying. While people may complain about the service provided by airlines, the one thing consumers have little to complain about is safety. Over the years airlines have proven that they successfully meet their foremost objective—safety. In the months after the September 11, 2001 terror attacks, passenger miles on the main US airlines fell by about 16 percent, while road use increased drastically. According to one study, the temporary increase in driving caused an extra 1,595 Americans fatality in the year after the attacks.[2]

Table 13.2   US Fatality Rate for Various Modes of Transportation

Years	Us Passenger Fatalities per 100 Million Passenger Miles			
	Autos	Buses	Passenger Trains	Airlines
1999–2010	0.72	0.05	0.05	0.01

Source: Compiled by the author using Airlines for America (A4A) data

> "…North America has an accident rate so far this year of 1.18 per million takeoffs against 1.51 in 2010, Europe has a rate of 1.39 per million takeoffs against 1.59, Asia-Pacific's rate per million takeoffs is 1.39 against 2.61, Latin America 4.57 against 6.85, and Africa has 6.34 against 17.11…"
>
> IATA, 2012.

The increase in aviation safety has been most dramatic over the past few decades. The Second World War was a significant event in aviation history as aircraft technology advanced rapidly to cope with wartime demand. Following the war, commercial aviation soared as the new technologies and aircraft developed during the war were applied to commercial applications. An excellent example of this technology transfer was the German Messerschmitt 262, which was the world's first operational jet-powered fighter aircraft. This aircraft's technology was subsequently used as a basis for future jet aircraft. And, although it took some time for jet aircraft to power the majority of commercial aircraft, the development of the jet engine marked a significant event in commercial

---

2   September 11's indirect toll: road deaths linked to fearful flyers.

aviation safety. By allowing aircraft to fly further, faster, and higher, the jet engine proved to be more reliable than the piston engine, thereby increasing safety. This increase in safety is displayed in Figure 13.1, which displays the number of accidents per million departures for commercial aircraft in the US. The evolution of the jet engine through the 1960s resulted in an accident rate that was decreasing in an exponential fashion, signaling a dramatic change in aviation safety.

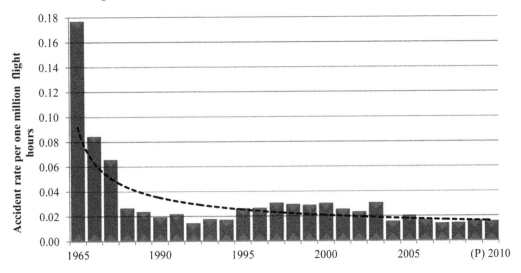

Figure 13.1   Commercial Aircraft Accidents per Million Flight Hours in the US

Source: Compiled by the author using Airlines for America (A4A) data

Other new technologies, such as fly-by wire, have also made commercial aviation safer. Advanced computer simulation and modeling have made it easier to design redundant and fail-safe components on aircraft. While all of these factors have played a role in improving safety, they have also been accompanied by improvements in pilot training. With advanced simulation training and research into crew resource management, this has reduced the number one cause of aviation accidents—pilot error. Because of these advancements, aviation has continually become safer and safer. However, in the past 20 years aviation safety has appeared to plateau. In general, the accident rate has hovered between 0.02 and 0.04 accidents per million flight hours since the mid 1980s. This is due in part to fewer advancements in aviation technology, but it also may well be that aviation safety has reached an economic equilibrium. This implies that accidents could conceivably be further reduced, but the costs of doing so may be excessively expensive.[3] Therefore, aviation safety may be reaching a point where the benefits of safety are approximately equal to the costs (the cost-benefit analysis of commercial aviation will be discussed in more detail later in this chapter). The plateau effect discussed above is displayed in Figure 13.2, which displays the number of aviation fatalities per 100 million aircraft miles in the US.

---

3   A good analogy here might be automobile safety. It would be possible to reduce automobile fatalities if every abutment on the interstate highway system was surrounded by crash absorbing material. However, it is clear that this would be too costly for the few fatalities that it might prevent.

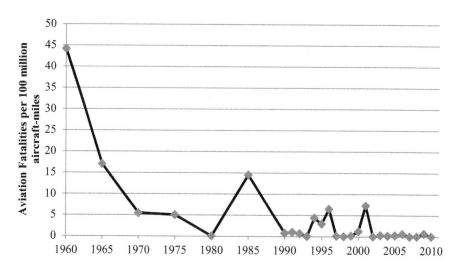

Figure 13.2   Number of Fatalities per 100 Million Aircraft-miles in the US

Source: BTS National Transportation statistics, Table 2.9: US Carrier Safety Data

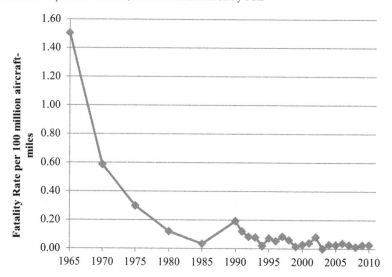

Figure 13.3   Commercial Aviation Fatal Accident Rate per 100 Million Aircraft-miles in the US

Source: BTS National Transportation statistics, Table 2.9: US Carrier Safety Data

As Figures 13.2 and 13.3 show (in comparison with Figure 13.1), fatalities and fatal accidents are more random than generic accidents, with significant fatality years followed by years with no fatalities. Even with this variability in the data, Figure 13.3 shows a significant downward trend in aviation fatalities, mirroring the downward exponential trend in aviation accidents. Today, the number of fatalities in commercial aviation is incredibly minute. In fact, based on 2006 data, it would take a passenger flying every single day, 43,720 years until they would experience a fatality in an aviation accident.[4] The probability of being killed in

---

4   In a similar study, Barnett (2000) examined the mortality risk of air travel and found it to be extremely

an aviation accident is less than a billionth of 1 percent. While any loss of life is a tragedy, the probability of being killed in an aviation accident is practically zero. In fact, many airlines such as JetBlue, Southwest Airlines, Virgin Atlantic, and Emirates have never experienced a fatal accident and many defunct airlines like Go, Laker Airways, Song, and MetroJet went through their whole operations without a fatal accident.

INCENTIVES FOR AVIATION SAFETY

While some observers (usually in sensationalized media stories) will assert that airlines occasionally cut corners that compromise safety in the interests of greater profit, we will argue in the paragraphs below that just the opposite is probably true. There are strong incentives for airlines to avoid any accidents or incidents so they are not likely to deliberately cut corners to compromise safety.

> "Regulation has gone astray… Either because they have become captives of regulated industries or captains of outmoded administrative agencies, regulators all too often encourage or approve unreasonably high prices, inadequate service, and anticompetitive behavior. The cost of this regulation is always passed on to the consumer. And that cost is astronomical."
>
> Senator Edward Kennedy, 6 February, 1975.

As the above data display, aviation is a tremendously safe mode of transportation. However, why do some people still believe that aviation is still unsafe? And why is the aviation industry continually focused on safety to the exclusion of many other considerations? The answers to these questions lie mainly (although there are other reasons covered below) in what might be termed asymmetrical media coverage of aviation accidents and incidents. While many more people in total are involved in automobile accidents and fatalities, aviation accidents typically involve more people in a single accident. Therefore, rightly or wrongly, the media sensationalize nearly every aviation accident, fatal or otherwise. Such extensive media coverage creates a situation where, if an airline has an accident, their logo may be emblazoned into the minds of consumers across the world for all the wrong reasons. While the media does not document the thousands of routine safe flights a day and the incredibly small probability of an aviation accident, any minor safety slip by an airline will draw extensive media coverage; this creates a climate where airlines have a very strong incentive to insure aviation safety.

However, media coverage is just one of many incentives that airlines have to provide safe and secure air travel. The other incentives for aviation safety can be grouped into five broad categories:

- passengers' reaction;
- labor reaction;
- financial concerns;
- insurance cost and liability risks;
- government enforcement.

---

small. He estimated a death risk per flight of one in 13 million for aircraft operated by countries that have a well-developed aviation industry. At this level of mortality risk, a passenger would have to take one flight per day for 36,000 years before having a fatal plane crash.

Passengers' reaction

Since consumers can choose amongst firms in the market, they can decide how important safety is to them. Therefore, an airline that is perceived to be less safe, as a result of an accident or investigation, is likely to see a decrease in demand compared to "safer" airlines. For passengers, the perception of the level of safety is ordinarily the key decision factor; this is opposed to the actual level of safety, since passengers usually know very little about the actual level of safety of their flight (Squalli and Saad, 2006). Perceptions of aviation safety are largely a result of accidents, particularly accidents that receive extensive media attention. Additionally, perceptions are difficult to change and may last for an extended period of time; this creates a situation where a single aviation accident may have a long-term impact on an airline's demand (Squalli and Saad, 2006). The ValuJet name was so tarnished by one accident that it was forced to be merged with AirTran Airways in 1997.

While logical economic reasoning implies that demand for a particular airline should be reduced when they have had an accident, empirical studies have had difficulty in proving a decrease in demand. Borenstein and Zimmerman (1988) conducted an extensive study of 74 accidents in the US from 1960 to 1985 and found that there was no statistically significant decrease in demand for the airline's services as a result of the accident.[5] Squalli and Saad (2006) did find a minor decrease in demand resulting from aviation accidents in the US, while Wong and Yeh (2003) estimated a 22.11 percent decline in monthly traffic lasting for 2.54 months resulting from an aviation accident in Taiwan. Part of the difficulty in statistically showing a decrease in demand is that following an aviation accident, airlines are likely to take competitive action to help offset a shift in the demand curve, such as lowering ticket prices. While not statistically significant, Borenstein and Zimmerman (1988) did find that consumers responded more adversely to aviation accidents post-deregulation than they did during regulation.

While it has been difficult to measure the decline in demand resulting from aviation accidents, economic logic suggests that, although the decline may not be large, it definitely exists. This is based on the simple fact that an accident on one carrier will undoubtedly cause some people (who are perhaps very risk adverse) to fly on another carrier. This may be particularly true of consumers who view aviation as unsafe in certain regions of the world, because there is a public perception that these regions have lower safety standards. While this may or may not be true, these so-called "unsafe" airlines will have difficulty in changing the public perceptions since consumers typically remember bad things before good things.

The fact that the demand for an airline's flights could be impacted severely provides an incentive for the airline to prevent accidents. Additionally, an airline that is perceived to be safer than the competition may receive a modest increase in demand if the consumers view safety as an important factor in their decision-making.

Labor Reaction

Similar in vein to passengers' reaction, the labor supply may also react adversely to an aviation accident. Employees do not want to work in an environment which they believe

---

5   Borenstein and Zimmerman (1988) had results that displayed a 4.3 percent reduction in consumer demand resulting from an accident during regulation and a 15.3 percent reduction in demand post-deregulation. However, neither value was statistically significant.

is unsafe. Therefore, an airline may experience two labor issues as a result of an accident: increased turnover and/or increased wage demands.

As a result of the perception or the reality of reduced safety, employees, particularly members of the flight crew, may feel that the airline is engaging in questionable safety practices. This could cause employees to leave the company or have the union enforce new safety measures. Moreover, an accident or series of incidents could make it more difficult for an airline to attract quality employees.

The other outcome affecting the labor supply resulting from decreased safety would be increased wages. Employees may demand better compensation for having to work in a less safe environment. In essence, this would be a form of "combat pay," where employees are compensated for working in an uncertain environment. While such demands may be difficult for employee groups to obtain in the short term, an airline's safety could become bargaining issues in the long term.

These impacts represent both explicit and implicit costs to the airline, and therefore provide strong incentives to avoid accidents and incidents. Hence, the airline has an economic incentive to continue operating safely in order to avoid the costs imposed by the market forces resulting from an accident.

Financial concerns

The stock market will always react negatively toward an airline that experiences an accident, particularly one that involves fatalities. The reason for this is the more or less obvious belief that such an accident will cause great uncertainty over the future of the airline in question. Borenstein and Zimmerman (1988) found that on average aviation accidents caused a 0.94 percent equity loss for the firm on the first day of trading, which was statistically significant at the 1 percent level. This value is slightly lower than two other studies that determined equity losses amounting to 1.18 percent and 1.19 percent on the first day of trading (Borenstein and Zimmerman, 1988). Mitchell and Maloney (1989) went one step further in analyzing the impact of an aviation accident on the firm's equity value in the long term. They found that if the accident was proved to be the airline's fault, then the equity value dropped by 2.2 percent (Mitchell and Maloney, 1989). However, if the accident was not deemed to be the airline's fault, equity value dropped by only 1.2 percent (Mitchell and Maloney, 1989). Regardless of who is to blame for the accident, the airline's equity will decline, which is another incentive to avoid aviation accidents and promote safety.

A loss of equity value for an airline will have other and possibly greater negative financial effects, namely, a large increase in the cost of capital. Because of the greater risk and uncertainty associated with the decline in equity value, the, airline will find it more expensive to raise capital; this can be quite serious since airlines are highly capital intensive. While a loss of equity through a decline in the stock price is not an explicit cost against the airline, an increase in the cost of capital is a direct cost to the airline, providing another major incentive to avoid accidents and incidents.

Insurance costs and liability risk

When an aviation accident occurs, airlines are usually fully indemnified from the losses through insurance. Insurance companies will pay out various liability and damage claims for the airlines, causing airlines little direct financial loss from an accident. However, as a result of an airline accident, particularly if the airline is determined to be liable to any extent, the airline's insurance premiums are likely to increase dramatically in the future. Like automobile insurance premiums, airlines will see insurance hikes if they experience an accident or incident. Moreover, the insurance rate hike does not occur for just one year, but lasts for several years. Such increases can have a significant effect on an airline's profit margins. This is true since, not only do airlines currently pay substantial insurance premiums, but it has been estimated that increases in insurance rates explain about 34 percent of equity loss (Wong and Yeh, 2003). Related to insurance premiums, airlines will find that their liability risks will increase substantially as a result of an accident. Therefore, the threat of increased liability risks provides airlines with one more economic incentive (from increased insurance premiums) to promote safety and avoid aviation accidents.

Government enforcement

The final major incentive for aviation safety is not a true market incentive; however, the threat of government penalties provides another real incentive. Similar to traffic laws, where the threat of a speeding ticket helps deter many from speeding, aviation regulations are designed to deter airlines from violating safety procedures. However, unlike traffic penalties, safety fines levied by the FAA can be substantial. For example, in 2008, the FAA levied more than $10 million in fines against Southwest Airlines for not inspecting cracks on dozens of its planes (Levin, 2008). The FAA argued that Southwest Airlines flew more than 60,000 flights with these planes without their required inspections. In another case in 2010, the FAA imposed a $24.2 million penalty on American Airlines charging that the airline made 14,278 flights on 286 MD-80 jets without making required upgrades to wiring during 2008 (Levin, 2010). Or, the $805,000 fine against United Airlines in 2002 for improper maintenance techniques (FAA, 2002). Clearly, fines levied by the FAA are substantial and provide an incentive for safety; however, the greatest threat posed to an airline would be the threat of a complete shutdown due to a severe violation in safety practices. The FAA has the authority to order an airline to cease operations; this would effectively cut off revenue while imposing sizeable costs and penalties. The fear of shutdown is one of the greatest threats to an airline and, while the FAA has rarely used its authority to temporarily shutdown an airline, the mere presence of this threat provides a tremendous incentive to adhere to FAA safety practices. For example, ValuJet, Kiwi Airlines, and Nation's Air all were shutdown by FAA for safety violations. The French authorities grounded Point Air Mulhouse back in 1988 subsequent to numerous and recurrent maintenance problems with the B-707 and DC-8. Therefore, the presence of the FAA and other aviation regulators provide airlines with a strong financial incentive to promote safe air travel.

CAUSES OF AVIATION ACCIDENTS

If you are looking for perfect safety, you will do well to sit on a fence and watch the birds; but if you really wish to learn, you must mount a machine and become acquainted with its tricks by actual trial.

Wilbur Wright, from an address to the Western Society of
Engineers in Chicago, 18 September, 1901

Aviation accidents occur for a variety of reasons and every accident is thoroughly investigated to help prevent future accidents. Accidents are rarely attributed to just one cause as a variety of factors must go wrong for the accident to occur. Understanding the nature of the accidents and how they occur is important to continually improve aviation safety and to help understand the economic principles of safety. The six major categories by which airline accidents can be categorized are:

- flight crew error;
- aircraft malfunction;
- weather related;
- airport/air traffic control;
- maintenance;
- miscellaneous/other.

Table 13.3 displays the number of worldwide fatalities resulting from aviation accidents between 2001 and 2010. The fatalities are classified according to Commercial Aviation Safety Team (CAST) and International Civil Aviation Organization (ICAO) principal categories agreed upon by air carriers, aircraft manufacturers, engine manufacturers, pilot associations, regulatory authorities, transportation safety boards, as well as members from Canada, the European Union, France, Italy, the Netherlands, the UK, and the US. Because of varying definitions, the cause of an aviation accident could be embedded amongst the different categories.

Table 13.3   Safety Record of US Air Carriers

Year	Total Accidents	Fatal Accidents
2000	49	2
2001	41	6
2002	34	0
2003	51	2
2004	23	1
2005	34	3
2006	26	2
2007	26	0
2008	20	0
2009	26	1
2010	26	0

Source: Compiled by the author using Airlines for America (A4A) data

For example, flight crew error could be categorized as CFIT (Controlled Flight Into Terrain) or LOC (Loss of Control), depending on the accident. Regardless of such cross-classifications, the ICAO taxonomy provides a standardized worldwide definition of aviation accidents that can be of use in safety research. Since CFIT and LOC are generally caused by crew mistakes, the table underscores the fact that the vast majority of aviation accidents are still caused by human error. As shown in Table 13.4, CFIT is the number one category of fatalities in aviation. CFIT can result from numerous issues; however pilot error is usually a central cause of CFIT accidents. LOC is the second major category, with LOC accidents being a result of numerous issues. While the ICAO classification is different, the major cause of aviation accidents remains roughly the same—human error.

Table 13.4   Fatalities by Occurrence Categories, Worldwide, 2001–2010

Cause	Number of Fatal Accidents	Fatalities
Loss of control—in flight	20	1,841
Controlled flight into terrain	17	1,007
Runway excursion (landing), undershoot/overshoot	17	783
Unknown	3	352
Midair/near midair collision	2	225
Non-powerplant failure or malfunction	1	225
Runway excursion (takeoff)	5	192
Other	4	125
Runway incursion—vehicle, aircraft, or person	1	118
Wind-shear or thunderstorm	1	97
Fuel related	1	23
Ground handling	9	9
Powerplant failure or malfunction	3	4
Fire/smoke	2	3
Evacuation	1	1

Source: Boeing Statistical Summary of Commercial Airplane Accidents, Worldwide Operations

Flight crew error

Flight crew error, or human error, is the number one cause of aviation accidents worldwide. While the period used is only from 2001–2010, human error has always been the primary cause of aviation accidents. New technology has helped make aviation safer, but it still cannot compensate for errors made by humans. Much research has been conducted into reasons why flight crews make errors, and while areas such as crew resource management have helped reduce human error, the fact is that as long as humans are in control of the aircraft, flight crew error will probably occur.

While aviation accidents are commonly a result of several contributory factors, most aviation accidents could have been avoided if the crew had done something differently.

For example, one of the worst aviation accidents in history was a result of human error. The 1977 PanAm/KLM accident in Tenerife was a result of the KLM pilot starting his takeoff role prior to receiving air traffic control clearance. The subsequent collision with a PanAm 747 killed 583 passengers in total. With that said, many other fatal aviation accidents have been avoided due to exemplary efforts by the flight crew. For example, an Aloha Airlines 737-200 was able to land safely (with only one fatality) after part of the fuselage was torn apart by a sudden decompression. Therefore, while human errors have caused accidents, flight crews have also saved numerous lives.

The development of realistic flight simulators has made it possible for pilots to experience a variety of problems without ever taking to the sky. Thus, while technology has successfully made aviation safer, future effort toward better crew training and management may result in fewer flight crew errors.

Aircraft malfunction

The second major determinant of an aviation accident is an airplane-related malfunction. Current aircraft are sophisticatedly designed with safety in mind; however, systems may still malfunction and this can ultimately cause a serious accident. Parts ranging from multi-million dollar engines to trivial items have all been the cause of serious aviation accidents. The Aloha Airlines flight highlighted above is an example of an airplane-related accident. In this case metal fatigue caused part of the fuselage to deteriorate and the aircraft experienced rapid decompression. Another example of an aircraft malfunction accident would be United Airlines Flight 232. This flight crash landed in Sioux City, Iowa, after one of the engines failed and thereby disabled the hydraulic systems on the aircraft. Most fatal aircraft malfunction accidents usually occur when the engines experience a problem; however, other systems can also cause fatal accidents.

Figure 13.4 overleaf displays a ratio of aircraft hull losses resulting from accidents per one million departures for the major aircraft manufacturers since 1959. While the number of aviation accidents per manufacturer includes accidents for a variety of reasons, a trend could possibly be extrapolated if it was assumed that the probability of an accident due to pilot error, weather, maintenance, or air traffic control is approximately the same for all manufacturers (this may not be a good assumption for the smaller companies that do not have their aircraft spread throughout the world). From this we can see that aircraft have become significantly safer as newer models of aircraft have been produced; for example, Boeing's 707 had a 8.87 hull loss rate per million departures whereas its latest model for which data was available, the 777, had a hull loss rate of only 0.18 per million departures.

Weather related

The third major cause of aviation accidents is the weather. Through the development of the jet engine, weather-related accidents have become less of a concern since jet engines enable aircraft to fly higher and avoid any troublesome weather. Additionally, the development of instrument landing systems allows aircraft to auto-land in adverse weather, reducing the chance of error during landing in bad weather conditions.

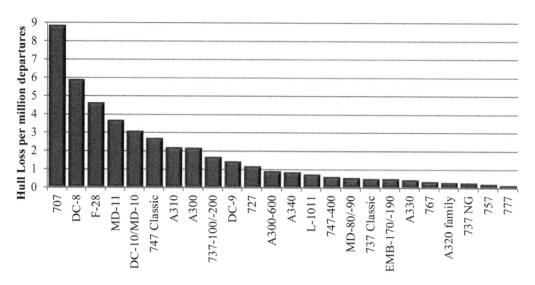

Figure 13.4   Commercial Aircraft Hull Loss by Aircraft per Million Departures between 1959 and 2010

Source: Compiled by the author using Boeing (2011) data[*]

[*] Note: The sample includes all commercial aircraft during the time period, except for regional and commuter aircraft; therefore, Bombardier's CRJ-200 aircraft and Embraer's ERJ aircraft are excluded

However, weather-related accidents still occur. For example, in 2005, an Air France A340 overran the runway at Toronto Pearson during adverse weather conditions (poor visibility and strong winds). Icing can also be a major problem for commercial aircraft, as evidenced by the American Eagle ATR-72 that crashed in 1994 while waiting to land at Chicago O'Hare. While the primary cause of the accident was icing on the wing, it was also determined that the aircraft type had poor de-icing equipment. As a result of the accident, modifications were made to the aircraft to reduce the risk of another accident of this type.

Airport/air traffic control

Air traffic control can also be prone to human error resulting in accidents. Air traffic controller fatigue and stress, as well as poor communication, are important human factors that contribute to accidents. Also, the air traffic control system generally dates from the mid-1960s and has had great difficulty in managing the recent huge increases in aviation traffic.

One of the more recent aviation accidents attributed to air traffic control was a 2002 midair collision between a Bashkirian Airlines Tupolpev 154 and a DHL 757 near the border of Germany and Switzerland. The air traffic controller ordered the TU-154 aircraft to disobey the aircraft's TCAS (Traffic Collision Avoidance System) warning, resulting in both aircraft descending and making contact. In this accident, a series of events led to the air traffic control system failing, resulting in the fatal crash. About 50 of the victims of the midair plane crash over Germany were Russian children going on holiday to Spain.[6]

---

6   CNN, Children's Holiday Party on Doomed Plane, July 02, 2002.

Air traffic controller fatigue has also been blamed for several aviation accidents in the US.[7] For example, although the 2007 Comair accident in Lexington, Kentucky was a result of several factors, the investigation found that the controller was working on just two hours of rest (Ahlers, 2007). Had the controller been better rested, it is possible that he may have noticed the CRJ aircraft beginning to takeoff on the wrong runway. Several other incidents may have resulted from air traffic controller fatigue, and regulations have been amended in order to help minimize human error by air traffic controllers (Ahlers, 2007).

Maintenance

The maintenance department of any airline is critical to ensuring that the airline operates safely. Aviation accidents sometimes occur as a result of maintenance being performed either incorrectly or not thoroughly. However, maintenance has accounted for only 4 percent of hull loss accidents worldwide from 1996 to 2005; this in itself is a strong testament to the generally high quality of work maintenance personnel do on a worldwide basis. Through domestic and international regulations and inspections, maintenance is usually performed to strict standards; unfortunately, there have been exceptions to this rule.

In 1985, a Japan Airlines 747 crashed outside of Tokyo killing 520 passengers and crew. The accident resulted from the aircraft losing its rear stabilizer and hydraulic systems due to an explosive decompression. The accident investigation determined that repairs performed by Boeing on an earlier tail strike of the aircraft were inadequate. Over time the repairs began to fatigue, and finally the fuselage cracked causing a massive depressurization in the rear of the aircraft.

Unfortunately, a similar accident occurred in 2002 when a China Airlines 747 flight from Taipei to Hong Kong crashed into the ocean killing all aboard. Once again the investigation uncovered the fact that the aircraft had experienced a tail strike over 20 years earlier, and the repairs were not made up to appropriate standards. Eventually metal fatigue caused rapid depressurization and the subsequent accident.

On May 11, 1996, a ValuJet aircraft crashed in the Florida Everglades and killed 110 passengers and crew. According to the NTSB the ValuJet crash was the result of failures by the airline, its maintenance contractor, and the FAA. Consequently, the Department of Transportation's Inspector General required that the FAA should be more proactive to monitor airline maintenance work performed by non-certified contractors. The ValuJet crash has led to changes at the FAA, including closer scrutiny of new carriers and more monitoring of their growth.

Miscellaneous/other

The final category of aviation accidents is miscellaneous/other which can include a variety of things, with hijackings representing the largest share. Unfortunately, commercial aircraft are still used for ulterior (usually political) motives, as the terrorist attacks of September 11, 2001 showed. Increased screening and security will assist in helping to prevent further terrorist attacks. However, more and more restrictive and onerous security regulations

---

7   According to National Transportation Safety Board (NTSB) four aviation mishaps, between 2001 and 2007, were contributed to by air traffic controller fatigue and the lack of sleep.

can rapidly become more detrimental to the traveling public than any small (to non-existent) increase in safety that they generate. Finally, the cause of some aviation accidents is simply unknown due to a lack of evidence or unusual issues. Air France Flight 447 on its way from Rio de Janeiro (GIG) to Charles De Gaulle (CDG) crashed into the Atlantic Ocean on 1 June 2009, killing all 216 passengers and 12 aircrew.[8] The investigation into the accident was hindered by the lack of evidence and the difficulty finding the aircraft's black boxes. Finally, the black boxes were located and recovered from the ocean floor in May 2011, nearly two years after the accident.[9] On April 28, 2012 a Somali Antonov 24 passenger plane sustained substantial damage in a landing because its tires blew out at Galkayo Airport (GLK).[10]

Classification of accidents by phase of flight

Another classification of aviation accidents is the phase of flight when the accident occurred. Figure 13.5 displays both the number of accidents and fatalities for the worldwide commercial jet fleet between 2001 and 2010 as categorized by the phase of flight. Looking at the figure, it is important to note the large disparities between accidents and fatalities for various phases of flight. Over one-third (36 percent) of worldwide aviation accidents occur during landing and 24 percent of fatalities occur during this stage. Conversely, 25 percent of aviation fatalities occur during climb, yet only 17 percent of accidents occur during this phase of flight. The statistics indicate that climb is the most dangerous phase of commercial flight (as far as fatalities are concerned), since engine failures in this particular phase of flight can easily result in a fatal accident. Cruise is the safest phase of

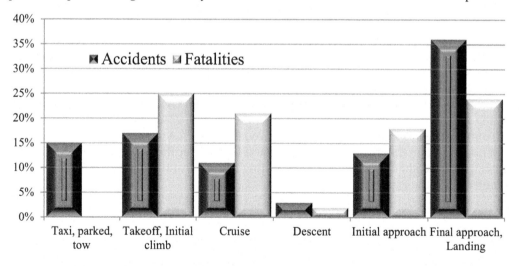

Figure 13.5   Worldwide Commercial Accidents and Fatalities between 2001 and 2010, Categorized by Phase of Flight

Source: Compiled by the author using Boeing (2011) data

---

8   Accident description F-GZCP. Flight Safety Foundation, June 1, 2009.
9   Flight AF 447 on 1 June, 2009. Bureau d'Enquêtes et d'Analyses (BEA), July 25, 2011.
10   Flight Safety Foundation, Aviation Safety Network, May 1, 2012.

flight, even though it takes up roughly 57 percent of a flight's duration. While accidents and fatalities have occurred at cruise, pilots generally have more time to react and to avoid more serious consequences. An example of a cruise incident that ended safely was the Air Transat A330 that ran out of fuel and glided to safety in the Azores in 2001. Had the aircraft run out of fuel at a lower altitude or during climb or approach, the result could have been catastrophic.

Classification of accidents by region

Another important classification of aviation safety is by region. Figure 13.6 displays hull loss rates per million flights for various regions worldwide in 2011.

As Figure 13.6 shows, aviation safety varies widely by region, with Africa being the worst region for aviation safety. These differences in aviation safety can be attributed to a variety of reasons; however, overall economic prosperity appears correlated with aviation safety. Poor regions such as Africa have neither the same level of safety oversight nor infrastructure, as more wealthier nations have developed. Air traffic control coverage can be sporadic across Africa and lack of instrument landing systems can make rough weather landings even more dangerous. Moreover, since the number one cause of aviation accidents is human error, training standards are extremely important and it is difficult to gauge the overall training standards in less-developed countries. Finally, many airlines in developing nations use older aircraft which may be more prone to accidents. Because of these reasons, one would expect the number of accidents in developing nations to be higher than developed nations; however, probably not to the extent that currently exists. Older airplanes have to meet strict safety requirements, and proper maintenance and supervision are essential to safe flight.

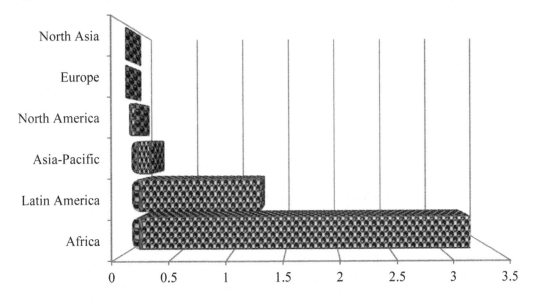

Figure 13.6   Hull Losses per Million Flights by Region

Source: Compiled by the author using IATA (2012) data

BASIC ECONOMICS OF SAFETY

To understand the economics of aviation safety, one needs to look at the industry from the macro-level, where the benefits of safety regulations to consumers and companies are weighed against the costs of imposing the regulations. Since the costs and benefits of safety are both explicit and implicit, it is sometimes difficult to fairly evaluate a regulation; this difficulty is one of the main reasons why many aviation safety regulations put into place by the government do not always make economic sense.

More specifically, some safety regulations enacted by governments are blanket responses to potential threats, or media-generated reactions that merely alleviate passenger concerns while not increasing safety in any substantial way. An example of this would be the requirement for all passengers to always take off their shoes while going through security screening. This regulation was created immediately following a potential terrorist threat, but the increased level of security this extra check provides is probably negligible.

This example highlights the fact that there is a strong probability that a number of aviation safety regulations likely generate more costs than benefits. Indeed, it is quite likely that some costly airline regulations, even if they provide some positive safety benefits to aviation, will actually decrease net safety in society! This follows because airline regulation will drive up prices for consumers, as well as sometimes making flying more cumbersome and time consuming, thus causing some who would have otherwise flown to instead travel by rail or, worse, drive. Since, as explained earlier, these other modes of transportation are far more dangerous than flying, lives are lost whenever costly regulations convert air travelers to ground travelers. Likewise, eliminating costly airline regulations that do little or nothing to improve safety would reduce ticket prices, draw people away from cars and rail and to aircraft, thus improving total travel safety. To some extent, we can increase total safety by decreasing airline safety.

The "Southwest rule" is a good example of a regulation that, if abolished, would probably increase total safety. In fact, it is likely the benefit of this regulation is about zero, since in all the years prior to the rule there were no injuries caused by the aircraft slowly taxiing toward takeoff position as passengers continued to settle into their seats. Ben-Yosef (2005, chapter 6) argues that the FAA's eventual 1996 decision to ground ValuJet Airlines, the low-cost leader in the eastern US at the time, also cost far more in increased highway fatalities than any conceivable benefit. Ben-Yosef also documents the general public's misperception that aircraft maintenance failures cause the most accidents, and, more broadly, the huge disconnect between public perception regarding the causes of airline crashes versus the reports of safety regulators and other industry experts. Breyer (1994) maintains that the interaction of public misperceptions and the political process produces an essentially random agenda. It is, of course, difficult for politicians to rise above the politics of safety and make decisions based on costs and benefits. The notion that airlines may well be too safe, forcing travelers into riskier transportation modes, appears to be too sophisticated to be effectively dealt with through the political process. Perhaps this will change in time with continued efforts to educate the public in this regard; however, such efforts have so far been remarkably ineffective and probably will not succeed in the future.

The very basic analysis provided above is an example of a simple economic cost-benefit analysis for aviation safety. Such an analysis could be conducted for almost all safety regulations to determine if the regulation is economically efficient. However, since it may be difficult to quantify all the benefits from improved safety, such analysis is rarely undertaken. Therefore, and as pointed out above, the political and bureaucratic process usually assumes, especially where aviation safety is concerned, that the benefits of almost any safety or security regulation outweigh the costs. Moreover, as mentioned earlier in the chapter, the reasons for this are easy to identify; although commercial aviation is a critical sector of the economy, the vast majority of people who fly do so only a few times a year. Therefore, the actual number of people who fly on a regular basis is only a small proportion of the population.[11] Add to this the media tendency to sensationalize all aviation-related accidents or incidents, and the natural inclination of regulators to avoid even the remote appearance of not being vigilant on safety, and one can readily see that even elementary cost-benefit analysis would be difficult to implement in this culture.

Moreover, it is also probably true that, if one were to perform the economic cost-benefit analysis for all safety regulations, many would pass due to the substantial benefits from improved safety. However, many regulations would also fail, largely because they were enacted in response to political pressure. Some safety decisions do not receive economic scrutiny because there are other competitive factors in play. For example, the regulation banning aircraft push back until everyone is seated was created in part as a response to other airlines lobbying against Southwest's practices. Therefore, for the reasons discussed earlier, it is probably true that, contrary to popular belief, aviation safety and security exceed the levels that might be considered economically efficient.[12]

POLITICS AND SAFETY REGULATION

Robert Poole (1981) has argued that airline safety regulation can be effectively privatized, thus driven by economic analysis rather than shallow politics. Poole envisions a system where, in essence, private insurers replace politicians as the ultimate safety authorities. Insurers have a vested interest in assuring that airlines do not take imprudent risks since they must pay for any damages caused by an accident. However, insurers do not have to explain their decisions to uninformed voters. Thus, insurance companies are unlikely to have any interest in continuing the "Southwest rule" or any other regulation that doesn't truly improve safety. Likewise, insurers would allow airlines to cancel safety programs that produce more costs than benefits; that is, an airline would be allowed to slightly increase risk as long as a higher insurance premium was paid to cover the slightly higher risk. This sort of behavior is observable now in that consumers can buy personal injury insurance, at a higher price, for a motorcycle or sub-compact car even though such vehicles are substantially less safe than standard size automobiles. Insurers know better than to try to eliminate all risk—any movement in that direction would result in customers leaving them for a more reasonable insurer. The fact is that insurers ban only imprudent risks and insist that customers pay more for any increase in prudent risks. In this setting, government could merely require that airlines purchase legitimate insurance, and then let insurers handle the details. The FAA and comparable regulators in other countries could continue their same basic mission but be converted into a private organization, paid by insurers rather than by taxpayers. The head of the FAA would, of course, no longer be a political appointee but would be a private manager, appointed by a board of stockholders, comparable to any other corporate CEO. Ideally, this privately reborn FAA would be driven by economic analysis, and able to maintain a more long-term focus on true accident risks, rather than being driven by the latest headlines and the whims of politics.

---

11   Again, this can be contrasted to the automobile where the ill-fated safety regulation tying the ignition of the car to a fastened seatbelt was quickly abandoned when a major proportion of the population (automobile drivers and voters) discovered what a nuisance this particular regulation would be in practice.

12   The opposite is probably true of automobile safety.

Although there is currently no private regulator of airline safety, Poole points to a number of examples of private safety regulation in other areas. Underwriters Laboratories, for example, sets safety standards for a number of electronic components. (It seems that many people have assumed that the company is some sort of public agency since its function is so commonly associated with government). Fire departments in the US, though usually government bureaucracies themselves are, in effect, regulated by a private insurance organization, the National Board of Fire Underwriters (NBFU). NBFU inspects fire departments to rate their response time, quality of equipment, staffing, and so on. NBFU does not, of course have authority to demand corrections at problem fire departments. However, if a local fire department is poorly rated then fire insurance premiums in the area are raised, immediately exposing inadequacies. Therefore, fire departments generally work with NBFU to correct problems and improve themselves as needed. NBFU began this function in 1890, stepping in to deal with the problem of widely varying quality in firefighting, and, after more than a century of experience, continues to operate without incident. The standards set by NBFU do not appear to have anything comparable to the Southwest rule or other inappropriate regulation.

More broadly, any independent private agency that provides product information and ratings is performing a function similar to the private airline regulation envisioned by Poole and other supporters. Just as consumers have difficulty judging if an airline is appropriately safe, they may have trouble judging the safety and general quality of many products. Government agencies sometimes provide such judgments to some extent and do not charge consumers for this service. However, government does not do this extensively enough to satisfy consumers; therefore, we have private companies like Consumer Digest and Consumer Reports that inspect products and make recommendations to consumers. Likewise, when investors want to know more about a company, they turn, not to the Securities and Exchange Commission, but to Moody's or Standard and Poor's.

Naturally, many people would question whether private regulation can really perform better than government. It is instructive to compare the incentives and operational nature of private and public entities in this area. Poole and other privatization proponents argue that private regulators have superior incentives and better flexibility to deal with problems. If the FAA were private, working under contract for insurers, it could be replaced, in part or in full, by another organization if it performed poorly. Basically, any entrepreneur, perhaps a former FAA employee, could approach an insurer and make a case that he or she could provide inspection in a given area more efficiently. Knowing it could be fired, the private FAA would seem to have a strong incentive to operate efficiently, appropriately monitor the competence and integrity of its workers, and so on. Likewise, being private would enable the FAA to more freely adjust policies or fire employees who weren't performing and more quickly promote those who were.

ACCIDENT PREVENTION

The ATR-72 airplane must be shown to be capable of continued safe flight and landing when operating in any weather conditions for which operation is approved, including icing conditions. As the FAA cannot issue a new or amended type certificate for an airplane with a known unsafe design feature. ATR must provide data to show that the [icing] problems experienced with the ATR-42 will not be present on the ATR-72.

FAA, 1989

The main reasons for the rapidly decreasing aviation accident rate since the 1950s are the various safety programs/inventions adopted by safety regulators, airlines, and aircraft manufacturers. All three groups have combined resources to make aviation the safest mode of transportation. This increase in safety has ultimately helped stabilize the industry and make it a more attractive transportation option for consumers. Based on the incentives described previously, methods to increase safety have sizeable economic benefits. Some examples of these are: aging aircraft regulations, collision avoidance systems, wind-shear detection, de-icing, and human factors.

As mentioned previously, aging aircraft can compromise aviation safety in some specific cases. In order to help minimize this problem, the FAA and the former Joint Aviation Authorities (JAA) require specific component overhauls at specified intervals (well ahead of the time the components would be expected to fail). The JAA was an associated body of the European Civil Aviation Conference (ECAC); an intergovernmental organization which was established by the ICAO and the Council of Europe. In 2002, the European Aviation Safety Agency (EASA) was created with the power to regulate civilian aviation safety and in 2008, also took over the functions of the JAA. Based upon a decision of ECAC's Director Generals in adopting the FUJA[13] II Report it was decided to disband the JAA system as of 30 June, 2009 and to keep the JAA Training Organization running. Some countries take aging aircraft regulations to an extreme, by not allowing airlines to operate commercial aircraft over a certain age. As pointed out in the previous section, this type of rather arbitrary safety regulation might improve safety somewhat, but it will also impose significant extra costs on the industry; these, of course, will ultimately be passed down to the passengers. From an economic standpoint, passengers may not be better off from such stringent aircraft age regulations (with no decrease in overall safety levels).

While midair collisions have never been the number one cause of aviation accidents, joint research by governments and industry resulted in the development and deployment of TCAS.[14] TCAS alerts pilots of a possible midair collision and provides instructions to help avoid a serious accident. The invention of TCAS has reduced the number of midair collisions; however, the system has not eliminated midair collisions all together, as in the case of the fatal accident involving the DHL and Bashkirian Airlines aircraft over Europe in 2002. Additionally, TCAS is not immune to human error, as pilots and air traffic controllers can still make mistakes and disobey TCAS warnings that may result in tragedy.

Wind-shear represents another significant threat to aircraft since it can cause an aircraft to become uncontrollable. Previously wind shear was undetectable; however, through government and industry research, warning devices have been created to alert pilots of possible wind-shear conditions. Based on the wind-shear warnings, regulations have been developed to help ensure aircraft do not fly during dangerous wind-shear conditions. While the American Eagle ATR-72 de-icing accident highlights the fact that fatal accidents still occur due to ice forming on the wings, advancements in anti-icing have significantly reduced the number of icing accidents. Aircraft manufacturers have designed aircraft with anti-icing boots, while chemical compositions have enabled de-icing to occur on the ground.

---

13   The Future of JAA (FUJA).
14   FAA Advisory Circular AC 20-151A (2009). Airworthiness Approval of Traffic Alert and Collision Avoidance Systems (TCAS II), Versions 7.0 and 7.1 and Associated Mode S Transponders.

SUMMARY

This chapter provides an overview of the state of aviation safety and security. The notion that there exists a perfectly safe aviation environment is discussed and critiqued from a more realistic economic benefit and cost approach. The very strong and ongoing economic incentives to pursue safety within the industry are discussed and contrasted with the sometimes irresponsible media coverage that is prevalent about this subject. The chapter analyzes the accident record of the aviation industry versus other modes of transportation. The various causes of aircraft accidents are then discussed and analyzed in some depth. In general terms, the chapter points out that safety and security in aviation have been highly effective from an economic point of view although there are probably numerous rules and regulations that could be relaxed with no decrease in overall safety. In fact, it is pointed out that stringent regulations for the introduction of new technologies and procedures probably act to decrease rather than increase overall safety in the industry. Future developments in this field will have to center on replacing human judgment with automated technologies. However, these developments have been and continue to be extremely difficult to implement due to bureaucratic and political inertia.

DISCUSSION QUESTIONS

1. What are some of the incentives for airlines to continuously improve safety?
2. What are the most common causes for airline accidents?
3. Over the past several decades, both the fatalities per million aircraft miles and accidents per million aircraft departures have decreased. Briefly explain some of the reasons for these improvements in aviation safety.
4. Explain how safety regulation may increase overall safety for travelers.
5. Explain how passengers and employees are able to influence the safety practices of an airline.
6. Evaluate the often heard statement that "airline safety must be preserved at any cost" from an economic cost-benefit perspective.
7. Who regulates the airline industry? What is the difference between the FAA and the EASA?
8. Which is the safest airline to fly?
9. Has the Airline Deregulation Act of 1978 improved airline safety?
10. What impact does the use of modern technologies such as jet engines, fly by wire programs, and flight simulators have on commercial aviation?

REFERENCES

Ahlers, M. M. (2007). NTSB: Air Controller Fatigue Contributed to 4 Mishaps. CNN, April 10.

Barnett, A. (2000). Air Safety: End of the Golden Age? *Blackett Memorial Lecture presented, November 27, 2000,* Royal Aeronautical Society. London, UK.

Ben-Yosef, E. (2005) *Evolution of the US Airline Industry.* The Netherlands: Springer.

Boeing. (2006). Statistical Summary of Commercial Jet Airplane Accidents, May.

Borenstein, S. and Zimmerman, M. B. (1988). Market Incentives for Safe Commercial Airline Operation. *The American Economic Review*, Vol. 78, No. 5, 913–935.

Breyer, S. (1994) Breaking the Vicious Circle: Toward Effective Risk Regulation, Harvard University Press, Vol 8, Issue 1, 127.

Federal Aviation Administration (FAA). (1989, Septermber 26). Final Special Condition. Docket No. NW-38; Special Conditions No. 25-ANM-30. Retrieved on May 14, 2012, from rgl.faa.gov/regulatory_and_guidance_library

Federal Aviation Administration (FAA). (2002). FAA Proposes $805,000 Fine against United Airlines, December 3.

IATA (2012, March 6) *Global Accident Rate Reaches New Low – Regional Challenges Remain.* Retrieved on April 22, 2012 from http://www.iata.org/pressroom/pr/pages/2012-03-06-01.aspx.

Levin, A. (2008, March 7) FAA Levels Record $10.2M fine against Southwest. *USA Today.* Retrieved on January 29, 2009 from http://www.usatoday.com/travel/flights/2008-03-06-fine_N.htm.

Levin, A. (2010, August 27) American Airlines tilts at record $24M fine. *USA Today,* Retrieved on August 22, 2012 from http://travel.usatoday.com/flights/2010-08-26-american-airlines-penalty_N.htm.

Mitchell, G. and Maloney, M. T. (1989). Crisis in the Cockpit? The Role of Market Forces in Promoting Air Travel Safety. *The Journal of Law and Economics*, Vol. 32, No. 6, 329–356.

O'Brien, B. (2006). *IOSA: IATA Operational Safety Audit.* Retrieved on August 23, 2012, from http://www.iata.org/SiteCollectionDocuments/Documents/ISMEd2Rev2.pdf.

Poole, R. (1981) *Instead of Regulation: Alternatives to Federal Regulatory Agencies.* Lanham, MD: Lexington Books.

Squalli, J. and Saad, M. (2006). Accidents, Airline Safety Perceptions and Consumer Demand. *Journal of Economics and Finance*, Vol. 30, No. 3, 297–305.

Vasigh, B. and Helmkay, S. (2000). An Empirical Examination of Airframe Manufacturers' Safety Performance: Boeing Versus Airbus. *Public Works Management & Policy*, Vol. 5, No. 2, 147–159.

Vasigh, B. and Helmkay, S. (2002). Airline safety: an application of empirical methods to determine fatality risk, in Jenkins, D. (Ed.), *Handbook of Airline Economics.* New York: McGraw-Hill, pp. 501–511.

Wong, J. T. and Yeh, W. C. (2003). Impact of Flight Accident on Passenger Traffic Volume of the Airlines in Taiwan. *Journal of the Eastern Asia Society for Transportation Studies*, Vol. 5, No. 5, 471–483.

14

An Overview of Macroeconomics for Managers

Economic depression cannot be cured by legislative action or executive pronouncement. Economic wounds must be healed by the action of the cells of the economic body—the producers and consumers themselves.

Herbert Hoover, State of the Union, December 2, 1930

Virtually all industries are affected by macroeconomic conditions—the ups and downs of the business cycle—but aviation and aerospace-related industries tend to be especially sensitive to economic conditions, often expanding faster during economic expansions but falling harder during recessions. Moreover, many management decisions are impacted by macroeconomic conditions. In a slow economic climate for instance, introducing discounted fares or sales promotions might be more effective strategies for generating demand than introducing a new luxury class. For aircraft manufacturers, macroeconomic variables become crucial in terms of predicting the demand for aircraft over the long term, since the financial health of airlines is highly dependent on economic cyclicality. Therefore, forecasting when the demand for aircraft is likely to fall off and pick up again will determine the timing of expected cash flows on aircraft, and consequently, the break-even year and quantity that will determine project success or failure.

Thus, it is especially crucial for managers in air transportation to have at least some fundamental knowledge about this complex topic. The following is designed to give readers a basic understanding of macroeconomic topics such as these: What factors determine economic growth and what do past trends suggest for the future? What is the likely impact of governments' sovereign debt problems and how much will these problems worsen? What are the key issues in monetary and fiscal policy? To what extent can these policies realistically smooth out the economy's ups and downs? We can start by considering three main goals for the economy—maximizing economic growth, attaining full employment, and promoting a stable currency (low inflation). We will begin with a consideration of the central issue, income growth.

This chapter will cover the following topics:

- Economic Growth
  - How can we best foster economic growth?
- Unemployment
- Inflation

ECONOMIC GROWTH

First, some definition of terms: The most widely followed economic statistic is Gross Domestic Product (GDP) which is a rough estimate of the total income of a nation. GDP can also be estimated from the viewpoint of total spending; since all spending is received by someone as income, spending and income are two sides of the same coin. GDP is not a precise calculation (no macroeconomic number is), but it is useful as a rough barometer of economic well-being. However, several factors can cause a diversion of GDP and well-being; in other words, there are some things which cause GDP to increase but harm the economy. Consider, for instance, inflation, which is a general rise in prices. (Inflation can also be thought of as a devaluation of the currency; higher prices mean paper money has less buying power.)

If your income doubles but prices of the things you buy also double, then you have not in any real sense enjoyed an increase in income. Thus we should, as best we can, subtract out the inflation component of any increase in GDP, so that we can see how *real* GDP changed. In fact, this is the very term that economists use—"real GDP," which is calculated by adjusting "nominal GDP" figures for inflation. The same adjustment for inflation applies to other data; a nominal variable is not adjusted for inflation while a real variable is inflation adjusted. Therefore, nominal wages, nominal interest rates, or any other nominal variable are not inflation adjusted. Real wages, real interests, or other real variables are inflation adjusted. (The term "constant dollar" is synonymous with *real*—so, GDP trends measured in constant dollars are the same as real GDP trends.)

When real GDP is increasing we say the economy is in an expansion, when real GDP is decreasing the economy is in recession. A severe recession is sometimes called a depression. Happily, since the industrial revolution, economies with substantial elements of free enterprise have, on net, expanded our real income, and our standard of living has grown dramatically over time. The transforming power of economic growth is most apparent when we compare the present to the distant past. In comparing economic well-being of, say, people in the US today versus people around 1900, it is clear that today's poor are in many ways better off than the middle class or even the wealthy in that era. In 1900, health care for example, was almost non-existent; thanks to numerous innovations, the poor have access to far better health care today than even the wealthy did then. Indoor plumbing was a rare luxury in 1900 as were automobiles. Today everyone in the middle class, and most of the poor, enjoys these things.

Figure 14.1 shows historical trends in real GDP from 1980–2011, with estimated projections through 2017. Purchasing power parity (PPP) means the real GDP figures reflect actual purchasing power as opposed to simple exchange rate conversion, which is often misleading. It is also interesting to look at real GDP per capita (per person) which provides a better picture of a country's standard of living as presented in Figure 14.2. A good contrast between GDP and GDP per capita is China and Japan. China recently surpassed Japan in total real GDP but still has a long, long way to go to raise per capita GDP to a level comparable to that of Japan and the world's other fully industrialized countries.

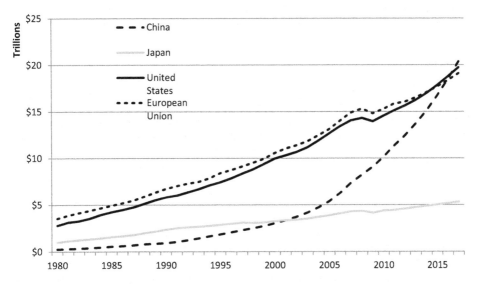

Figure 14.1   Gross Domestic Product Purchasing Power Parity, 1980–2017

Source: International Monetary Fund, World Economic Outlook Database, April 2012

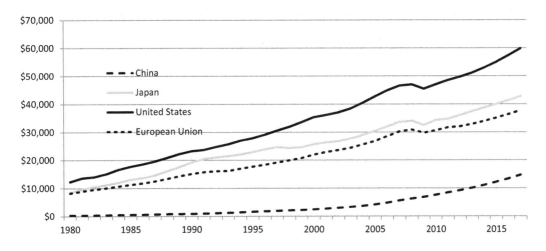

Figure 14.2   Gross Domestic Product per Capita Purchasing Power Parity, 1980–2017

Source: International Monetary Fund, World Economic Outlook Database, April 2012

Some critics argue that economists' focus on real GDP growth reflects excessive materialism. Although this may be true, the fact is that economic growth frees people to better pursue whatever goals they choose, material or otherwise. We have far more leisure time to be with loved ones today than we did in 1900 when six-day work weeks were common. A strong desire to protect the environment has emerged in every wealthy economy (and has not significantly emerged in any country not blessed by strong economic growth). We have already mentioned how economic growth has transformed the lives of the poor. Note that economic growth vastly outweighs any beneficial impact of private charity or government redistribution, though economists generally agree we need some of these things as well. Economic growth empowers us to pursue whatever interests us, including the most noble of goals as well as the most shallow.

How can we best foster economic growth?

Growth is powered largely by a sort of "round about" production process by investments in physical and human capital. Any production tool is an example of physical capital. We can, for example, catch far more fish if we first build a tool, such as a net, that will enable us to be extremely productive, as opposed to just wading into the water and trying to grab fish with our hands. We can make automobiles far more efficiently if we first produce a modern production plant with assembly lines featuring robotics. Likewise, we can be far more productive business managers if we first improve our human capital, if we first study management principles and related disciplines such as accounting and economics rather than rushing into the job market right after high school.

The question then becomes, how do we encourage people to invest the time and resources necessary to complete college or build a modern production plant? One key is that the people who make these investments must be allowed to reap rewards from their investments. People who live in war-torn countries are unlikely to attempt to build factories requiring three years to completion when they know the plant will almost certainly be bombed to rubble before it's completed. Likewise, few people will be willing to pursue years and years of education if in the end taxes are so high that their take home pay is hardly higher than those with little or no education. In economic language, income growth will only flourish if private property rights are well-protected. People must be protected from losing their property to foreign invaders or criminals. They must also, as we saw in the economic freedom data presented in earlier chapters, be protected from excessive taxation or regulation by their own government. At what point do taxes or regulations become excessive, strangling the golden goose of economic growth? How much government is too much government?

The answer is not always clear, or even objective. Economists agree that certain regulations, present to some degree in virtually all countries, are destructive—such as typical price controls or barriers to free international trade. However, other regulations create trade-offs that cannot be objectively weighed. Environmental regulation by definition harms the economy by forcing firms away from lowest-cost production. But if, say, an environmental regulation slows economic growth by 10 percent but makes the air 10 percent cleaner is it worth it? Economics cannot answer such a question since the answer depends on how much an individual values the competing interests of protecting the environment versus improving people's economic lives. Poorer people desperate for economic growth are likely to side with economic well-being while wealthy, satisfied people are more often willing to accept economic sacrifice to serve environmental interests. This is why the wealthiest countries have the most stringent environmental regulations.

Income redistribution also harms economic growth, but is subjectively viewed as a good thing by many. Taxing the wealthy and middle class more may reduce their incentive to be productive, while the lower-income people who receive the redistribution have less incentive to be productive themselves. But economists disagree as to how seriously incentives are harmed. At one end of the spectrum, some economists think these effects are usually small; substantial redistribution will only slightly harm economic growth. At the other end, some think the damage is so extensive that even the poor themselves are routinely harmed by government redistribution because the economy is so weakened. Subjectivity on some issues and inability to precisely measure damages on others leaves us without a clear consensus on what the exact role of government should be.

However, there is some consensus on the broad issues. It is clear that too much government—too much regulation and taxation, strangles economic growth, and the historical data on economic freedom presented in prior chapters confirms that the world's governments are more likely to be too large than too small.

UNEMPLOYMENT

> True individual freedom cannot exist without economic security and independence. People who are hungry and out of a job are the stuff of which dictatorships are made.
>
> Franklin D. Roosevelt, State of the Union, January 11, 1944

Using US Bureau of Labor and Statistics classifications, a person is considered unemployed if they do not have a job, have actively looked for work in the prior four weeks, and are currently available for work. Persons who were not working and were waiting to be recalled to a job from which they had been temporarily laid off are also included as unemployed. The unemployment rate represents the number unemployed as a percent of the labor force. The labor force is the number of persons in the economy who have jobs or are seeking a job, are at least 16 years old, are not serving in the military, and are not institutionalized. In other words, the labor force is all people who are eligible to work in the everyday US economy. Assume for example, that 9,000 people are unemployed in a small city with a total labor force of 100,000. Then the unemployment rate is 9,000/100,000 = 9 percent.

Figure 14.3 shows historical trends in unemployment rates, with some esitmated projections for the years beyond 2011. The ups and downs of the unemployment rate reflect the state of the economy, roughly mirroring the ups and downs of real GDP.

It might seem that 0 percent unemployment would be ideal but this is not the case. Economic growth will inherently destroy some jobs as technology advances, causing some people to be unemployed at least temporarily. When technology allows us to increase output per worker in a given industry it often happens that we then need fewer workers in that industry—some people lose jobs. Two centuries ago, for example, most people were employed in agriculture; as technology improved vast numbers of farm jobs were "destroyed" to the point that very few people in the US work on farms today. Technology allows us to get more food with fewer people producing food. Since it takes so little labor to make food we have people and resources freed up to now make airplanes,

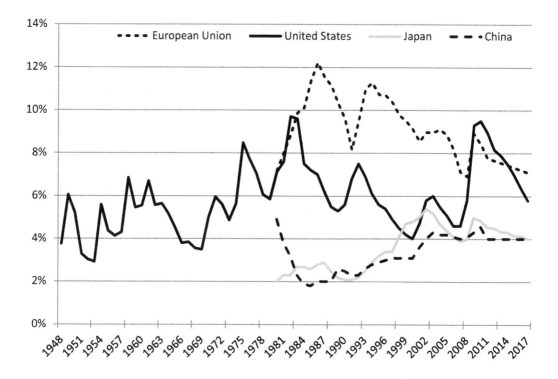

Figure 14.3   Historical Unemployment Rates, 1948–2017

Source: Bureau of Economic Analysis (US 1948–2012); World Bank, World Development Indicators Databank (1961–1980);
International Monetary Fund, World Economic Outlook Database (1981–2017)

cars, computers, and many other products. Indeed, the essence of an advancing economy is that there must be more output per worker in order for there to be a higher real wage, thus creating more consumption per worker and a higher standard of living.

As such, our employment goal is not to try to protect people from losing jobs—this would prohibit virtually any technological progress, but rather to promote a growing economy where there are new, generally improved jobs available to those whose previous jobs are obsolete. Historically, the solution for blacksmiths who lost their jobs as autos replaced horses was not to try to slow progress to save their old jobs, but to become auto workers, construction workers, or something else in the new, advanced economy. Of course, sometimes a worker displaced by technological and economic progress may have great problems in landing a worthwhile new job, may require private charity or public assistance to avoid great hardship in transition. Though there might be considerable quibbling about details, economists would generally support some sort of assistance for such cases—we want to protect people struggling in transition but we do not normally want to protect inefficient jobs. Happily, such transitional assistance is usually not necessary; historically the vast majority of US workers who have lost jobs were able to obtain a new job paying as well or better within eight weeks (Trading Economics, 2012).

There are three different types of unemployment. Unemployment triggered by technological advances is classified as "structural unemployment," a category that also includes job loss from major changes in consumer tastes. "Frictional unemployment" includes job seekers just entering the work force, such as new college graduates, as well as people who are between jobs but will have no significant problem finding work. An example of the latter would be a worker who quit a job in one city in order to relocate to another. Together, structural and frictional unemployment are referred to as "natural unemployment." As the name implies, it is natural for a healthy, dynamic economy to have substantial amounts of frictional and structural unemployment. Although it is difficult to put a precise number on how much unemployment is natural, the Congressional Budget Office forecasts the natural unemployment rate of the US to be 5 percent through 2017 (Brauer, 2007). In other words, if this figure is correct, the US would be considered to be at "full employment" if the unemployment rate is 5 percent.

The third category, "cyclical unemployment," is not considered natural, and is inherently problematic. This is unemployment caused by a recession, a general downturn in the economy where real GDP shrinks. Why do recessions occur? Economists do not agree on a clear answer. Broadly speaking, there are two different schools of thought on this. Some believe that recessions occur naturally in a free enterprise-oriented economy in response to large, unanticipated shifts in supply or demand. Others maintain that a truly free market economy would have few, if any, recessions, and that errant government policies generate the boom-bust business cycle. We will consider this debate later but for now, let us consider less controversial aspects of unemployment.

In many ways the labor market functions like the product markets already discussed; the normal laws of supply and demand apply and determine the level of employment and the pay rate, as shown in Figure 14.4. Conventionally, the price of labor is referred to as the wage, though the term can include salaried workers as well as those paid by the hour. The demand for labor by firms is a normal, downward sloping curve, meaning that, other things equal; a higher wage will result in fewer people being hired. Two separate effects cause employment to fall if wage rises:

1. Higher wages raise production costs, triggering higher prices for output. In turn, higher product prices cause consumers to buy less; with less output produced fewer workers are needed. In other words, if aircraft manufactures must pay more for labor then aircraft will be more expensive, fewer of them will be sold, so firms will employ fewer workers to make less planes.
2. The more expensive labor becomes, the more firms will substitute other inputs for labor, mainly production capital. In other words, higher wages will cause more of a shift toward robotics and other automated production techniques. Thus, higher wages result in fewer aircraft being produced and less labor used per unit of production.

The shape of the labor supply curve is intuitive—the higher the wage the more people want to work. In general, the higher the wage in a particular profession the more people want to work in that profession. Together, supply and demand determine the wage and number of people employed, as shown in Figure 14.4. At W* the market clears; the number of qualified people willing to work equals the number of people employed by firms, L*. This is a graphical presentation of the idea of full employment; at W* and L* everyone qualified and willing to work is able to find work in a reasonable time period.

Many factors influence unemployment rates but in a mechanical sense unemployment is caused by the wage being too high, above the equilibrium level. In this regard, labor markets are different from most output markets; wages have some tendency to get stuck above the equilibrium whereas product prices normally fall to equilibrium. If, for example, demand for air travel plummets then the price plummets, too. Airlines will

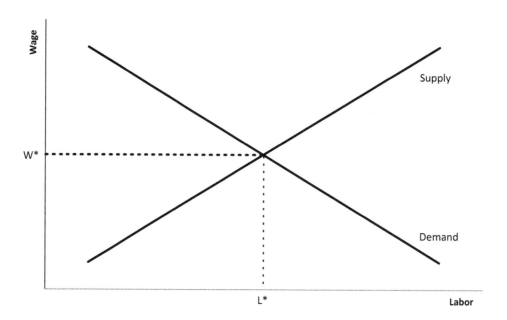

Figure 14.4   Supply and Demand for Labor

not get a "good" price but will be able to sell at the "going" price, the new equilibrium. However, a similar drop in the demand for pilots will not result in a large, immediate drop in their salaries; instead, the wage is stuck well above equilibrium, as in Figure 14.5. Some unemployed pilots will be equally qualified as currently employed pilots and happy to accept the current pay but will find that "no one is hiring." Serious unemployment problems result from this tendency for wages to be "sticky," to remain above equilibrium for many months, possibly years.

Of course, those who remain employed are pleased that their wage did not abruptly fall, but for the overall economy it would be beneficial if wages were more flexible. To understand this, consider the impact on real GDP. Clearly, real GDP will be lower if many people are not working, not producing anything. If the wage falls to allow more people to get back to work then we will have more total output, a higher real GDP. So, whatever is happening to nominal wages, real income unambiguously rises. A more flexible wage would reduce unemployment and make recessions less severe. So, it is worth considering policy changes that would make wages less rigid, which would encourage workers to accept lower wages and encourage firms to offer lower pay rather than so much reliance on layoffs.

Consider government regulation that explicitly keeps wages from falling; for example, the minimum wage. In essence, the minimum wage keeps the wage perpetually above equilibrium, as in Figure 14.5, so that maximum employment and real GDP can never be achieved. Most economists agree that there are better ways to supplement income for low-wage workers that desperately need assistance; since minimum wage regulations tend to reduce employment and, hence, real income. For low earners who do need help we might better supplement their income with tax credits or food stamps. Such programs will cost something but will be balanced by overall economic gains if there are more people working productively rather than remaining unemployed. Abolishing the minimum wage, replaced by income supplements where needed, would also allow the creation of

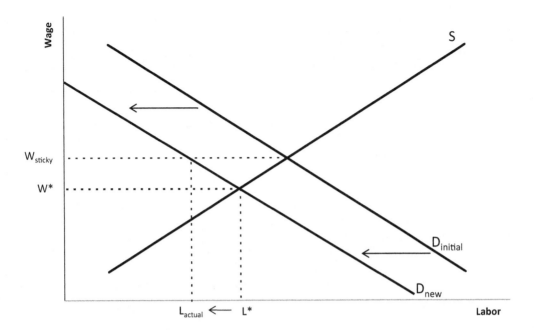

Figure 14.5   **Wage is "Stuck" above Equilibrium; Labor is Less than Full Employment**

more entry level, trainee-type jobs for young people and others with little experience and/ or lower skill levels—younger, lower-skill workers tend to suffer disproportionately high unemployment rates (BLS, 2011).

Another source of wage stickiness is a low savings rate. If workers are living paycheck to paycheck then even a modest pay cut will likely disrupt their lives, and perhaps affect their productivity at work. Alternatively, a high-saving worker can readily absorb temporary pay cuts in rough times. No one is happy about a wage cut, even if temporary and reasonable in a tough economy, but if a person normally saves 20 percent per paycheck then any pay cut of 20 percent or less will have little or no impact on current consumption and suffer no immediate hardship. Indeed, if a worker has accumulated savings then even a cut over 20 percent may not seriously reduce current consumption since the worker can draw funds from savings. A simple way to increase saving is to stop taxing it, or at least to reduce taxes on saving. Saving encompasses anything that is not current consumption; this includes buying stocks and bonds, for instance, as well as putting funds in a bank savings account. Thus, lower taxes on interest income, corporate income, capital gains, and dividends will increase savings and thereby make wages more flexible and reduce unemployment. Of course, higher savings rates also enables faster capital accumulation; the more savings available, the lower interest rates and the cost of capital, the easier it is for firms to borrow or sell stock to finance expansion. Naturally, this increases economic growth; suggesting higher savings rates will improve growth as well as increase stability and reduce unemployment. On the other hand, a common political objection to tax cuts on savings is that wealthier people tend to save more and would thus directly benefit more from such tax cuts. However, everyone benefits from lower unemployment and a stronger economy. Moreover, cutting or even abolishing taxes on savings can be offset by higher taxation of consumption so that the relative tax burdens of the wealthy and middle class do not change. In other words, it is possible to have pro-savings tax reform without changing the relative tax burden (Seidman, 2003).

Other policy changes that would reduce unemployment create some trade-offs and would be more controversial. For instance, the higher unemployment benefits offered to those out of work, and the longer those benefits last, the higher unemployment rates will tend to be. This is not to say that every unemployed person will lazily sit back and collect benefits until they run out, though some probably will. But even conscientious workers will be choosier and take more time when benefits are generous (Branek, 2011). Of course, few societies would want to scrap all welfare programs in order to maximize incentives to work; we face some trade-offs in economic growth and employment versus income redistribution, as mentioned earlier. It has been argued that any regulation intended to help workers will increase employers' costs, pushing some marginally profitable businesses to shut down and leading stronger businesses to reduce growth and job creation. Some labor regulations may still be worth their weight, but it is important to weigh costs and benefits carefully given that even the best regulations will virtually always have costs that increase unemployment.

Some economists believe that inflation, though it generates significant costs and risk, can sometimes be used to stimulate employment. Before considering that possibility let us begin with a basic discussion of inflation.

INFLATION

Inflation is a general rise in prices of goods and services in an economy over time, the implication being that a unit of currency has less purchasing power. In order to measure inflation, the Bureau of Labor Statistics (BLS) uses several standards, but most commonly the Consumer Price Index (CPI). The CPI is a measure of the average change over time in the prices paid by urban consumers for a market basket of consumer goods and services.

Figure 14.6 depicts historical inflation rates. Notice the surge in inflation in the 1970s in the US, Europe, and Japan. In 1971 the industrialized world abandoned the last official link between paper currencies and gold, the goal of which was to let exchange rates be decided by the market, just as it decides the price and quantity of nearly everything else. But by doing so, this policy effectively freed governments to create as much paper money, and resulting inflation, as they chose. The dominant economic theory of the time saw inflation as a possible stimulant of economic growth. Although there is still considerable support for this theory, the results of this 1970s inflation experiment were very disappointing and attitudes toward inflation adjusted accordingly. This is reflected in the changes in monetary policy since, resulting in much lower inflation rates in the industrialized world beginning in the 1980s and continuing from the 1990s to present times.

Although many factors can influence the rate of inflation, a main factor is the supply of paper currency and other liquid assets such as cash deposited in checking accounts—the money supply. It is important to point out that in common conversation, the terms money and income are sometimes used interchangeably but they are very different things. Inflation is not caused by rising income but it is commonly caused by governments "printing too much money." (They generally create money electronically rather than literally printing it but the result is the same.) Let us consider in more detail exactly how governments control the money supply.

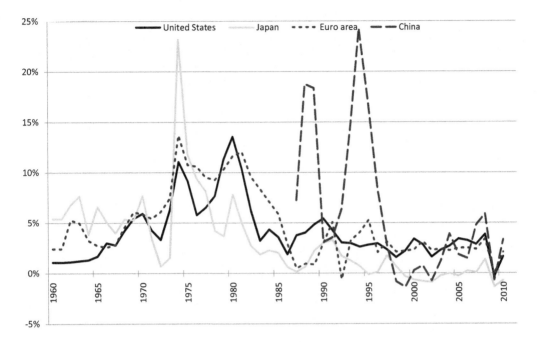

Figure 14.6   Annual Percent Change in CPI, 1961–2011

Source: Bureau of Economic Analysis (US); World Bank, World Development Indicators Databank

Money creation by government central banks

Achieving price stability is not only important in itself, it is also central to attaining the Federal Reserve's other mandate objectives of maximum sustainable employment and moderate long-term interest rates.

> Ben Bernanke, Testimony to the Committee on Financial Services,
> US House of Representatives, February 15, 2006

Government "central banks" often have a hybrid structure that is technically a mix of private and government control, such as that of the Federal Reserve System — commonly referred to as "the Fed" in the US. However, any "independence" from political control on the part of these institutions is generally more style than substance. The Fed is controlled by political appointees and all its "profits" are turned over to the US Treasury; similar arrangements exist at other countries' central banks. To put it simply, central bank policies are generally driven by the politicians in power.

The Fed and other government central banks have three basic methods of changing the money supply: open market operations, the 'discount rate," and banks' required reserve ratios. Open market operations are the closest to the euphemism of printing money. To increase the money supply in this case, the central bank creates new money and then buys some financial assets, most commonly government bonds, thus releasing the newly printed money into the market. Let us take a moment to explain in more detail. To increase debt, governments sell securities which are basically IOUs. If, for example, you bought a ten-year bond from the US Treasury they would promise to send you interest payments over those ten years and would pay back the principle at the end of the ten years; you would receive the amount you lent plus interest. If you buy such a bond you do not have to hold it to maturity; you can at any time sell it in the open market to whoever wants to buy it. Thus, when the Fed creates new money and then uses it to buy existing US bonds in this market, the bond sellers walk away with newly created money, and the money supply is increased. Of course, if the Fed wants to decrease the money supply, a much less common occurrence, then it would take previously purchased bonds and sell them back to the open market. The Fed and other central bank counterparts may buy and sell various other assets at times, such as foreign currency and various mortgage-backed securities, depending on the particulars of politics and economic policy.

The "discount rate" is the interest rate at which the central bank will lend money to private banks. A lower discount rate encourages more borrowing and thus puts more cash into circulation in the same general way as if the Fed bought bonds. Likewise, if the discount rate is raised, banks will borrow less cash thus slowing the growth of money. In the US the discount rate is not commonly used as a main instrument of monetary policy; the Fed usually acts as a "lender of last resort" and prefers banks to attract private capital. In theory, this can help inject some degree of market discipline, discouraging imprudently risky bank practices. However, this impact is dampened by the fact that central banks often do stand ready to lend to banks that find themselves in financial trouble, especially those deemed "too big to fail."

Lastly, the central bank has regulatory control over reserve requirements for private banks. Lower required cash reserves essentially allow banks to take more cash out of vaults and put it back into circulation in the economy by having more money available to lend to borrowers. In short, lower required reserves increases money supply while higher required reserves decrease money supply. However, this is rarely used to influence the money supply because it is too powerful to easily calibrate; wild swings in bank lending and the money supply can be induced by relatively minor changes in required reserves. The US Fed learned this lesson the hard way in 1937 when it abruptly raised required reserves, causing lending activity to plummet, and plunging an economy that had been recovering back into depression.

To summarize, the government can increase the money supply through an open market purchase, using newly created money to buy some asset, usually an existing government bond. Alternatively, the government could increase the money supply by reducing the discount rate or by reducing required bank cash reserves, though the latter is riskier and seldom used. Governments do not have to literally print more money to increase the money supply but the effects are the same. Let us turn to an examination of those effects.

Costs of inflation

Economists agree that excessive inflation—certainly anything in the double digit range (10 percent or more), can wreck an economy. However, the impact of moderate inflation is more controversial. In fact, some believe that moderate doses of inflation at the right time can actually stimulate economic growth. Let us begin, though, with the damage done by inflation where there is broad agreement in general terms.

One category of inflation costs are known as "menu costs," which are the *physical* costs a firm incurs as a result of changing prices, such as a restaurant needing to print new menus.

The more inflation there is the more firms must print new catalogues, change billboards, and commercials that mention price, and so on. It is important to note that menu costs are not zero in a zero inflation economy; things like seasonality and technological change will cause some prices to change although menu costs are minimized with zero percent inflation. Zero inflation does not mean that no price is changing; it simply means that the overall value of the currency is stable, with any price increases being offset by price decreases. Although menu costs can be significant they are unlikely to cause great harm in and of themselves.

Potentially far more damaging is the fact that inflation arbitrarily redistributes income. This in turn makes contract and long-run planning more expensive and, perhaps, even impossible. For example, consider the interest rate on a 30-year mortgage, a key interest rate for the economy. Suppose both lender and borrower expect inflation over the life of the loan to be 2 percent per year. Accordingly, the mortgage contract is made with a 4 percent nominal rate (the rate actually charged) which means both parties estimate the real interest rate to be 2 percent (the nominal rate minus inflation). Suppose instead that inflation surges to 10 percent per year for 30 years. This means the real interest rate turns out to be -6 percent (4 percent–10 percent). In other words, the lender suffers a huge loss even if the borrower fulfills the contract; the lender gets 6 percent more dollars than was lent but since each dollar is worth 10 percent less the lender receives far less real value back than was lent. As with any zero-sum transaction, the lender's loss is the borrower's gain, exemplifying an artificial redistribution of income.

Typically, wages may not be instantly adjusted for inflation but the lag is seldom more than a year. As such, even if a worker's real wage is constant this means that his nominal wages will also rise about 10 percent annually. At that rate, his nominal wage will double about every seven years! When this is the case, loan payments, like those in the previous example, are easier and easier to afford because in real terms the payments are worth less and less; they are a smaller and smaller percentage of the borrower's income. In itself, this complete distortion of the original contract, this redistribution of income from lender to borrower, nets out to zero, whatever the borrower gains the lender loses.

However, the story does not end there. Because lenders face this risk from inflation they will charge a higher interest rate, a risk premium to cover the risk of higher inflation. With higher interest rates, fewer people can afford to buy houses and cars, and fewer businesses can afford to expand. Resultantly, economic growth gets slower and slower the more the risk of higher inflation rises. Note in this scenario it is not necessary that inflation be high, just significant risk of future inflation can devastate economic growth.

Another problem with inflation, potentially even more damaging than the inhibition of contracts from inflation risk, is that inflation distorts the price system leading to misallocation of resources. The phrase "misallocation of resources" sounds much more mild than the economic devastation that can result from this problem. The "great recession" triggered by the collapse of the housing market beginning in 2008 is an example of just how serious mistakes in allocating resources can be. There were multiple causes of the housing collapse but most, though not all, economists agree that distortion caused by monetary inflation played a significant role. To understand this best, it will be useful to revisit the basic consequences of money creation by governments.

When government prints "too much" paper money the resulting inflation does not necessarily spread evenly over the economy. In other words, 2 percent inflation can be comprised of some prices surging far more than 2 percent while other prices do not rise or even fall. Sometimes, the initial inflation can be very concentrated in a particular sector of the economy, helping to create a "bubble" in the price of a certain asset. The term bubble is used to convey an artificial rise in price that is analogous with someone blowing a bubble bigger and bigger. As with any bubble, there is a point where it cannot sustain its size and growing internal pressure eventually forces it to burst. Relating this metaphor to economics, when an artificially inflated price reaches its maximum the next step is a sudden collapse in price. This, of course, is what we saw in the housing market in the 2000s.

The basic economics of this relate to confusion between nominal and real prices, confusion over what is being driven by real forces and what is driven by a general inflation of the currency. Specifically, when producers believe that an increase in demand for their product is driven merely by a general inflation that happens to have hit their industry earlier than others, they raise prices but do not produce more output. In other words, the appropriate response to 10 percent inflation for the average business is to raise their own prices 10 percent, and leave real variables like output and employment unchanged. Producers should only increase output if demand for their particular product has risen relative to other products. A good rule of thumb for managers to follow would be to increase output only if real demand increases and leave output unchanged only if nominal, overall demand changes.

However, it is not always easy to tell whether rising demand is driven by the early stages of general inflation or driven by real forces. Furthermore, firms strive to respond quickly to real changes in demand, in part because they want to increase supply ahead of competitors. If firms mistakenly interpret an early rise in inflation for a rise in real demand for their product then they respond by hiring more people, buying more raw materials and other inputs, and investing in more capital, such as more aircraft or more production plants. To finance all this they often borrow heavily, partly to take advantage of government tax systems that encourage debt financing. Thus, inflation can, in this situation, trigger a temporary boom in production and employment. But this type of boom is a sort of castle in the air; it has no rational foundation and is based on mistakes and misallocation of resources. Such booms tend to turn into busts; the bubble eventually bursts and collapses.

These booms usually turn into busts because misallocations are not easily undone. If for example, a developer caught up in this artificial boom contractually commits to building a large number of houses, he cannot readily change course once he realizes his error. He must either breach the contract or eventually be stuck with a large inventory of homes worth less than they cost to produce. In either case the business will incur major losses, often enough to trigger bankruptcy; especially if the builder went into debt to finance this project. The sudden collapse can cause massive layoffs and job loss. Because there has been overproduction with inventories now artificially high, the number of jobs lost from the bust will typically be larger, often much larger, than the jobs gained during the boom. All the workers that were mistakenly drawn into construction during the boom are suddenly thrown out of work. Problems of this ilk are probably the most severe of those caused by inflation.

Again, the above housing example is for illustration; we do not wish to imply that excessive money printing was the sole, or even necessarily the principal cause of the recent housing debacle. Many factors played a part—including ill-conceived government guarantees on over 70 percent of high-risk mortgages, wildly optimistic risk assessments by credit rating agencies, and rigid assumptions by government regulators and most private parties held on the belief that mortgage-backed securities were inherently low risk.[1] And although most economists do see "easy money" as a substantial part of the problem it should be noted that some prominent economists, such as Federal Reserve President Alan Greenspan, and his successor, Ben Bernanke deny that monetary policy played a significant role in the crisis at all.

In any event, the key point to be made is that inflation can cause major problems in the economy. The damage done by higher menu costs, inhibition of long-term contracts and price distortions that misallocate resources is only a portion of all the harm that inflation can cause and not an exhaustive list. However, this is hopefully enough to demonstrate the serious risk that inflation poses to an economy. The possibility of financial bubbles and subsequent collapse and crisis is especially dangerous.

MONETARY POLICY

After contemplating the above discussion on the ills of inflation it may seem strange that many—though certainly not all—economists support the use of moderate inflation as a temporary stimulant. The theory that inflation can be beneficial was popularized by J. M. Keynes beginning in the 1930s, and current supporters of his theories are often referred to as "Keynesians." For Keynesians, the economic impact of inflation is roughly analogous to the health impact of a powerful drug, such as morphine. Just as too much morphine can harm, even kill the patient, Keynesians agree that too much inflation can wreck an economy. However, just as moderate doses of morphine can help a very sick patient, moderate inflation might help a weak economy. Morphine creates an artificial high in the patient where serious pain is relieved even though the underlying malady is unchanged. But the reduction in pain helps the patient sleep and allows the body to heal itself. Likewise, printing lots of money and dumping it into the economy will, at best, produce a sort of artificial growth that cannot be sustained in the long run. But, if all goes well, by the time the artificial stimulus wears off, natural forces such as technological advances and accumulation of physical and human capital, will kick in to keep the economy expanding.

We just discussed an example of how inflation might temporarily stimulate economic growth, essentially by getting people to mistake a general rise in prices with a favorable change in relative or real prices, as perhaps, happened in the manic production of houses phase before its collapse. From the Keynesian perspective, an economy can generally enjoy the boom without suffering the bust *if* moderate inflation policies are implemented only when the economy is weak. Recall, economies with substantial elements of free enterprise have normally grown as population grows, as the stock of human and physical capital grows, and as technology advances. For the US, the summation of these elements has equated into about 3 percent per year real GDP growth over the long term.

In a recession, growth is temporarily negative but with time positive growth resumes. Suppose that, during a recession, instead of just waiting for the economy to recover on its own, the Fed increases the money supply to cause moderate inflation that tricks the economy into expanding. According to Keynesians, as long as we do not increase the money supply too much for too long, the natural forces of growth will keep the economy going; we only get the boom/bust scenario if have too much monetary stimulus for too long.

---

1   A good, comprehensive study of the housing crisis can be found in Sowell (2009). For a more concise overview see Wallison (2010) or White (2008).

To return to the housing example, the Fed aggressively expanded the money supply in the 2001 recession, which was short and mild, in part because the housing market stayed buoyant throughout the recession (Ferguson, 2004). In reality, the Fed kept the printing presses going aggressively for several more years but suppose, instead, the Fed had begun to gradually reduce monetary stimulus in 2002 and imagine that by 2003 the Fed pursued a more neutral, steady monetary policy. It may well be that this would have been a great success in Keynesian monetary policy; perhaps the housing boom through 2006 and subsequent collapse would never have happened, or at least been substantially more moderate. We might have enjoyed a shortening of the 2001 recession without any future "great recession." Again, we use the 2000s housing market as a timely, historically important illustration; we do not mean to imply that the housing fiasco sprang exclusively or even necessarily principally from excessive money creation. The broader point is that monetary stimulus when there is plenty of "slack" in the economy can be effective and may not trigger the boom/bust cycle so greatly feared.

In conclusion, inflation used judiciously may help reduce the length and severity of recessions. On the other hand, there are serious costs of inflation and the attempt at stimulus may backfire and make things ultimately worse. As, perhaps, happened with the monetary stimulus of the 2000s. So, is inflation stimulus worth the risk? Keynesians would say that it is, would argue that the economy is prone to get stuck in recession for long periods if stimulus is not applied. They would agree there are risks to inflation stimulus and it can backfire but see these risks as being less than the risk of leaving the economy to its own devices.

Critics of the Keynesian view

Critics of the Keynesian view — most famously, perhaps, Milton Friedman and Frederick Hayek, believe the economy would perform best if we simply pursued zero inflation at all times and abandoned any idea of monetary stimulus. One problem they stress is that it is not so easy to time Keynesian monetary policy because of the "recognition lag" and "impact lag." The recognition lag refers to the fact that we cannot accurately see exactly what state the economy is in; we only see where the economy was a few months ago. Consider that the monthly change in real GDP is a fraction of 1 percent. In good times the monthly growth is positive but tiny, averaging something like 0.25 percent. In a recession, real GDP is contracting but again by a small amount; roughly 0.08 percent monthly in the 2001 recession and still only about 0.4 percent in the great recession of 2007–2009. Since the economy is such a large, complex system with relatively small month-to-month aggregate changes, it is very hard to tell whether the economy is growing reasonably well or not. If we assume we are in a recession but in reality reasonable growth has already resumed than monetary stimulus is likely to cause net harm, maybe even a disastrous boom/bust cycle. Adding to this problem is the fact that today's "money printing" will not impact aggregate demand much for at least three months and sometimes much longer, up to about 14 months; this is known as the impact lag. The impact lag follows from the fact that new money must be spent and re-spent many times before the effects ripple throughout the economy strongly enough to stimulate aggregate demand and inflation.

Some have likened monetary policy to driving a car with the front and side windows covered so that you can only see out the back window, where you were. Then, when you decide to apply the brakes or gas the car responds about three to 14 months later! Of course, if Keynesians are correct, once the car is stuck in the mud you can be fairly sure it is still stuck so you should go ahead and hit the gas, albeit not too hard for too long! Economists simply do not understand the business cycle and the effects of money supply well enough to reach a unanimous conclusion on monetary policy. There is a consensus that inflation is dangerous and, if it is to be used in an attempt at stimulus, it should be used sparingly when, we think, the economy is still weak.

FISCAL POLICY

In addition to monetary policy, fiscal policy is another government option for influencing the macro economy. Fiscal policy concerns government spending, taxing, and borrowing. Though the two often coordinate, fiscal policy is completely separate from the money supply. An increase in government spending, for example, is funded by either more tax revenues or more government borrowing — government spending does not change the total money supply; it just moves existing money around. Likewise, an increase in government debt does not increase the money supply, though, politically, massive debt increases often are correlated with massive increases in the money supply. When a government runs a deficit this means that spending is greater than taxes collected, debt must be accumulated to make up the difference, and bonds are sold as a way of borrowing some of the existing money in circulation. A government surplus, comparatively rare, occurs when government spending is less than taxes collected, in which case the government buys back some of its own bonds and retires outstanding debt.

As with monetary policy, there is a difference of opinion as to whether or not governments can effectively use fiscal policy to manage aggregate demand and smooth out some of the ups and downs of the economy. Keynesians, again, favor active government intervention depending on the state of the economy. Essentially, Keynesians believe fiscal policy should be unstable in order to somewhat counter the natural instability of the economy. Specifically, their policy calls for intentionally running large deficits (government spending way above taxes collected) when the economy is weak, running surpluses (spending less than taxes collected) when economic growth is unusually and unsustainably strong, and a roughly balanced budget in normal times. In their view, a budget deficit has the same basic effect as an increase in the money supply; it drives up spending and demand, another way of causing inflation.

Underlying the Keynesian view is the assumption that the economy is naturally prone to some boom and bust instability. Keynesians believe, for example, that any drop in the economy will be magnified by a multiplier effect. If, for instance, one firm lays off a large group of workers then those workers will likely reduce spending which could trigger layoffs at other firms. However, non-Keynesians point out that any major decline in the economy will tend to reduce prices and, often, interest rates, creating bargains that tend to drive up spending, which can partially offset the initial decline. In other words, rather than being cursed with a multiplier that causes any problem to snowball, the economy may be protected by a dampener that naturally buoys the economy when it tends to weaken. For example, when major airlines reduce routes, lay off workers, and sell aircraft this creates opportunities for upstart airlines to expand, and makes their expansion more affordable since aircraft and workers will likely be available at discounted rates. Thus, the decline by the major airlines is partially offset by the expansion of the newcomers; the net loss to the economy is less than the initial drop.

We can see historical examples of a dampener effect but we can also see examples of a multiplier effect. The question is: Which is more typical? Is the overall economy generally burdened by a multiplier or protected by a dampener? Unfortunately, the economy has thus far proved to be too complex for us to definitively answer this question and debates in this area continue. While we cannot resolve such issues we can explain major aspects of these opposing viewpoints. Let us begin with a consideration of the Keynesian theory; then we will consider the opposing view.

A basic Keynesian model

In the Keynesian fiscal policy model, the government can use its massive capacity to spend, tax, and borrow to manipulate aggregate demand. Aggregate demand is the total demand for goods and services in any economy. It encompasses the demand generated by all segments of the economy to include consumers, firms, and the government. As with all demand curves, the aggregate demand curve slopes downward and for the same reason. As the aggregate price level, including interest rates, goes down, people feel wealthier and are willing to spend more on goods and services. In addition, firms are more likely to undertake profitable investments with lower interest rates. Aggregate supply (AS) is the total supply of goods and services within the economy; sloping upward as in Figure 14.7 below. Coordination, on the other hand, is simply the matching of aggregate demand to aggregate supply. From a macroeconomic point of view, the trouble that is perceived in the economy during a recession is that there may be a persistent lack of coordination in the economy between aggregate demand and supply. That is, there might be an imbalance between the specific goods and services demanded and those supplied because of problems from sticky wages and prices. To clarify, during periods of cyclical

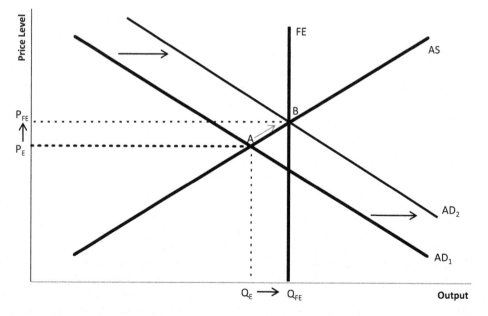

Figure 14.7   **Keynesian Model of Aggregate Supply and Demand Below Full Employment**

unemployment both employment and output are below full employment equilibrium. Thus, the AS-AD_1 equilibrium (point A) in this example is below full employment output (Q_{FE}).

In this circumstance, Keynesian fiscal policy calls for the government to intentionally run a large deficit, accomplished by a large increase in government spending, a large tax cut, or some combination of the two. Theoretically this will shift aggregate demand to the right (from AD_1 to AD_2 in Figure 14.7) and, if all goes ideally, boost the economy back to full employment. Additionally, the increase in AD will increase output from Q_E to Q_{FE}, but will also be accompanied by an increase in overall price level from P_E to P_{FE} resulting in a new equilibrium at point B in Figure 14.7. This increase in price level could be a natural response to the increase in aggregate demand, but it may also simply take the form of artificial inflation if the price increase is not equally matched with an increase in output as demonstrated below in Figure 14.8. Figure 14.8 depicts a case where the economy is actually operating above full employment, where AD_1 intersects AS. This is the situation discussed earlier where inflation has essentially "tricked" people into expanding the economy to an unsustainable level, as, perhaps, happened in the housing industry in the 2000s. While this may not seem so bad to have "too much" output, employment, and economic growth, the longer this sort of artificial boom continues the worse the collapse tends to be when the bubble bursts. This situation calls for restraint rather than stimulus, a government surplus rather than a deficit. By raising taxes and/or decreasing government spending, if all goes ideally, AD_1 will shift down to AD_2 and the economy will return to a sustainable full employment.

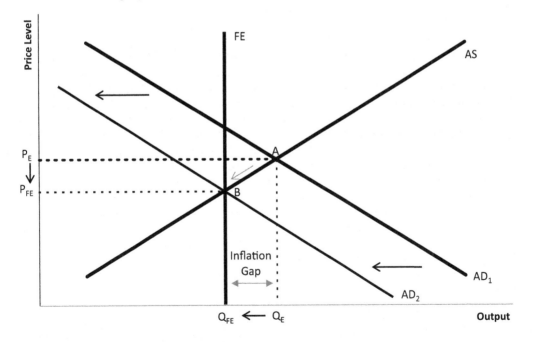

Figure 14.8   Keynesian Model of Aggregate Supply and Demand Below Above Full Employment

The question then becomes should the policy be implemented by an increase in government spending or tax cuts to stimulate private spending by consumers and businesses? Advocates of direct government spending argue that government spending through the construction of roads, the defense budget, and other government-funded projects, will offset decreased private spending. The theory is that government spending is a more direct and immediate stimulus to the economy than tax cuts, or, for that matter, monetary policy. Advocates for decreased taxation point out that since the problem is one of coordination, there is little likelihood that government spending will affect the correct areas of the economy and a strong likelihood that it will only benefit politically favored sectors. On the other side, a general tax cut will provide immediate incentives to consumers to purchase products and firms to invest in response to the added consumption. Firms are then more likely to produce those goods and services more highly valued in the economy thereby acting to alleviate the coordination problem. Also, reductions in tax rates will increase aggregate supply by improving incentives to work, save, and invest. Economists are unsure as to how powerfully aggregate supply is stimulated by cuts in tax rates but all agree there is a positive effect.

Why might the Keynesian approach fail?

Opponents see several problems with the Keynesian approach. For starters, it is possible that whatever fiscal policy undertaken may not move aggregate demand throughout the economy (the AD curve). For example, if the government borrows from Peter to send a "stimulus" check to Paula it stands to reason that Paula's spending may increase but it seems that Peter's spending will go down. When government borrows to finance its spending the lenders have less left over to spend themselves. Thus, government spending tends to *crowd out* private spending. Keynesians agree that in normal times, an increase in deficit spending will tend to cause an off-setting decrease in private spending. However, they argue that when the economy is weak, this crowding out is much less than 100 percent so that total spending does rise. Some of the funds Peter lends to the government would have been kept in cash reserves rather than spent by him; so when government borrows his funds and then spends them there is a net increase in overall spending.

Evidence on this point is not conclusive but let us assume that deficit spending is indeed a stimulus. The next problem, as with monetary policy, is timing. In fact, this problem is much worse for fiscal policy because, in addition to the recognition and impact lags already cited there is also a serious *implementation lag*. Major changes in fiscal policy require that new laws be enacted, which typically takes many months, often even years to achieve. As a result, there is significant risk that by the time a law is passed and then later has its impact on the economy, it will be exactly the wrong policy for the economy's current state. For example, a major stimulus that hits when the economy is already growing rapidly with rising inflation may very well lead to the boom/bust scenario we discussed earlier.

Another problem with Keynesian policy is that people may try to anticipate government policy rather than passively allowing themselves to be manipulated. Consider, for instance, the common policy of offering investment tax credits for businesses that invest in plants and equipment when the economy is weak. This policy could end up destabilizing the economy as businesses come to anticipate it. Suppose the economy shows slight signs of slowing but is actually still growing solidly. However, when business leaders see the preliminary signs of weakness suppose many decide to temporarily delay investments they were planning, thinking that if they wait then maybe the economy will become weak enough that government will give them a new tax credit for their investments. The problem, of course, is that if many businesses cut back their investment spending at the same time, waiting for tax credits, this can in itself bring on a recession!

The last challenge for Keynesian policy we will mention is the impact of real world politics. Recall, ideal Keynesian policy would have large deficits only when the economy is weak, with at least roughly balanced budgets in normal times, and surpluses in unusually good years to help offset the deficits of the bad years. In reality, no government on earth has come remotely close to this ideal. In the real political world, most governments engage in deficit spending almost all the time, with a rare balanced budget or slight surplus occurring now and then, mostly by accident. It is true, however, that governments do generally run much larger deficits than usual when the economy is weak, so the policy may still offer some stimulus when most needed.

Because of experience with the problems cited above, there has not been a strong consensus in support of Keynesian policy since the late 1960s or early 1970s. Still, many modern economists are basically moderate Keynesians, believing that government intervention, however imperfect and problematic, is still better than the alternative of leaving the economy to its own devices—which, in their view, would increase the average severity of recessions. Critics, however, maintain that Keynesian policy has actually made things worse and that the theory that government deficits can be a stimulus has helped make continuous deficit spending politically acceptable. In other words, the practical political impact of Keynesian theory in the long run may be to drown economies in a tidal wave of government debt.

THE SOVEREIGN (GOVERNMENT) DEBT CRISIS

How European sovereign debt became the new subprime is a story with many culprits, including governments that borrowed beyond their means, regulators who permitted banks to treat the bonds as risk-free and investors who for too long did not make much of a distinction between the bonds of troubled economies like Greece and Italy and those issued by the rock-solid Germany.

The New York Times, November 11, 2011

Although it is debatable as to how much proponents of Keynesian theory are responsible, there is no doubt that massive increases in government debt have occurred and pose a serious threat to virtually all the world's economies as some have already begun to experience. When governments are overwhelmed by debt the typical results are either default (not fully paying back lenders) or massive money creation that pays the debt off but tends to derail the economy through hyperinflation. Of course, default on a large scale also tends to wreck the economy since many private parties holding the government debt will be driven into bankruptcy themselves when government fails to repay. As this text is being written, Greece is already in turmoil because of excessive government sovereign debt, with Spain, Italy, Ireland, and Portugal, among others, on the brink of serious problems. If present trends continue, virtually every government will eventually find itself in a sovereign debt crisis with the sort of economic decline that Greece has been experiencing. Figure 14.9 shows historical trends in government central debt as a

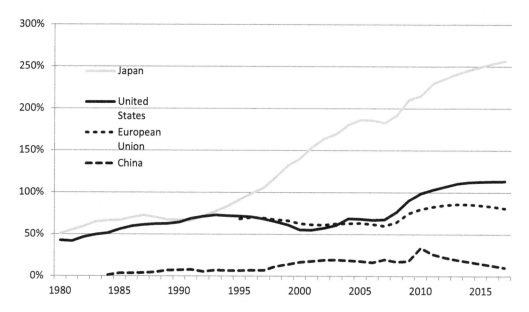

Figure 14.9   Central Government Debt as Percent of Gross Domestic Product

Source: International Monetary Fund, World Economic Outlook, April 2012

percentage of GDP[2] from 1980 through 2011, with forecasts beyond that time. Note that these figures do not include debt from lower government levels—states, provinces, or cities—which is also considerable. The forecasts of national government debt assume a return to solid real GDP growth in the US and EU, an assumption that looks more and more likely to be over-optimistic.

Although details differ slightly across countries, the key problems are basically the same. Spending on government health care and pension programs has surged and is set to explode as the "baby boomer" generation enters retirement. A key fiscal flaw in these government programs is that they are basically financed on a "pay as you go" basis where there is no real savings mechanism. In the US, for example, social security taxes paid in by current workers are paid out to current retirees; they are not deposited in some account for that taxpayer's future retirement, as they are in a typical private pension.[3] This was not an obvious flaw, at least to the rationally ignorant, as long as there were many more taxpayers than retirees collecting benefits. Unfortunately, those days are drawing to a close; with the ratio of workers to retiree falling now, and set to plummet in the future, the program is essentially a ticking debt time bomb. Analogies to pyramid or ponzi schemes are not so far off. Figure 14.10 shows historical and future trends in the proportion of

---

2   This is the best way to consider the burden of debt, analogous to an individual considering debt as a percentage of her income.
3   Until recently, because there were so few retirees, there actually was more social security taxes collected than payments made to retirees. However, this surplus was generally more than offset by deficit spending elsewhere. Moreover, since any surplus in social security was used to buy US bonds, the Social Security Trust Fund has nothing but these bonds, government IOUs, in it. While administrators can resell these bonds for cash this is not really any different than the Treasury selling new bonds. In other words, the system was unable to directly save for the future; at best it could only accumulate US bonds, liabilities to the US not assets. However, even this was not achieved since the surpluses were smaller than general deficits in other government spending.

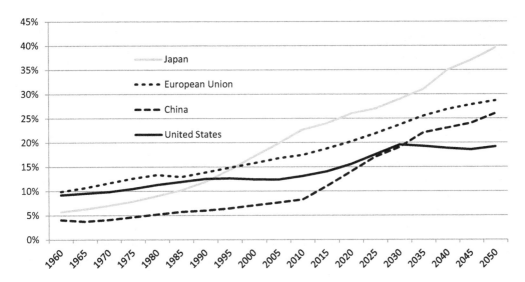

Figure 14.10   Percent of Population over 65 Years of Age

Source: US Census Bureau, Eurostat, Statistics Bureau of Japan

the population over 65 years old, a good proxy for the problems posed by the defective government fiscal mechanism that politicians have burdened economies with. Problems in the US are projected to be at their worse depths in about 2030. Comparatively, much of Europe, China, and Japan will be in even worse shape.

The situation is similar for separate government pension plans for government workers; workers have been promised very generous pensions, often in six figures, with nowhere near enough funds set aside to help fund these promised benefits (Buttonwood, 2011). The financial situation is typically even worse for government-provided health care; Medicare in the US, for instance, is in far worse financial condition than even Social Security (Lotterman, 2011). A full discussion of the pros and cons of socialized medicine is beyond our scope but suffice it to say that socialized medicine is least problematic when there are many healthy, young taxpayers to fund if for relatively few elderly and sick people. But, as with government retirement systems, the best days of socialized medicine are behind us. Again, the recent turmoil in Greece is likely to be only a mild foreshadowing of the fiscal pain awaiting them and most of the world's other economies, unless there are dramatic changes.

CAN THE US AVOID GREECE'S FATE?

The US has a few advantages over most countries in regard to our mounting debt problems. Culturally, the US may be more open than most societies to cuts in government spending that shift more responsibility to individuals to pay more for their own health care and retirement. This is not certain since few if any cuts have yet emerged but there are a few encouraging signs. For instance, the American Association of Retired Persons (AARP), arguably the most powerful lobbying force in US politics, has historically been staunchly opposed to any reduction in Social Security payments but recently admitted that the program was essentially insolvent and required such reform (Peterson, 2011).

A plan to partially convert Medicare into a program where participants would receive a set amount of funds to buy private health insurance, rather than the tradition of being eligible for limitless funds from a government run system, was passed by the US House of Representatives in 2012, but there is no immediate chance of this plan becoming law.

The US can also readily increase its tax base by allowing more immigration of young workers with the education and skills to be very productive. This is, of course, not a permanent solution, since young, healthy taxpayers eventually become less healthy retirees drawing government benefits, but it would help buy some time for more structural reform. Increases in tax rates, at least for the middle class where most of the income is, may be more problematic. There seems to be little political support for higher taxes on the US middle class, though this group is taxed lightly compared to their counterparts in most countries.[4] Higher tax rates on the wealthy are probably more politically viable but it is economically more difficult to get much more revenue from this group. Higher tax rates will increase revenue extraction as long as taxable income does not fall too much but, for the wealthy, significant decreases in taxable income are very possible. With higher tax rates, there is more incentive to find legal loopholes, to find illegal tax dodges, to work less, take early retirement (the wealthy can readily afford this), or to migrate to a different country that is a tax haven such as the Cayman Islands. Because of these effects, historically very high tax rates on the wealthy have tended to produce less revenue (The Economist, 2011). The wealthy in the US already pay a greater portion of income taxes than the wealthy in any other industrialized country. The top 10 percent of US income earners paid 45.1 percent in the mid-2000s, compared to the OECD-member average of 31.6 percent.[5]

While there are economic problems with raising tax rates on the wealthy and political problems with raising them on the middle class, tax reform that reduces rates but also closes loopholes holds substantial promise. Suppose, for example, an individual currently faces a marginal tax rate of 35 percent, meaning 35 percent of each additional dollar earned goes to the goverment. Suppose that the individual uses various legal tax shelters and deductions so that the actual taxes paid are 20 percent of his total income. For the simplest illustration, suppose we were to completely eliminate the various shelters and loopholes and cut his tax rate to 20 percent of all income. Then he would pay the same total tax bill if his income were the same—however, average income increases with this sort of reform, even if it is not as complete as eliminating all loopholes. With a lower marginal tax rate there is more incentive to work, save, and invest, less incentive to cheat on taxes or move to a tax haven. There are also gains from not having to take the time and expense to keep records and fill out complicated tax forms, known as compliance costs, which were estimated to be slightly over $400 billion in 2012 (Kroll, 2012). Moreover, all economic decisions regarding labor and investment become more efficient, decisions are made more on the basis of economic efficiency rather than on tax consequences. Thus, economic growth increases, as do government tax revenues. The stimulation to economic growth can also be enhanced by moving to a system that taxes consumption more and savings less. This sort of approach, reducing tax rates but increasing tax revenues through improving incentives to work and save, and closing loopholes, does have substantial political support, and was in fact recommended by the bipartisan Commission on Fiscal Responsibility and Reform known as Simpson-Bowles (Tax Policy Center, 2010).

---

4   See: Worldwide-Tax.com.
5   See OECD Income Distribution, 2008.

WHAT IS THE FUTURE OF THE WORLD ECONOMY?

First, the bad news: sovereign debt crises stemming from unfunded obligations in government health care and retirement programs are likely to retard economic growth in much of the world, particularly in Europe. Such fiscal crises in individual countries have occurred fairly routinely throughout history and, though painful, have not been cataclysmic; in some sense this is business as usual—governments are almost universally prone to eventually overspend their way into insolvency (The Center for Small Government, 2011). However, so many countries facing such serious problems at the same time is more historically unique. The timeliness of political responses will be a good indicator of how severe this impending crisis will be. If governments seriously address these problems before they become acute then all this government debt may be just a bump in the road. Though tax increases may sometimes be part of the mix, effective reform will require some true spending restraint, and, probably, tax and regulatory reform to better promote economic growth. Remember, the faster economic growth is, the easier it is to deal with any problem.

The coming surge in retirees in addition to contributing to debts finally coming due in socialized health care and retirement will also directly reduce economic growth. Furthermore, not only will total number of workers rapidly decline but, other things equal, the average productivity of workers will tend to also fall because older, more experienced workers who remain healthy tend to be highly productive. An ironic silver lining: the debt crisis is likely to fuel some cuts in retirement benefits that may help motivate healthy seniors to keep working longer than they otherwise would, mitigating these problems somewhat.

The fate of future economic performance will also be revealed in coming policy choices regarding the environment. The unfortunate reality, again, is that there is an unavoidable trade-off; the more stringent environmental regulation is the worse the economy will perform. As of this writing, it seems that economic concerns are likely to be paramount. People, for example, seem reluctant to embrace expensive measures to combat global warming once they realize the economic costs are considerable (Fankenhauser, 2000). If events do unfold in this manner then environmentalists may be displeased but the debt crises will be substantially less severe, allowing societies to properly revisit the issue afterward.

Another indicator of future growth possibilities will be tax policy. As discussed earlier, reforms that lower marginal tax rates, even if rendered "revenue neutral" by the closing of loopholes, tend to accelerate growth even though the magnitude of this effect is difficult to gauge. Reform that removes tax burdens on savings and investment is also helpful to growth. A number of regulatory reforms can accelerate economic growth; reductions of international trade barriers are a prime example. In recent years, the movement to more free trade seems to have slowed; the future is uncertain with some danger of increased protectionism.

A brighter spot, the collapse of communism, and more broadly, the whole idea that government command and control is a useful economic system, continues to pay dividends. This is most obvious, perhaps, in the rapid economic growth of China and India following their embrace of substantial economic reform. Although much of their growth is attributable to them being able to readily import technology much more advanced than their own, the end of this sort of growth is nowhere in sight since these two still have, comparatively, very low-tech economies with very low per capita real GDP. In other words, once a developing country substantially catches up to state-of-the-art technology, rapid growth is not so easy—superior technology must be invented instead of merely imported, but it will be many decades before India and China reach that point. Likewise, many of the economies in Eastern Europe, Central and South America and, even more so, Africa, have made strides in market reforms and have tremendous growth potential. For example, over the last 15 years Poland has enjoyed 4.5 percent annual real GDP growth, while Uganda has grown 4.1 percent per year (World Bank, 2011).

The stronger economic growth in the developing world is likely to only partially offset the declining growth of the wealthier economies. Most forecasters expect slower growth in the world economy (Hannon, 2011). But growth in the developing world may be especially strong in aviation-related industries. Many of these countries do not have the infrastructure needed for advanced ground transportation; expanding airports may be the most efficient way to transport people and products. The future of aviation and aerospace has its share of challenges but there is some promise and reason for optimism as well.

SUMMARY

The key goal of macroeconomic policy is to maximize economic growth and to raise the standard of living so that we have more resources to address whatever goals we choose. This is not necessarily an overly materialistic focus, as seen in the fact that only well-developed economies tend to be consistently concerned with protecting the environment. Full employment is another key goal, achieved not when unemployment is zero, but when the unemployed can find a job commensurate with their skills reasonably quickly. A stable currency (zero, or at least, low inflation) is the other main macroeconomic goal. Although inflation can ruin an economy many economists believe it is possible that moderate inflation, properly timed, can actually stimulate economic growth and full employment. Keynesian theory advocates a policy of intentional inflation when the economy is weak. This can be done through monetary policy or, possibly, through fiscal policy. Keynesian fiscal policy calls for aggressive government deficit spending during a recession. However, this policy may be useless if deficit spending simply crowds out private spending, leaving aggregate spending unchanged. Also, Keynesian policy can be destabilizing because of complexities in trying to hit the economy with just the right amount of inflation at just the right time. Economists disagree as to whether Keynesian policy has been useful or destructive. The modern trend toward perpetual deficit spending, driven largely by government health care programs and government retirement systems has resulted in large government indebtedness that poses a serious threat to economic stability and growth, as seen already in Greece. Slow population growth has exacerbated this problem as the ratio of working people to retired people is set to drop dramatically in the future. Unless there is major reform to control government spending and stimulate economic growth it is very likely that the sort of problems observed in Greece will become much worse and spread through most of the world.

DISCUSSION QUESTIONS

1.	Explain the difference between real and nominal variables.
2.	What are three possible ways a central bank could increase the money supply?

3. Suppose a government raises tax rates in order to redistribute income. How would this likely impact economic growth?
4. Summarize the various ways inflation can harm an economy.
5. At the equilibrium wage there is full employment, though some people are unemployed. Explain.
6. According to Keynesian theory, what should fiscal policy be if the economy is in recession? In an unsustainable growth spurt with rising inflation? In a time of normal growth?
7. Why might Keynesian policy fail, or even make things worse?
8. What are the key spending programs driving the impending crisis in government sovereign debt?
9. How do population demographics relate to the sovereign debt crisis and future economic growth?
10. What are the likely trends in future economic growth?

REFERENCES

Beranek, W. (2011). Unemployment Benefits and Unemployment, *Modern Economy* 2, 800-803.

Bureau of Labor Statistics (BLS). (2011). Feburary 25. Retrieved May 2012 from http://www.bls.gov/cps/minwage2010.htm.

Buttonwood. (2011). Finance and Economics: Pensions, Ponzis and Pyramids. *The Economist*, 88.

Brauer, D. (2007). *The Natural Rate of Unemployment, Congressional Budget Office*, Washington DC.

Fankenhauser, S. (2000). How Much Damage Will Climate Change Do? Recent Estimates. *World Economics*, 1(4): 179-206.

Ferguson, R. W. (2004). *A Retrospective on Business-cycle Recoveries: Are "Jobless" Recoveries the New Norm?*, At the Exchequer Club of Washington Luncheon, Washington, D.C., July 21, 2004. Retrieved on March 26, 2012 from http://www.federalreserve.gov/boarddocs/speeches/2004/20040721/default.htm.

Hannon, P. (2011). World News: New Indicators Suggest Growth Will Slow in Most Major Economies. *The Wall Street Journal*, June 15.

Kroll, K. M. (2012). Can a Fairer Tax Code Create Jobs? *Business Finance*, May 8.

Lotterman, E. (2011). *Medicare Hurting a Lot Worse than Social Security*. St Paul. MN: Saint Paul Pioneer Press.

Peterson, K. (2011). Seniors' Groups Slam AARP for Shift on Social Security. *The Wall Street Journal*, June 17.

Seidman, L. (1997). *The USA Tax: A Progressive Consumption Tax*. Cambridge, MA: The MIT Press.

Tax Policy Center (2010) *The Bowles-Simpson "Chairmen's Mark" Deficit Reduction Plan*. Retrieved on April 4, 2012 from http://www.taxpolicycenter.org/taxtopics/Bowles_Simpson_Brief.cfm.

The Center for Small Government (2011) *Deficit or Overspending: The Difference One Word Makes*. Retrieved on March 18, 2012 from http://www.centerforsmallgovernment.com/small-government-news/deficit-or-overspending-the-difference-one-word-makes/.

The Economist (2011). Diving into the Rich Pool, September 24.

The World Bank (2011). *GDP Growth (Annual %)*. Retrieved on May 6, 2012 from http://data.worldbank.org/indicator/NY.GDP.MKTP.KD.ZG.

Trading Economics. (2011). Retrieved May 2012 from http://www.tradingeconomics.com/united-states/median-duration-of-unemployment-weeks-m-sa-fed-data.html.

Airport and Airline Codes

AIRPORT CODES

Code	Name	City	Country
AKL	Auckland International Airport	Auckland	New Zealand
AMS	Schiphol Airport Amsterdam	Amsterdam	The Netherlands
ATL	Atlanta International Airport	Atlanta	USA
BOS	Boston Logan International Airport	Boston	USA
BWI	Baltimore/Washington International Airport	Baltimore	USA
CDG	Charles De Gaulle International Airport	Paris	France
CLT	Charlotte/Douglas International Airport	Charlotte	USA
CVG	Cincinnati/Northern Kentucky International Airport	Cincinnati	USA
DCA	Ronald Reagan Washington National Airport	Washington	USA
DEN	Denver International Airport	Denver	USA
DFW	Dallas/Fort Worth International Airport	Dallas	USA
DMK	Don Muang International Airport	Bangkok	Thailand
DTW	Detroit Metro Airport	Detroit	USA
EWR	Newark Liberty International Airport	Newark	USA

Code	Name	City	Country
FCO	Rome Leonardo da Vinci Airport	Rome	Italy
FLL	Fort Lauderdale Hollywood International Airport	Fort Lauderdale	USA
FRA	Frankfurt International Airport	Frankfurt	Germany
HKG	Hong Kong International Airport	Hong Kong	Hong Kong
HNL	Honolulu International Airport	Honolulu	USA
IAD	Washington Dulles International Airport	Washington	USA
IAH	Bush Intercontinental Airport	Houston	USA
ICN	Incheon International Airport	Seoul	Korea
JFK	John F. Kennedy International Airport	New York	USA
LAS	Mccarran International Airport	Las Vegas	USA
LAX	Los Angeles International Airport	Los Angeles	USA
LHR	London Heathrow	London	United Kingdom
LGW	London Gatwick	London	United Kingdom
HND	Tokyo International Airport	Tokyo	Japan
LGA	La Guardia Airport	New York	USA
MAD	Madrid-Barajas Airport	Madrid	Spain
MCO	Orlando International Airport	Orlando	USA
MDW	Chicago Midway Airport	Chicago	USA
MIA	Miami International Airport	Miami	USA
MSP	Minneapolis/St. Paul International Airport	Minneapolis	USA
MXP	Milan Malpensa Airport	Milan	Itay

Code	Name	City	Country
NRT	Narita International Airport	Tokyo	Japan
OAK	Oakland International Airport	Oakland	USA
ORD	Chicago O'Hare International Airport	Chicago	USA
PEK	Beijing Capital International Airport	Beijing	China
PHL	Philadelphia International Airport	Philadelphia	USA
PHX	Phoenix Sky Harbor International Airport	Phoenix	USA
PIT	Pittsburgh International Airport	Pittsburgh	USA
SAN	San Diego International Airport	San Diego	USA
SEA	Seattle/Tacoma International Airport	Seattle	USA
SEL	Seoul Gimpo International Airport	Seoul	Korea
SFO	San Francisco International Airport	San Francisco	USA
SIN	Changi International Airport	Singapore	Singapore
SLC	Salt Lake City International Airport	Salt Lake City	USA
STL	Lambert-St. Louis International Airport	Saint Louis	USA
TPA	Tampa International Airport	Tampa	USA
YVR	Vancouver International Airport	Vancouver	Canada
YYZ	Toronto Pearson International Airport	Toronto	Canada
ZRV	Zurich International Airport	Zurich	Switzerland

AIRLINE CODES

Airlines	Country	Code
Aeroflot-Russian Airlines	Russia	SU
Aeromexico	Mexico	AM
AHK Air Hong Kong Ltd	Hong Kong	LD
Air Caledinie International	Caledonia	SB
Air Canada	Canada	AC
Air China Ltd	China	CA
Air France	France	AF
Air New Zealand Ltd	New Zealand	NZ
Air Nippon	Japan	EL
Air Niugini Pty Ltd	Niugini	PX
Air Pacific	Fiji	FJ
Air Tahiti NUI	Tahiti	TN
Air-India Ltd	India	AI
Alitalia-Linee Aeree Italiane	Italy	AZ
All Nippon Airways	Japan	NH
American Airlines	USA.	AA
Asiana Airlines Inc.	Korea	OZ
Austrian Airlines	Austria	OS
British Airways Plc	UK	BA
Cargolux Airlines Int'l SA	Luxemburg	CV
Cathay Pacific Airways	China	CX
China Airlines Ltd	Taiwan	CI
China Eastern Airlines	China	MU
China Southern Airlines	China	CZ
Dalavia Far East Airways	Russia	H8
Delta Air Lines, Inc.	USA	DL
EgyptAir	Egypt	MS
Emirates Sky Cargo	UAE	EK
EVA Airways Corp.	Taiwan	BR
Fedex	USA	FX
FinnAir O/Y	Finland	AY
Garuda Indonesia	Indonesia	GA
Hong Kong Dragon Airlines Limited.TD.	Hong Kong	KA

Airlines	Country	Code
Iran-Air	Iran	IR
Japan Airlines Co., Ltd	Japan	JL
Japan Asia Airways Co., Ltd	Japan	EG
Jetblue	USA	B6
KLM Royal Dutch Airlines	Netherlands	KL
Korean Air Lines Co.,Ltd	Korea	KE
Lufthansa Cargo AG	Germany	LH
Malaysia Airlines System Berhad	Malaysia	MH
Miat-Mongolian Airlines	Mongolia	OM
Nippon Cargo Airlines	Japan	KZ
Pakistan Int'l Airlines	Pakistan	PK
Philippine Airlines, Inc.	Phillipines	PR
Polar Air Cargo Inc.	USA	PO
Qantas Airways Ltd	Australia	QF
Royal Nepal Airlines Corp.	Nepal	RA
Scandinavian Airlines System(SAS)	Sweden	SK
Shanghai Airlines Co., Ltd	China	FM
Singapore Airlines	Singapore	SQ
Southwest Airlines	USA	WN
Spirit Airlines	USA	NK
Srilankan Airlines Ltd	Sri Lanka	UL
Swiss Int'l Air Lines	Switzerland	LX
Thai Airways Int'l Public Co.,Ltd	Thailand	TG
Turkish Airlines	Turkey	TK
United Airlines, Inc.	USA	UA
United Parcel	USA	5X
Uzbekistan Airways	Uzbekistan	HY
Vietnam Airlines	Vietnam	VN
Virgin Atlantic	UK	VS

List of Countries by GDP (2012)

	Country	GDP in Thousands of $US
1	European Union	18,084,866
2	United States	15,880,207
3	People's Republic of China	7,209,418
4	Japan	5,920,556
5	Germany	3,599,981
6	France	2,834,353
7	United Kingdom	2,602,487
8	Brazil	2,576,244
9	Italy	2,245,905
10	Russia	2,197,710
11	India	1,858,969
12	Canada	1,809,315
13	Spain	1,524,063
14	Australia	1,470,027
15	Mexico	1,231,642
16	South Korea	1,201,535
17	Indonesia	908,125
18	Turkey	876,583
19	Saudi Arabia	606,016

	Country	GDP in Thousands of $US
20	Switzerland	602,593
21	Sweden	591,544
22	Republic of China (Taiwan)	545,450
23	Poland	535,270
24	Argentina	526,299
25	Belgium	518,073
26	Norway	494,242
27	Iran	450,945
28	Austria	419,945
29	South Africa	402,493
30	United Arab Emirates	384,196
31	Thailand	367,875
32	Denmark	350,875
33	Colombia	331,856
34	Greece	313,856
35	Venezuela	299,932
36	Nigeria	288,822

Traffic Movements (2012)

Rank	Airport	Location	Code (IATA/ICAO)	Total Movements
1	Atlanta International Airport	United States	ATL/KATL	950,119
2	O'Hare International Airport	United States	ORD/KORD	882,617
3	Los Angeles International Airport	United States	LAX/KLAX	666,938
4	Dallas/Fort Worth International Airport	United States	DFW/KDFW	652,261
5	Denver International Airport	United States	DEN/KDEN	630,063
6	George Bush Intercontinental Airport	United States	IAH/KIAH	531,347
7	Charlotte/Douglas International Airport	United States	CLT/KCLT	529,101
8	Beijing Capital International Airport	China	PEK/ZBAA	517,584
9	McCarran International Airport	United States	LAS/KLAS	505,591
10	Paris-Charles de Gaulle Airport	France	CDG/LFPG	499,997
11	Frankfurt Airport	Germany	FRA/EDDF	464,432
12	Philadelphia International Airport	United States	PHL/KPHL	460,779
13	London Heathrow Airport	United Kingdom	LHR/EGLL	454,883

Rank	Airport	Location	Code (IATA/ICAO)	Total Movements
14	Detroit Metropolitan Wayne County Airport	United States	DTW/KDTW	452,616
15	Phoenix Sky Harbor International Airport	United States	PHX/KPHX	449,351
16	Minneapolis-Saint Paul International Airport	United States	MSP/KMSP	436,625
17	Barajas Airport	Spain	MAD/LEMD	433,683
18	Toronto Pearson International Airport	Canada	YYZ/CYYZ	418,298
19	Newark Liberty International Airport	United States	EWR/KEWR	403,880
20	Amsterdam Airport Schiphol	Netherlands	AMS/EHAM	402,372
21	John F. Kennedy International Airport	United States	JFK/KJFK	399,626
22	Munich Airport	Germany	MUC/EDDM	389,939
23	San Francisco International Airport	United States	SFO/KSFO	387,248
24	Miami International Airport	United States	MIA/KMIA	376,208
25	Phoenix Deer Valley Airport	United States	DVT/KDVT	368,747
26	Salt Lake City International Airport	United States	SLC/KSLC	362,654
27	LaGuardia Airport	United States	LGA/KLGA	362,137
28	Logan International Airport	United States	BOS/KBOS	352,643
29	Tokyo International Airport	Japan	HND/RJTT	342,804
30	Mexico City International Airport	Mexico	MEX/MMMX	339,898

Passenger Traffic (2010)

Rank	City (Airport)	Code	Total Passengers
1	Atlanta GA, US	ATL	89,331, 622
2	Beijing, CN	PEK	73, 948, 113
3	Chicago IL, US	ORD	66, 774, 738
4	London, GB(LHR)	LHR	65, 884, 143
5	Tokyo, JP	HND	64, 211, 074
6	Los Angeles CA, US	LAX	59, 070, 127
7	Paris, FR	CDG	58, 167, 062
8	Dallas/Fort Worth TX, US	DFW	56, 906, 610
9	Frankfurt, DE	FRA	53, 009, 221
10	Denver CO, US	DEN	52, 209, 377
11	Hong Kong, HK	HKG	50, 348, 960
12	Madrid, ES	MAD	49, 844, 596
13	Dubai, AE	DXB	47,180, 628
14	New York NY, US	JFK	46, 514, 154
15	Amsterdam, NL	AMS	45, 211, 749
16	Jakarta, ID	CGK	44, 355, 998
17	Bangkok, TH	BKK	42, 784, 967
18	Singapore, SG	SIN	42, 038, 777
19	Guangzhou, CN	CAN	40, 975, 673

Rank	City (Airport)	Code	Total Passengers
20	Shanghai, CN	PVG	40, 578, 621
21	Houston TX, US	IAH	40, 479, 569
22	Las Vegas NV, US	LAS	39, 757, 359
23	San Francisco CA	SFO	39, 253, 999
24	Phoenix AZ, US	PHX	38, 554, 215
25	Charlotte NC, US	CLT	38, 254, 207
26	Rome, IT	FO	36, 227, 778
27	Sydney, AU	SYD	35, 991, 917
28	Miami FL, US	MIA	35, 698, 025
29	Orlando FL, US	MCO	34, 877, 899
30	Munich, DE	MUC	34, 721, 605

Index